CURVATURE AND VARIATIONAL MODELING IN PHYSICS AND BIOPHYSICS

To learn more about AIP Conference Proceedings, including the
Conference Proceedings Series, please visit the webpage
http://proceedings.aip.org/proceedings

CURVATURE AND VARIATIONAL MODELING IN PHYSICS AND BIOPHYSICS

Santiago de Compostela, Spain
17 – 28 September 2007

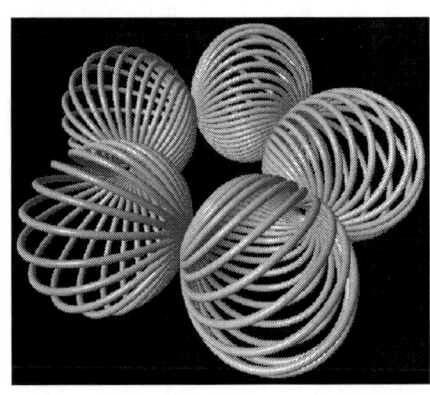

EDITORS

Óscar J. Garay
University of País Vasco
Bilbao, Spain

Eduardo García-Río
Ramón Vázquez-Lorenzo
University of Santiago de Compostela
Santiago de Compostela, Spain

SPONSORING ORGANIZATIONS
Institute of Mathematics of the University of Santiago de Compostela
Faculty of Mathematics of the University of Santiago de Compostela
Research Project Consolider-Ingenio i-MATH CSD2006-00032 03 (Ministry of Science
    and Education, Spain)
Research Project MTM2004-04934-C04-03 (Ministry of Science and Education, Spain)
Research Project IT-256-07 (Gobierno Vasco)
Research Project MTM2006-01432 03 (Ministry of Science and Education, Spain)
Research Project PGIDIT06PXIB207054PR (Xunta de Galicia)

Melville, New York, 2008
AIP CONFERENCE PROCEEDINGS ■ VOLUME 1002

Editors:

Óscar J. Garay
Departamento de Matemáticas
Facultad de Ciencia y Tecnología
Universidad del País Vasco
Aptdo. 644
48014 Bilbao
Spain

E-mail: oscarj.garay@ehu.es

Eduardo García-Río
Ramón Vázquez-Lorenzo

Facultad de Matemáticas
Universidad de Santiago de Compostela
15782 Santiago de Compostela
Spain

E-mail: xtedugr@usc.es
        ravazlor@usc.es

L.C. Catalog Card No. 2008924318
ISBN 978-0-7354-0521-9
ISSN 0094-243X
Printed in the United States of America

CONTENTS

Preface vii
Committees ix
Acknowledgments xi

COURSES

Lectures on Elastic Curves and Rods 3
    D. A. Singer
Variational Problems Which are Quadratic in the Surface Curvatures 33
    B. Palmer
Simple Geometrical Models with Applications in Physics 71
    M. Barros
The Geometry and Topology of Liquid Crystals 114
    C. Santangelo
**Semiflexible Polymers and Filaments: From Variational Problems to
Fluctuations** 151
    J. Kierfeld, K. Baczynski, P. Gutjahr, and R. Lipowsky
An Experimental Trip to the Calculus of Variations 186
    J. Arroyo
Computational Methods in Mathematics 213
    J. C. Díaz-Ramos

List of Participants 253
Author Index 255

PREFACE

This volume contains a series of lectures given in the International School "Curvature and variational modelling in Physics and Biophysics" held in Santiago de Compostela, Spain. They were grouped in six mini-courses and two computer labs, offered in two weeks, from September 17th to September 28th, 2007.

The idea behind this event organization was to promote the cooperation between the mathematical world and the physical and life sciences. Historically, institutions of higher learning have treated them as separate spheres. However, there is at least an emerging rhetoric concerning the importance of acknowledging and promoting the reciprocal relationship between these fields and we have bet to foster the cooperation between Mathematics and the "outer" world. Our aim thus was to explore and stimulate connections among Differential Geometry, Physics and Biology.

Probably, the most important concept in Riemannian Geometry is Curvature. On the other hand, a basic tool in modern Physics is the Least Action Principle, and Soft Matter is one of the most active research fields in the physical and life sciences. Combining these ingredients we have obtained the fundamental topics covered in this school.

The school was designed as an open post-graduate school. That is, although our main concern was to provide qualified postgraduate students with a starting research point, interdisciplinary researchers belonging to related fields were also welcome. Our aim was to furnish these groups with an intensive mathematical experience in the hope that interaction among people with different backgrounds and professional needs would increase each participant's appreciation of the mathematical world as a whole, as well as of the relevance of mathematical modelling in real life sciences.

Guest lecturers were selected among leading specialist in the world not only because their expertise in the area but also because their teaching skills. We are deeply indebted to them for their generous effort and for their participation in this meeting. Professors M. Barros, R. Lipowsky, J. Kierfeld, B. Palmer, C. Santangelo and D. Singer, were in charge of the morning courses while Professors J. Arroyo and J.C. Díaz-Ramos conducted the computer labs in the afternoon.

In this volume, lectures have been grouped in seven blocks which give a good approximation to the school courses contents. The first two blocks are purely mathematical papers dealing with variational problems having strong connections to Physics and Biophysics, which are, in turn, the natural areas for the fourth and fifth articles where the authors deal with important physical problems modelled by means of geometric variational tools. The third paper locates at the interface between Differential Geometry and Mathematical Physics and it is a good transition from the mathematical formalism of the previous papers to the more experimental flavour of the next ones. The last two papers show a brief description of the two computer labs offered in the school. They focused in computational and visualization aspects of curves and surfaces and related variational problems. Slides of some courses are available at the school's webpage
`http://xtsunxet.usc.es/schoolsantiago2007/`

CURVATURE AND VARIATIONAL MODELLING IN PHYSICS AND BIOPHYSICS

September 17-28, 2007, Santiago de Compostela, Spain

ORGANIZING COMMITTEE

- Oscar J. Garay (University of País Vasco, Spain)
- Eduardo García-Río (University of Santiago de Compostela, Spain)
- Ramón Vázquez-Lorenzo (University of Santiago de Compostela, Spain)

SCIENTIFIC COMMITTEE

- Manuel Barros (University of Granada, Spain)
- Marisa Fernández (University of País Vasco, Spain)
- Ángel Ferrández (University of Murcia, Spain)
- Oscar J. Garay (University of País Vasco, Spain)
- Eduardo García-Río (University of Santiago de Compostela, Spain)
- R. Kamien (University of Pennsylvania, USA)
- R. Lipowsky (Max Planck Institüt of Colloids and Interfaces, Potsdam, Germany)
- U. Seifert (Institut für theoretical Physik, Universität Stuttgart)

ACKNOWLEDGMENTS

We are grateful to the Institute of Mathematics and the Faculty of Mathematics of the University of Santiago de Compostela where this School was held. Sponsorship from different Organizations was very welcome. In particular, the School was supported by the Ministerio de Educación y Ciencia under the research project CONSOLIDER INGENIO-MATHEMATICA (i-MATH) CSD2006-00032. Also financial support by projects MTM2004-04934-C34-03 and Gobierno Vasco IT-256-07, as well as by projects MTM2006-01432 and PGIDIT06PXIB207054PR is gratefully acknowledged.

Special thanks are given to the Scientific Committee and to L.A. Cordero (Univ. of Santiago de Compostela), M. Barros (Univ. of Granada), M. Fernández (Univ. Pais Vasco), A. Ferrández (Univ. Murcia), J. J. Nieto (Univ. of Santiago de Compostela), M. Elena Vázquez-Abal (Univ. of Santiago de Compostela) and J.J. Mencía (Univ. Pais Vasco) for their collaboration. Finally, we wish to show our gratitude to all the speakers and attendees for their participation and their ability to create a pleasant atmosphere which contributed greatly to the school success.

Santiago de Compostela, February 2008

The Editors:

Oscar J. Garay

Eduardo García-Río

Ramón Vázquez-Lorenzo

COURSES

Lectures on Elastic Curves and Rods

David A Singer

Dept. of Mathematics
Case Western Reserve University
Cleveland, OH 44106-7058

Abstract. These five lectures constitute a tutorial on the Euler elastica and the Kirchhoff elastic rod. We consider the classical variational problem in Euclidean space and its generalization to Riemannian manifolds. We describe both the Lagrangian and the Hamiltonian formulation of the rod, with the goal of examining the (Liouville-Arnol'd) integrability. We are particularly interested in determining closed (i.e., periodic) solutions.

Keywords: elastica, Kirchhoff rod
PACS: 02.30.Xx,02.30.Yy

1. CLASSICAL BERNOULLI–EULER ELASTICA

The classical curve known as the *elastica* is the solution to a variational problem proposed by Daniel Bernoulli to Leonhard Euler in 1744, that of minimizing the bending energy of a thin inextensible wire (See, e.g., [27], §263). The mathematical idealization of this problem is that of minimizing the integral of the squared curvature for curves of a fixed length satisfying given first order boundary data. In this lecture, we will use the classical techniques of the calculus of variations to derive the equations of the elastica.

Consider regular curves (curves with nonvanishing velocity vector) in Euclidean space defined on a fixed interval $[a_1, a_2]$:

$$\gamma: [a_1, a_2] \longrightarrow \mathbb{R}^3, \quad \|\gamma'(t)\| = \frac{ds}{dt} = v \neq 0.$$

We will assume (for technical reasons) that the (geodesic) curvature k of γ is nonvanishing. This will allow us to define the *Frenet Frame* along the curve.

The Frenet frame $\{T, N, B\}$ is orthonormal and satisfies

$$\gamma' = vT, \quad \frac{dT}{ds} = kN, \quad \frac{dN}{ds} = -kT + \tau B, \quad \frac{dB}{ds} = -\tau N.$$

The elastica minimizes the bending energy

$$\mathscr{F}(X) = \int_\gamma k(s)^2 ds = \int_{a_1}^{a_2} k(t)^2 \, v dt,$$

with fixed length and boundary conditions. Accordingly, let α_1 and α_2 be points in $\mathbb{R}^3$ and α_1' and α_2' nonzero vectors.

We will consider the space of smooth curves

$$\Omega = \{\gamma | \gamma(a_i) = \alpha_i, \gamma'(a_1) = \alpha_i'\},$$

CP1002, *Curvature and Variational Modeling in Physics and Biophysics*
edited by O. J. Garay, E. García-Río, and R. Vázquez-Lorenzo
© 2008 American Institute of Physics 978-0-7354-0521-9/08/$23.00

and the subspace of unit-speed curves

$$\Omega_u = \{\gamma \in \Omega | \, \|\gamma'\| \equiv 1\}.$$

Later on we need to pay more attention to the precise level of differentiability of curves, but we will ignore that for now.

$\mathscr{F}^\lambda : \Omega \longrightarrow \mathbb{R}$ is defined by

$$\mathscr{F}^\lambda(\gamma) = \frac{1}{2} \int_\gamma \|\gamma''\|^2 + \Lambda(t) \left(\|\gamma'\|^2 - 1 \right) dt.$$

One version of the Lagrange multiplier principle says a minimum of $\mathscr{F}$ on Ω_u is a stationary point for $\mathscr{F}^\lambda$ for some $\Lambda(t)$. ($\Lambda(t)$ is a *pointwise* multiplier, constraining speed.) The name $\mathscr{F}^\lambda$ for the function will be justified later, when we will see that $\Lambda(t)$ depends on a constant λ.

Assume that γ is an extremum of $\mathscr{F}^\lambda$. Then if W is a vector field along γ, that is, an infinitesimal variation of the curve, we have

$$\partial \mathscr{F}^\lambda(W) = \frac{\partial}{\partial \varepsilon} \mathscr{F}^\lambda(\gamma + \varepsilon W)|_{\varepsilon=0} = 0,$$

$$
\begin{aligned}
0 &= \frac{1}{2} \frac{\partial}{\partial \varepsilon} \int_{a_1}^{a_2} \|(\gamma + \varepsilon W)''\|^2 + \Lambda \|(\gamma + \varepsilon W)'\|^2 - \Lambda(t) dt \\
&= \int_{a_1}^{a_2} \gamma'' \cdot W'' + \Lambda(t) \gamma' \cdot W' dt.
\end{aligned}
$$

Integrating by parts,

$$
\begin{aligned}
0 &= \int_{a_1}^{a_2} -\gamma''' \cdot W' - (\Lambda \gamma')' \cdot W dt + (\gamma'' \cdot W' + \Lambda \gamma' \cdot W) \Big|_{a_1}^{a_2} \\
&= \int_{a_1}^{a_2} \left[\gamma'''' - (\Lambda \gamma')' \right] \cdot W dt \\
&\quad + (\gamma'' \cdot W' + (\Lambda \gamma' - \gamma''') \cdot W) \Big|_{a_1}^{a_2} \\
&= \int_{a_1}^{a_2} E(\gamma) \cdot W dt + (\gamma'' \cdot W' + (\Lambda \gamma' - \gamma''') \cdot W) \Big|_{a_1}^{a_2},
\end{aligned}
$$

where $E(\gamma) = \gamma'''' - \dfrac{d}{dt}(\Lambda \gamma')$.

The elastica must satisfy

$$E(\gamma) = \gamma'''' - \frac{d}{dt}(\Lambda \gamma') \equiv 0,$$

for some function $\Lambda(t)$.

Integrating,

$$\gamma''' - \Lambda(t)\gamma' \equiv J,$$

4

for J a constant vector.

This equation can also be derived from Noether's Theorem: If γ is a solution curve and W is an infinitesimal symmetry, then $\gamma'' \cdot W' + (\Lambda\gamma' - \gamma''') \cdot W$ is constant. In particular, for a translational symmetry, W is constant; so

$$(\Lambda\gamma' - \gamma''') \cdot W = \text{constant.}$$

Letting W range over all translations, we get

$$\Lambda\gamma' - \gamma''' = C,$$

for C some constant field.

Now it is helpful to assume γ is parametrized by arclength s. Then $\gamma' = \gamma_s = T$, $\gamma'' = kN, \gamma''' = -k^2 T + k_s N + k\tau B$, so

$$\gamma''' - \Lambda(s)\gamma' = (-k^2 - \Lambda(s))T + k_s N + \tau B = J.$$

Differentiate J to get

$$0 = J_s = (-3kk_s - \Lambda_s)T + (k_{ss} - k^3 - \Lambda k - k\tau^2)N + (k\tau_s + 2k_s\tau)B.$$

From this it follows that $\Lambda(s) = -\frac{3}{2}k^2 + \frac{\lambda}{2}$ for some constant λ.

The vector field $J(s) = \frac{k^2-\lambda}{2}T + k_s N + k\tau B$ is constant along the curve. Thus it is the restriction of a translation field to the curve. From $J_s = 0$ we get the equations

$$k_{ss} + \frac{1}{2}k^3 - k\tau^2 - \frac{\lambda k}{2} = 0, \tag{1}$$

and

$$k\tau_s + 2k_s\tau = 0. \tag{2}$$

We may use Noether's theorem to derive additional first integrals of these equations. Recall that for any variation W,

$$0 = \int_{a_1}^{a_2} E(\gamma) \cdot W \, ds + (\gamma'' \cdot W' - J \cdot W)\Big|_{a_1}^{a_2}.$$

If W is a symmetry, we have

$$\gamma'' \cdot W' - J \cdot W = \text{constant.}$$

Now let W be the restriction of a rotation field:

$$W = \gamma \times W_0, \quad W_0' = 0.$$

Differentiating,

$$kN \cdot T \times W_0 - J \cdot \gamma \times W_0 = \text{constant,}$$

or

$$(kN \times T - J \times \gamma) \cdot W_0 = \text{constant.}$$

5

Since this works for any W_0, we get

$$kB + J \times \gamma = A,$$

for A a constant vector. So the vector field

$$I = kB = A + \gamma \times J,$$

is the restriction of an isometry to the curve.

The vector fields J and I play an important role in the integration of the equations of the elastica. A *Killing field* along a curve is a vector field along the curve which is the restriction of an infinitesimal isometry of the ambient space. If γ is an elastica in $\mathbb{R}^3$, then we have two Killing fields along γ:

$$J = \frac{k^2 - \lambda}{2} T + k_s N + k\tau B,$$

and

$$I = kB = A + \gamma \times J,$$

where A and J are constant fields.

$$k^2 \tau = I \cdot J = A \cdot J = c \tag{3}$$

is constant, as is

$$4\|J\|^2 = (k^2 - \lambda)^2 + 4k_s^2 + 4k^2\tau^2 = a^2. \tag{4}$$

Observe that equation (1) integrates to equation (4) and equation (2) integrates to equation (3). Now we can integrate these equations as follows:

Eliminating τ and replacing k^2 by u:

$$(u - \lambda)^2 + \frac{(u_s)^2}{u} + \frac{4c^2}{u} = a^2,$$

or

$$(u_s)^2 = P(u).$$

for P a cubic polynomial.

Solving this differential equation gives

$$k^2 = u = k_0^2 \left(1 - \frac{p^2}{w^2} \operatorname{sn}^2\left(\frac{k_0}{2w}s, p\right)\right) = \frac{c}{\tau}, \tag{5}$$

with $\operatorname{sn}(x, p)$ the *elliptic sine with parameter* p, and with $p, w,$ and k_0 parameters.

The parameters $p, w,$ and k_0 are related to the constants λ and c by

$$2\lambda = \frac{k_0^2}{w^2}(3w^2 - p^2 - 1),$$

$$4c^2 = \frac{k_0^6}{w^4}(1 - w^2)(w^2 - p^2).$$

6

A planar curve has $\tau = 0$, so $c = 0$. Thus either $w = 1$ or $w = p$. The parameter k_0 determines the maximum curvature.

Up to similarity, every solution corresponds to a point in the triangle $0 \le p \le w \le 1$. The planar curves correspond to two of the three edges of the triangle. The third edge of the triangle, $p = 0$ is made up of curves of constant curvature and torsion (helices).

What do the three-dimensional solutions look like? We can use the Killing fields to construct a preferred cylindrical coordinate system. Choose coordinates in $\mathbb{R}^3$ so that

$$J = \begin{pmatrix} 0 \\ 0 \\ \frac{a}{2} \end{pmatrix}, \quad A = \begin{pmatrix} 0 \\ 0 \\ b \end{pmatrix}, \quad b \ge 0.$$

The second equation is achieved by replacing γ by $\gamma - K$ for some K (translation). Since $I = \gamma \times J + A$, we have

$$c = I \cdot J = A \cdot J = \frac{ab}{2}.$$

$$I - \frac{I \cdot J}{J \cdot J} J = I - A = \gamma \times J.$$

We have the coordinate fields

$$\partial z = \frac{2}{a} J,$$

$$\partial \theta = \frac{2}{a} \gamma \times J = \frac{2}{a}\left(I - \frac{4c}{a^2} J\right),$$

$$\partial r = \frac{\partial z \times \partial \theta}{\|\partial z \times \partial \theta\|} = \frac{J \times B}{\|J \times B\|}.$$

Writing $T = \frac{dr}{ds}\partial r + \frac{dz}{ds}\partial z + \frac{d\theta}{ds}\partial \theta$, we get

$$\frac{dr}{ds} = T \cdot \partial r, \quad \frac{dz}{ds} = T \cdot \partial z, \quad \frac{d\theta}{ds} = \frac{T \cdot \partial \theta}{\|\partial \theta\|^2},$$

and the equations for γ can be integrated explicitly.

Using equation 4, the differential equation for r can be seen to be

$$\frac{dr}{ds} = \frac{2k_s}{\sqrt{4k_s^2 + (k^2 - \lambda)^2}} = \frac{2kk_s}{\sqrt{a^2 k^2 - 4c^2}}.$$

This integrates to

$$r = \frac{2}{a^2}\sqrt{a^2 k^2 - 4c^2}.$$

So r has the same periodicity and critical points as k. The elastica lies between two concentric cylinders (the inner one perhaps degenerating to a line) around the z - axis. The maxima of the curvature occur on the outer cylinder and the minima on the inner cylinder.

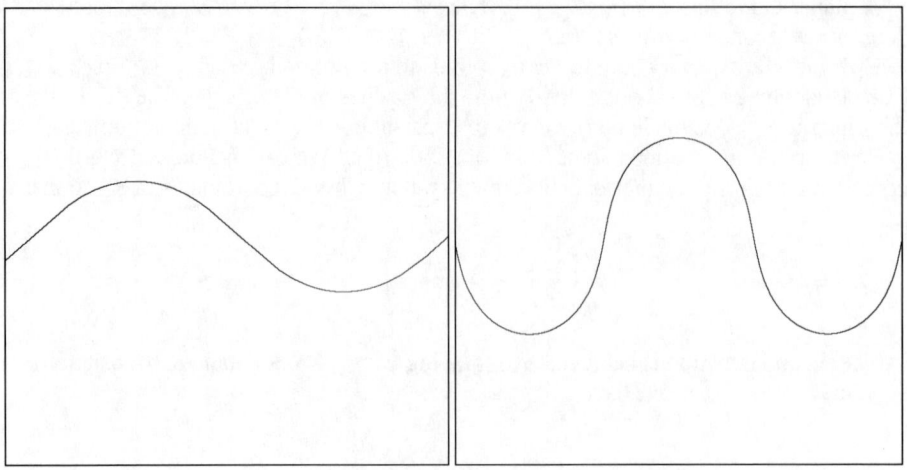

FIGURE 1. Wavelike elastica

In the 2-dimensional case $(c = 0)$, the curve lies in a strip parallel to the $z-axis$. In this case the formula for r, the distance from the axis, simplifies to

$$r = \frac{k}{a}.$$

When $w = p$, $k^2 = k_0^2 \left(1 - \text{sn}^2 \left(\frac{k_0}{2p}s, p\right)\right)$ so

$$k = k_0 \, \text{cn} \left(\frac{k_0}{2p}s, p\right).$$

The curvature oscillates between $-k_0$ and $+k_0$. We call such a curve a "wavelike" elastica. (Figure 1 and Figure 2)

When $w = 1$, $k^2 = k_0^2 \left(1 - p^2 \, \text{sn}^2 \left(\frac{k_0}{2}s, p\right)\right)$ so

$$k = k_0 \, \text{dn} \left(\frac{k_0}{2}s, p\right),$$

(where dn(x) is the elliptic delta), and k is non-vanishing. We call such a curve "orbit-like". (Figure 3a)

The borderline case, $p = w = 1$, (Figure 3b) has non-periodic curvature:

$$k = k_0 \text{sech} \frac{k_0}{2}s.$$

For $p = 0$ we get the helices:

$$k \equiv k_0, \qquad \tau \equiv \tau_0 = \frac{c}{k_0^2}.$$

8

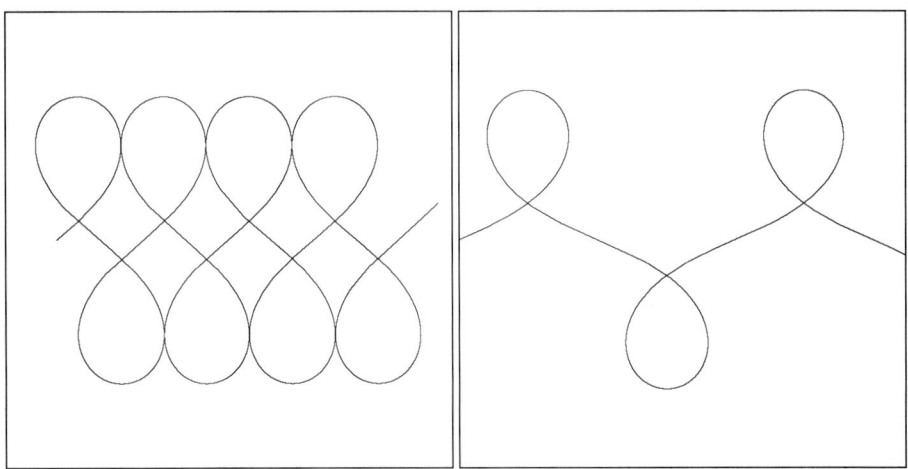

FIGURE 2. Wavelike elastica

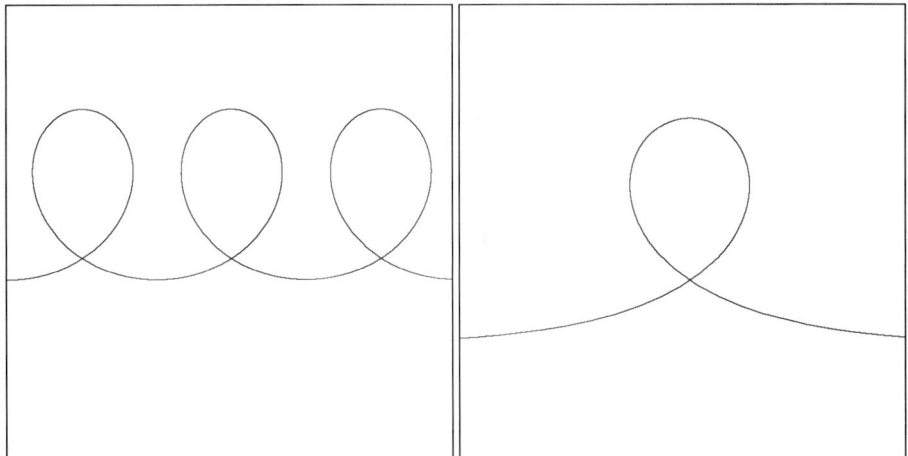

FIGURE 3. a) Orbitlike elastica b) Borderline elastica

The curves with $0 < p < w < 1$ are non-planar. For such curves, $0 < k < k_0$ and the curve has non-vanishing curvature and torsion.

When is the curve *closed*? Its curvature (and its torsion) is always periodic except in the case of a borderline elastic curve ($p = 1$), which is obviously not closed. We need the coordinates of γ to be periodic.

9

To answer this we must briefly review the properties of elliptic functions. (See [6] for exhaustive details.) The complete elliptic integrals $E(p)$ and $K(p)$ are given by:

$$K(p) = \int_0^{\frac{\pi}{2}} \frac{1}{\sqrt{1 - p^2 \sin^2 \phi}} \, d\phi,$$

and

$$E(p) = \int_0^{\frac{\pi}{2}} \sqrt{1 - p^2 \sin^2 \phi} \, d\phi.$$

The elliptic function $\mathrm{sn}(x, p)$ is an odd function with $\mathrm{sn}(x + 2K(p), p) = -\mathrm{sn}(x, p)$. So the curvature of an elastica is periodic with period $2wK(p)/k_0$.

If $\triangle z$ and $\triangle \theta$ represent the change in z and θ over one period of k, then γ is a smooth closed curve if and only if $\triangle z = 0$ and $\triangle \theta$ is rationally related to 2π.

$$
\begin{aligned}
\triangle z &= \int_0^{\frac{2w}{k_0} K(p)} \langle \partial z, T \rangle \, ds = \frac{2}{a} \int_0^{\frac{2w}{k_0} K(p)} (k^2 - \lambda) ds \\
&= \frac{4w}{ak_0} \int_0^{K(p)} (k_0^2 - \lambda) - k_0^2 \frac{p^2}{w^2} \mathrm{sn}^2 (x, p) \, dx.
\end{aligned}
$$

The closure condition may be written:

$$\triangle z = 0 \Longleftrightarrow 1 + w^2 - p^2 - 2 \frac{E(p)}{K(p)} = 0.$$

There is one closed planar curve (besides the circle): It requires

$$w = p, \quad 2E(p) = K(p) \Longleftrightarrow p \approx .82.$$

We have seen that J is a constant field, the restriction of a translation field to the curve. From the formula

$$J = \frac{k^2 - \lambda}{2} T + k'N + k\tau B,$$

it is clear that the curvature achieves its extrema at places where $N \cdot J = 0$.

The second condition for closure is that the θ coordinate be periodic. Let $\triangle \theta$ denote the increase in the θ coordinate in one period of the curvature function. Then it is necessary that $\triangle \theta$ be a rational multiple of 2π. The formula for $\triangle \theta$ in terms of the parameters p and w is quite complicated; see [19] for details. The essential fact is that as the point on the curve $\triangle z$ varies, the value of $\triangle \theta$ varies monotonically between 0 and $-\pi$. It therefore takes on every rational multiple of pi between those two values. This leads to the following theorem:

Theorem 1 (Langer and Singer, 1983 [19]) $\triangle \theta$ *is monotonically decreasing from π to 0 along $\triangle z = 0$. Thus there are infinitely many closed elastic curves which are nonplanar. All such elastica are embedded, lie on embedded tori of revolution, and represent (m,n) - torus knots, one for each $m > 2n$.*

FIGURE 4.   The figure-eight elastica

Figure 5 allows one to visualize the space of elastic curves in $\mathbb{R}^3$. Three parameters control the size and shape of an elastica. One parameter, k_0, gives the maximum curvature of the curve; dilations of $\mathbb{R}^3$ will adjust this. So we are reduced to two parameters, p and w, with $0 \leq p \leq w \leq 1$, and the parameter space is a triangle. Two of the three sides of the triangle correspond to planar elastic curves, while the third side corresponds to helices. The three vertices of the triangle correspond to borderline elastica, circle, and straight line. The interior of the triangle consists entirely of nonplanar curves. The curve $\triangle z = 0$ contains all of the closed elastic curves. The line $p^2 + w^2 = 1$ contains those (nonplanar) elastic curves that cross the axis of symmetry. Finally, there is a line along which are the solutions of the *free elastica* problem: for these curves there is no length constraint. We will have more to say about the free elastica later.

2. THE ELASTICA IN A RIEMANNIAN MANIFOLD

In this lecture we will formulate a generalized variational problem, that of the elastica in a Riemannian manifold. By this we mean a curve which is an extremal for the integral of the squared (geodesic) curvature among curves with specified boundary conditions. Here we summarize the machinery needed for calculations.

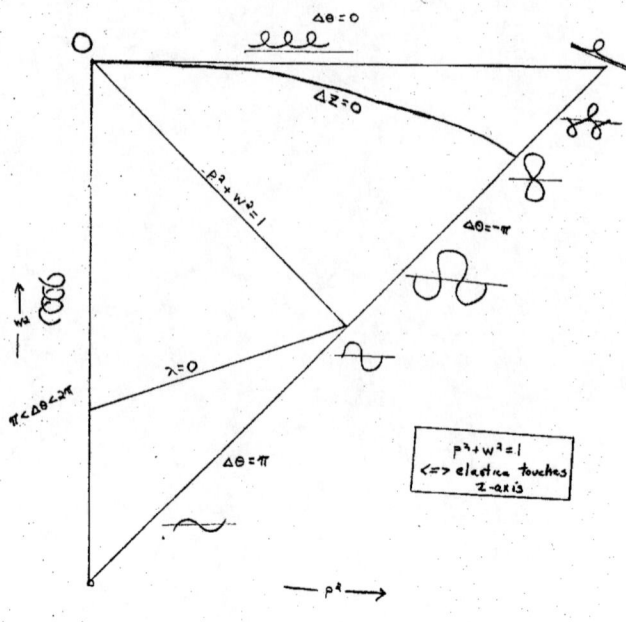

FIGURE 5. Parameter space of elastica in $\mathbb{R}^3$

In what follows, M is a smooth Riemannian manifold, with Riemannian metric $g(X,Y) =< X,Y >$, that is, a positive-definite symmetric bilinear form on tangent vectors X and Y at each point. The ordinary derivative is replaced by the covariant derivative $\nabla_X Y$, which measures the derivative of a vector *field* Y in the direction of a vector X.

For vector fields X and Y the equality of mixed partial derivatives is replaced by the bracket formula:

$$\nabla_X Y - \nabla_Y X = [X,Y] = XY - YX.$$

Ex: If $X = \frac{\partial}{\partial x}$, and $Y = x\frac{\partial}{\partial y}$ in local coordinates, then

$$\nabla_X Y - \nabla_Y X = \frac{\partial}{\partial y} - 0 = \left[\frac{\partial}{\partial x}, x\frac{\partial}{\partial y}\right].$$

If $\gamma(t)$ is an immersed curve in M then it has velocity vector $V = vT$ and squared geodesic curvature

$$k^2 = \|\nabla_T T\|^2.$$

The Frenet equations for a curve are

$$\gamma' = vT, \quad \nabla_T T = kN, \quad \nabla_T N = -kT + \tau B, \quad \nabla_T B = -\tau N. \tag{6}$$

For a family of curves $\gamma_w(t) = \gamma(w,t)$ we will write

$$W = W(w,t) = \frac{\partial \gamma}{\partial w},$$

$$V(w,t) = \frac{\partial \gamma}{\partial t} = v(w,t)T(w,t).$$

So V is velocity and $v = \frac{ds}{dt}$ is speed, and W represents an infinitesimal variation of the curve. s is the arclength parameter along a curve.

The basic formulas needed in calculating the Euler equations are as follows:

$$0 = [W,V] = [W,vT] = W(v)T + v[W,T]. \tag{7}$$

So $[W,T] = -\frac{W(v)}{v}T = gT$.

$$2vW(v) = W(v^2) = 2\langle \nabla_W V, V \rangle = 2\langle \nabla_V W, V \rangle = 2v^2 \langle \nabla_T W, T \rangle. \tag{8}$$

So $W(v) = -gv$, $g = -\langle \nabla_T W, T \rangle$.

$$W(k^2) = 2\langle \nabla_T \nabla_T W, \nabla_T T \rangle + 4gk^2 + 2\langle R(W,T)T, \nabla_T T \rangle. \tag{9}$$

Here the curvature tensor R is given by

$$R(X,Y)Z = \nabla_X \nabla_Y Z - \nabla_Y \nabla_X Z - \nabla_{[X,Y]}Z.$$

Proof of (9):

$$\begin{aligned}
W(k^2) &= 2\langle \nabla_W \nabla_T T, \nabla_T T \rangle \\
&= 2\langle \nabla_T \nabla_W T + \nabla_{[W,T]}T + R(W,T)T, \nabla_T T \rangle \\
&= 2\langle \nabla_T \nabla_T W + \nabla_T(gT) + \nabla_{gT}T \\
&\quad + R(W,T)T, \nabla_T T \rangle \\
&= 2\langle \nabla_T \nabla_T W, \nabla_T T \rangle + 2\langle R(W,T)T, \nabla_T T \rangle \\
&\quad + 4g\langle \nabla_T T, \nabla_T T \rangle.
\end{aligned}$$

In what follows, $\gamma : [0,1] \to M$ is a curve of length L. Now for fixed constant λ let

$$\mathscr{F}^\lambda(\gamma) = \frac{1}{2}\int_0^L k^2 + \lambda \, ds = \frac{1}{2}\left(\int_0^L k^2 ds + \lambda L\right)$$

$$= \frac{1}{2}\int_0^1 (\|\nabla_T T\|^2 + \lambda)v(t)dt.$$

13

For a variation γ_w with variation field W we compute

$$
\begin{aligned}
\frac{d}{dw}\mathscr{F}^{\lambda}(\gamma_w) &= \frac{1}{2}\int_0^1 W(k^2)v + (k^2+\lambda)W(v)\,dt \\
&= \frac{1}{2}\int_0^1 W(k^2) - (k^2+\lambda)g\,ds \\
&= \int_0^1 \langle \nabla_T\nabla_T W, \nabla_T T\rangle + 2gk^2 \\
&\quad + \langle R(W,T)T, \nabla_T T\rangle - \tfrac{1}{2}(k^2+\lambda)g\,ds.
\end{aligned}
$$

One of the symmetries of the curvature tensor allows us to replace $\langle R(W,T)T, \nabla_T T\rangle$ with $\langle R(\nabla_T T, T)T, W\rangle$. Now integrate by parts, using $g = -\langle \nabla_T W, T\rangle$

$$
\begin{aligned}
\frac{d}{dw}\mathscr{F}^{\lambda}(\gamma_w) &= \int_0^1 \langle \nabla_T\nabla_T W, \nabla_T T\rangle - \langle \nabla_T W, 2k^2 T\rangle \\
&\quad + \langle R(\nabla_T T, T)T, W\rangle + \frac{1}{2}\langle \nabla_T W, (k^2+\lambda)T\rangle\,ds \\
&= \int_0^L \langle E, W\rangle\,ds \\
&\quad + \left[\langle \nabla_T W, \nabla_T T\rangle + \langle W, -(\nabla_T)^2 T + \Lambda T\rangle\right]_0^L,
\end{aligned}
$$

where

$$
E = (\nabla_T)^3 T - \nabla_T(\Lambda T) + R(\nabla_T T, T)T,
$$

and

$$
\Lambda = \frac{\lambda - 3k^2}{2}.
$$

When M is a manifold of constant sectional curvature G, the formula for E can be simplified to

$$
E = (\nabla_T)^3 T - \nabla_T(\Lambda_G T),
$$

where

$$
\Lambda_G = \frac{\lambda - 2G - 3k^2}{2}.
$$

Now by using the Frenet equations (6) we compute

$$
\begin{aligned}
E &= \nabla_T\left(\nabla_T kN - \frac{\lambda - 2G - 3k^2}{2}T\right) \\
&= \nabla_T\left(\frac{k^2 - \lambda + 2G}{2}T + k_s N + k\tau B\right) \\
&= \frac{2k_{ss} + k^3 - \lambda k + 2Gk - k\tau^2}{2}N + (2k_s\tau + k\tau_s)B.
\end{aligned}
$$

The equations $E = 0$ for the elastica become:

$$2k_{ss} + k^3 - \lambda k + 2Gk - k\tau^2 = 0, \tag{10}$$

and

$$2k_s\tau + k\tau_s = 0. \tag{11}$$

The second equation integrates to

$$k^2\tau = c.$$

Eliminating τ from the first equation and integrating:

$$k_s^2 + \frac{k^4}{4} + (G - \frac{\lambda}{2})k^2 + \frac{c^2}{k^2} = A.$$

Letting $u = k^2$ this becomes

$$u_s^2 + u^3 + 4(G - \frac{\lambda}{2})u^2 - 4Au + 4c^2 = 0.$$

This has the following solutions:

1. $u = k^2$=constant, τ=constant   ("helices" and circles)
2. $k = k_0 \operatorname{sech} \left(\frac{k_0}{2w}s \right), \tau = 0$ ("borderline elastica")
3. $k = k_0 \operatorname{cn} \left(\frac{k_0}{2w}s, p \right), \tau = 0$ ("orbitlike elastica")
4. $k = k_0 \operatorname{dn} \left(\frac{k_0}{2w}s, p \right), \tau = 0$ ("wavelike elastica")
5. $k^2 = k_0^2 \left(1 - \frac{p^2}{w^2} \operatorname{sn}^2 \left(\frac{k_0}{2w}s, p \right) \right)$

where $4G - 2\lambda = \frac{k_0^2(1+p^2-3w^2)}{w^2}$ and $0 \le p \le w \le 1$.

2.1. Example: Closed elastic curves on the 2-sphere

The two-sphere $\mathbb{S}^2$ with constant Gauss curvature G provides an interesting and re-markably complex illustration of the theory of elastic curves.[1] Obviously the compact-ness of the sphere insures that there will be a more interesting collection of closed elastic curves than the plane, but also fact that there are no dilations suggests that the structure is richer. To determine the conditions for closedness, we again rely on the notion of Killing fields.

Proposition 2 *Let M be a (simply-connected) manifold with constant sectional curva-ture G, and let γ be an elastica in M. Then the vector fields $J = \frac{k^2-\lambda}{2}T + k_sN + k\tau B$ and $I = kB$ along γ extend to Killing fields (infinitesimal isometries) on M.*

---

[1] For proofs of results in this subsection, see [21]

Idea of proof: Verify that when $W = I$ or $W = J$, then W preserves arclength parameter, curvature, and torsion of γ. Since the isometry group of the sphere is three-dimensional, a dimension count shows that vector fields which satisfy these conditions must be the restrictions of Killing fields.

For arclength, one checks that $\langle \nabla_T W, T \rangle = 0$. For curvature, use the formula (9) for $W(k)$. For torsion, use the formula:

$$W(\tau^2) = 2\left\langle \frac{1}{k}(\nabla_T)^3 W - \frac{k_s}{k^2}(\nabla_T)^2 W + \left(\frac{G}{k} + k\right)\nabla_T W - \frac{k_s}{k^2} GW, \tau B \right\rangle.$$

In the two-sphere, the Killing field $J = \frac{k^2 - \lambda}{2} T + k_s N$ must be the restriction to γ of a rotation field. By choosing coordinates x, y of longitude and latitude on the sphere, we may assume that

$$\frac{\partial}{\partial x} = aJ,$$

where a is a constant chosen so that J has unit length on the equator. Since

$$\|J\|^2 = \frac{(k^2 - \lambda)^2}{4} + k_s^2 = A + \frac{\lambda^2}{4} - Gk^2,$$

it is clear that the norm of J is maximized where k^2 is minimized. So if $k = k_0 \operatorname{cn}\left(\frac{k_0}{2w} s, p\right)$, then k vanishes at the maxima of $\|J\|$. Since $\langle N, J \rangle = k_s \neq 0$ when $k = 0$, the curve γ is transverse to the coordinate curves $y = \text{const.}$ at these points. It follows that the curve is crossing the equatorial curve $y = 0$ at the inflection points. The normalizing constant a is precisely $\sqrt{A + \frac{\lambda^2}{4}}$.

Theorem 3 *If γ is a wavelike elastica on a two-dimensional space-form, then the inflection points of γ all lie on a geodesic (the "axis" of the curve).*

A wavelike elastic curve on $\mathbb{S}^2$ is now seen to oscillate across a great circle. Define the *wavelength* Λ of a wavelike elastica to be the amount of progress it makes along its axis in one complete period of k, as measured by arclength along the geodesic. If $\frac{\Lambda}{\pi\sqrt{G}}$ is rational, then the elastica will close up. To determine the closed elastic curves, then, it is necessary to study the dependence of Λ on parameters.

When the parameter λ is fixed at a value less than $2G$, the non-geodesic elastic curves are wavelike, with curvature given by

$$k(s) = k_0 \operatorname{cn}(\frac{k_0 s}{2p}, p),$$

where the maximum curvature is given by

$$k_0 = \frac{p\sqrt{4G - 2\lambda}}{\sqrt{1 - 2p^2}} \quad (0 \leq p^2 < \frac{1}{2}).$$

16

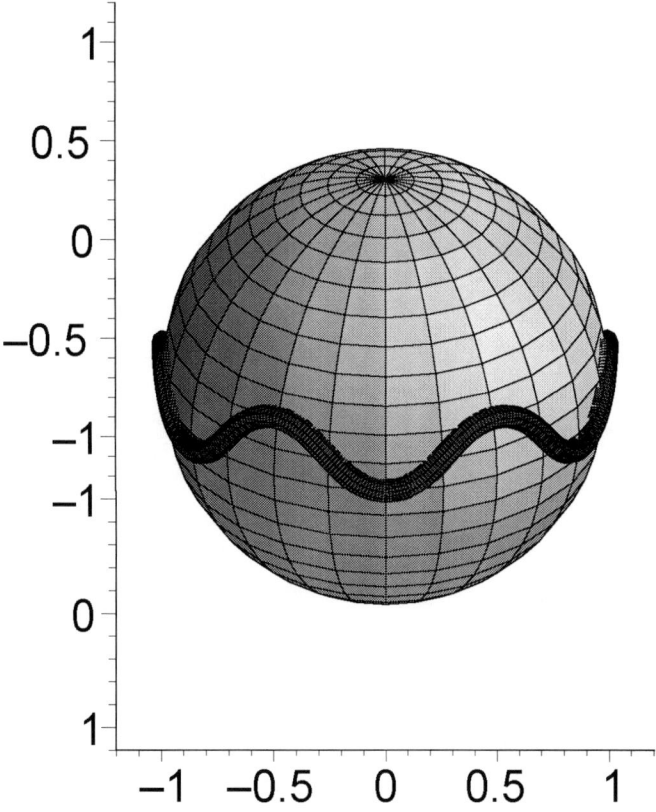

FIGURE 6. A wavelike elastica in $\mathbb{S}^2$

The significance of the quantity $4G - 2\lambda$ can (partly) be accounted for as follows: By analyzing the *second variation* formula along a closed geodesic, it is possible to determine that the $n - fold$ geodesic is a stable critical point of $\mathscr{F}^\lambda$ if and only if

$$\frac{2n-1}{n^2} \geq \frac{\lambda}{2G}.$$

For large λ the length penalty causes the closed geodesic to cease to be a minimum. Indeed, when $\lambda > 2$ the small circle of curvature $\sqrt{\lambda - 2G}$ is a stable equilibrium. At these values, *orbitlike* elastic curves appear, whose curvature is nonvanishing.

Now suppose λ is some fixed positive constant less than $2G$. As a specific example, consider $\lambda = 1$. For this value, the equator, the double-covered equator, and the *triple-covered* equator are stable equilibria of $\mathscr{F}^\lambda$. As we will see (in lecture 5), the *minimax principle* is valid in this situation. Therefore, since the single and triple equators are regularly homotopic (that is, it is possible to smoothly deform the single circle to the triple circle through regular curves), we expect to find another elastica, a minimax

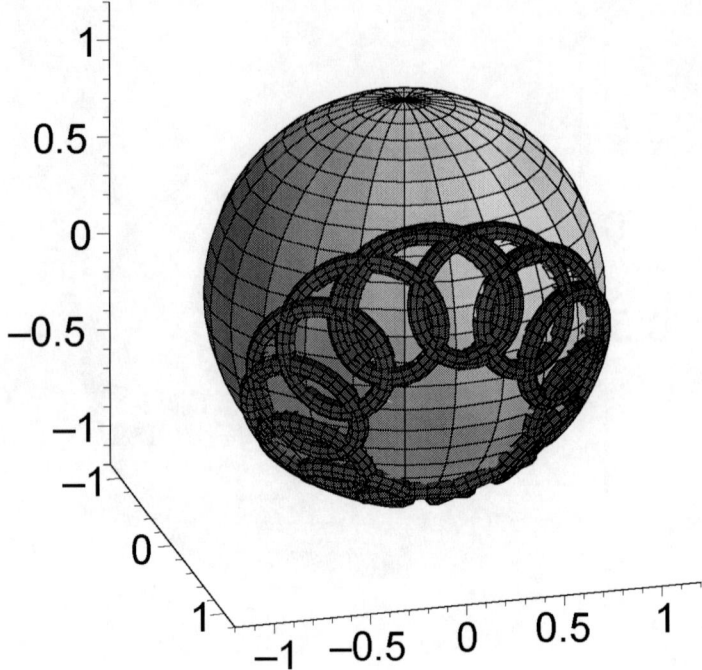

FIGURE 7. An orbitlike elastica in $\mathbb{S}^2$

critical point. In fact, using the fact that the regular homotopy can be chosen to preserve a four-fold symmetry, we expect to find an elastica whose wavelength is exactly $\pi\sqrt{G}$, which looks something like the seam on a tennis ball. And indeed, such an elastica does exist.

The full story turns out to be surprisingly subtle. The existence and uniqueness of such minimax critical points turns out to hold for $0 \leq \lambda \leq \frac{8G}{7}$.

Theorem 4 (L-S, 1987 [21]) *Let λ be a fixed constant with $0 \leq \frac{8}{7}G$. Then for each pair of positive integers m,n with*

$$\frac{m}{n} < 1 - \frac{\sqrt{G}}{\sqrt{4G-2\lambda}},$$

there is a unique elastica $\gamma_{m,n}^{\lambda}$ (up to congruence) which closes up in n periods while traversing the equator m times. All such curves are unstable (minimax) critical points for $\mathscr{F}^{\lambda}$.

18

3. THE KIRCHHOFF ELASTIC ROD

In the first two lectures we have considered an essentially one-dimensional object, a wire of negligible thickness. In this lecture, we will expand our horizon to include some consideration of the effect of thickness on elastic energy. We consider a thin elastic rod with circular cross-section and uniform density – the uniform symmetric (linear) Kirchhoff rod.[2] For a thorough treatment of the elastic theory of rods, see [1]. The configurations of the rod are described abstractly using *adapted framed curves*:

$$\Gamma = \{\gamma(s); T, M_1, M_2\},$$

where the *centerline* of the rod $\gamma(s)$ is a unit speed curve in $\mathbb{R}^3$, and the *material frame* $(T(s), M_1(s), M_2(s))$ is a positively oriented orthonormal frame. These are related by the condition:

$$\gamma'(s) = T(s).$$

We say that the frame is *adapted* to the curve.

The rotation of the material frame can be described by the *Darboux vector*

$$\Omega = mT - m_2 M_1 + m_1 M_2,$$

and the equations

$$
\begin{array}{rclcll}
T' & = & \Omega \times T & = & & m_1 M_1 \quad + m_2 M_2, \\
M_1' & = & \Omega \times M_1 & = & -m_1 T & \quad\quad\quad + m M_2, \\
M_2' & = & \Omega \times M_2 & = & -m_2 T & - m M_1.
\end{array}
$$

The *Total Elastic Energy* of a framed curve is given by

$$E(\Gamma) = \frac{1}{2} \int \underbrace{\alpha_1 (m_1)^2 + \alpha_2 (m_2)^2}_{bending} + \underbrace{\beta m^2}_{twisting} \, ds,$$

where α_1, α_2 and β are material constants.

We will be concerned with the *symmetric* case:

$$\alpha = \alpha_1 = \alpha_2.$$

In this case, the bending energy is $\frac{\alpha}{2} \int k^2 \, ds$.

To understand the behavior of the rod, we need to define another frame, called the *inertial frame* along a rod by the equations:

$$
\begin{array}{rcll}
T' & = & & k_2 U \quad + k_3 V, \\
U' & = & -k_2 T, \\
V' & = & -k_3 T.
\end{array}
$$

The inertial frame has no *twist*, i.e., U and V "have no T-component to their angular velocity." That is, the Darboux vector is $\Omega = -k_3 U + k_2 V$. The quantities k_2 and k_3 are

---

[2] [23] is the main reference for this section

19

the *natural curvatures* of γ. Note that they are related to the geodesic curvature by the equation

$$k^2 = k_2^2 + k_3^2.$$

To more easily understand the inertial frame, define an angle ϕ to be the angle between N and U. That is, write

$$\begin{aligned}
N &= U\cos\phi + V\sin\phi, \\
B &= -U\sin\phi + V\cos\phi.
\end{aligned}$$

Differentiating the second equation

$$-\tau N = B' = \phi'(-U\cos\phi - V\sin\phi) + (k_2\sin\phi - k_3\cos\phi)T.$$

Comparing, we derive the relationships:

$$\phi' = \tau, \quad k_2 = k\cos\phi, \quad k_3 = k\sin\phi.$$

The natural frame is uniquely determined once its initial value is specified; if we replace ϕ by $\phi+$ const. we get an equivalent frame. Unlike the Frenet frame, the inertial frame only depends on two derivatives of the curve. Since the curvature and torsion are determined by the natural curvatures, the curve itself is uniquely determined up to congruence by its curvatures. The natural frame has the further technical advantage that we need not require k to be nonvanishing.

Now to relate the natural frame to the material frame, assume $U(0) = M_1(0)$. Then we can measure the twisting of the rod by looking at the angle $\theta(s)$ between $M_1(s)$ and $U(s)$.

Write $M_1 = U\cos\theta + V\sin\theta$. Then

$$m = M_1' \cdot M_2 = \theta'(-U\sin\theta + V\cos\theta) \cdot M_2.$$

That is,

$$m = \theta'.$$

We can describe the rod by $\{\gamma, k_1, k_2, \theta\}$. The energy is

$$E = \frac{1}{2}\int \alpha k^2 + \beta(\theta')^2 \, ds.$$

An elastic rod is an equilibrium configuration for the energy with appropriate boundary conditions (usually, having each end fixed in position and clamped.) Note that we can change the rod without changing its centerline by changing θ. We can fix the ends of the rod and its centerline and reduce the energy by minimizing the second term. Thus we have

Proposition 5 *For a rod in equilibrium, θ' is constant. Thus $\theta(s) = ms$, for some fixed constant m.*

If instead of clamping the ends, we held them in collars (so the ends could not change direction but were free to twist), then the energy would be reduced by untwisting until $m = 0$. This shows that an untwisted rod is a minimizer of bending energy – an elastica.

Now assume the rod is closed of length L, and (for convenience) that the material frame M satisfies $M(0) = M(L)$. That is, we take a rod, 'twist' it n times and weld the ends together. The Frenet frame automatically closes up, but the natural frame need not. (More generally, we could assume the angle between $M(0)$ and $M(L)$ is a fixed value. This would lead to a similar conclusion, except that n would not be an integer.)

Let $\psi = \phi - \theta$ be the angle between the material frame and the Frenet frame. Then our assumption is that $\frac{\psi(L) - \psi(0)}{2\pi}$ is an integer. This leads to:

$$2\pi n = \psi(L) - \psi(0) = \int_0^L \phi'(s) - \theta'(s)ds$$
$$= \int_0^L \tau(s)ds - mL.$$

The previous calculation says that an elastic rod centerline has total torsion $\int_0^L \tau(s)ds$ given by the quantity $2\pi n + mL$. Using this, it is possible to formulate a variational problem whose solutions are exactly the elastic rod centerlines.

Theorem 6 *For a curve $\gamma(s)$ define*

$$\mathscr{F}(\gamma) = \lambda_1 \int_\gamma ds + \lambda_2 \int_\gamma \tau ds + \lambda_3 \int_\gamma k^2 ds,$$

with k and τ the curvature and torsion and $\lambda_3 \neq 0$. Then an extremal of $\mathscr{F}$ is an elastic rod centerline.

Clearly, when $\lambda_2 = 0$, this is an elastic curve.

For a Kirchhoff elastic rod centerline γ, there are two Killing fields: for constants α and σ:

$$J = \frac{\alpha k^2 - \lambda}{2}T + \alpha k'N + k(\alpha\tau - \sigma)B,$$

and

$$I = \sigma T + \alpha k B.$$

Theorem 7 *For rod centerline γ there is an 'associated' elastic curve γ_0 whose curvature is k and torsion is $\tau - \frac{\sigma}{2\alpha}$, where k and τ are the curvature and torsion of γ.*

Recall that for elastic curves, there are embedded nonplanar elastic curves, each lying on an embedded torus of revolution, representing an (m,n) - torus knot, one for each $m > 2n$. In the case of rods, there is an extra parameter, allowing for many more closed curves. For rods, the result is:

Theorem 8 (Ivey - Singer, 1998 [12]) *Every torus knot type is realized by a smooth closed elastic rod centerline. For any pair of relatively prime positive integers k and* *n there is a one-parameter family of closed elastic rods forming a regular homotopy between the k-times covered circle and the n-times covered circle. The family includes exactly one elastic curve, one self-intersecting elastic rod, and one closed elastic rod with constant torsion.*

4. HAMILTONIAN THEORY

In this lecture, we will investigate the remarkable fact that the equations of the elastica in Euclidean space, and indeed, in space forms, can be completely integrated. The key to understanding this is through the Hamiltonian approach to the variational problems of curves and rods. What follows is a brief survey of the ideas in geometric variational problems, using the machinery of Optimal Control Theory.[3] A different approach to this type of variational problem, using the theory of exterior differential systems, is developed in [5].

We begin with an n-dimensional smooth manifold E, called *Configuration Space*. Let $H : T^*E \to \mathbb{R}$ be a smooth function on the cotangent bundle (or *Phase Space*); H is the *Hamiltonian*.

In canonical coordinates $(p^1, \ldots, p^n, q^1, \ldots, q^n)$ H defines a time-independent Hamiltonian system:

$$\dot{q}^i = \frac{\partial H}{\partial p^i}, \quad \dot{p}^i = -\frac{\partial H}{\partial q^i}.$$

However, for certain problems, such as those considered here, a different type of coordinate will be more useful.

If $F, G : T^*E \to \mathbb{R}$, then the *Poisson bracket* is

$$\{F, G\} = \sum_{1}^{n} \frac{\partial F}{\partial q^i} \frac{\partial G}{\partial p^i} - \frac{\partial F}{\partial p^i} \frac{\partial G}{\partial q^i}.$$

$\{\bullet, H\}$ acts like a derivative:

$$\{FG, H\} = F\{G, H\} + G\{F, H\}.$$

In terms of the Poisson bracket, Hamilton's equations can be written:

$$\dot{q}^i = \{q^i, H\}, \quad \dot{p}^i = \{p^i, H\}. \tag{12}$$

If $\gamma(t) = (p^i(t), q^i(t))$ is a solution curve of equation (12), then we can differentiate any quantity along γ by $\frac{dF}{dt} = \{F, H\}$. In particular, if $\{F, H\} = 0$, then F is a conserved quantity along the integral curves of (12), or a *first integral* of H.

H is *Liouville integrable* if there are functions $F_1, \ldots, F_n$ with all $\{F_i, F_j\} = 0, F_n = H$. The F_i are n constants of motion in involution. The Liouville-Arnol'd theorem says that the trajectories of H can be found by quadratures.

---

[3] The book [14] by Velimir Jurdjevic is an excellent reference for this lecture. See [13] and [22] for details.

Now let $M = \mathbb{R}^3, \mathbb{S}^3$, or $\mathbb{H}^3$. Let $E = FM$ be the space of positively-oriented orthonormal frames $f = (X; f_1, f_2, f_3)$ on M. It is helpful to think of f as a linear map from $\mathbb{R}^3$ to the tangent space at X, taking the standard orthonormal basis (e_1, e_2, e_3) to the orthonormal vectors (f_1, f_2, f_3), where $f_3 = f_1 \times f_2$.

So configuration space E is a 6-dimensional manifold.

The group $SO(3)$ of rotations acts on E on the *right* by rotating frames: If $g : \mathbb{R}^3 \to \mathbb{R}^3$ is a rotation, then $(f, g) \mapsto f \circ g = R_g(f)$. The space E is the total space of a principal fiber bundle over M whose fiber is $SO(3)$.

$$E \longleftarrow SO(3)$$
$$\downarrow \pi$$
$$M$$

The standard basis for the Lie Algebra $\mathfrak{so}(3)$ determines three *fundamental vector fields* on E, as follows:

$$\text{Let } \alpha_1 = \begin{pmatrix} 0 & 0 & 0 \\ 0 & 0 & 1 \\ 0 & -1 & 0 \end{pmatrix}, \quad \alpha_2 = \begin{pmatrix} 0 & 1 & 0 \\ -1 & 0 & 0 \\ 0 & 0 & 0 \end{pmatrix}, \quad \alpha_3 = \begin{pmatrix} 0 & 0 & -1 \\ 0 & 0 & 0 \\ 1 & 0 & 0 \end{pmatrix}.$$

Then $[\alpha_1, \alpha_2] = \alpha_3$, $[\alpha_2, \alpha_3] = \alpha_1$, and $[\alpha_3, \alpha_1] = \alpha_2$.

Each α_i gives rise to a one-parameter subgroup of $SO(3)$ and by the right action a one-parameter group of diffeomorphisms of E with the vector field A_i as infinitesimal generator.

The vectors $A_1(f), A_2(f), A_3(f)$ span the *vertical* subspace $\mathbf{V}(f)$ of the tangent space at each point f of E. That is, $\pi_*(A_i(f)) = 0$, and the vectors are linearly independent at each point f.

Note: we may identify E with the Lie Group $\mathscr{G}$ of isometries of M. From that point of view, A_i becomes a *left invariant* vector field.

M	$E \equiv \mathscr{G}$	Matrix description
$\mathbb{R}^3$	$E(3)$	$\begin{pmatrix} 1 & 0 \\ v & R \end{pmatrix}, v \in \mathbb{R}^3, R \in SO(3)$
$\mathbb{S}^3$	$SO(4)$	$A^T A = I = \begin{pmatrix} 1 & 0 & 0 & 0 \\ 0 & 1 & 0 & 0 \\ 0 & 0 & 1 & 0 \\ 0 & 0 & 0 & 1 \end{pmatrix}$
$\mathbb{H}^3$	$SO(3,1)$	$A^T J A = J, J = \begin{pmatrix} -1 & 0 & 0 & 0 \\ 0 & 1 & 0 & 0 \\ 0 & 0 & 1 & 0 \\ 0 & 0 & 0 & 1 \end{pmatrix}$

The Riemannian metric defines the horizontal subspace $\mathbf{H}(f)$ of the tangent space at f. Roughly speaking, given any curve in M passing through $X = \pi(f)$, there is a unique lift of the curve to E achieved by parallel transport of the frame f along the curve. The

23

tangent vector at f to this lift is horizontal. A *basic vector field* $B(\xi)$ is a horizontal field (using the Riemannian connection) such that $\pi_*(f)(B(\xi)) = f(\xi)$, where ξ is any vector in $\mathbb{R}^3$. In particular, let $B_i = B(e_i)$.

It can be shown that $B(\xi)$ satisfies the equivariance property: $(R_g)_* B(\xi) = B(g^{-1}(\xi))$. (See [15], Chapter 2.) Again, viewing E as a Lie group $\mathscr{G}$, the fields B_i are left invariant vector fields.

More generally, if $\mathscr{G}$ is the isometry group of a Riemannian manifold M, then $\mathscr{G}$ acts on the space $\mathscr{E}$ of orthonormal frames on the *left*. If $\mathscr{I}$ is an isometry, then $d\mathscr{I}(x) : M_x \to M_x$ is an isometry of the tangent space at x, and takes frame f to $d\mathscr{I}(x) \circ f$.

$$\mathbb{R}^3 \xrightarrow{g} \mathbb{R}^3 \xrightarrow{f} M_x \xrightarrow{d\mathscr{I}(x)} M_x.$$

This diagram shows how the isometries of M act (on the left) and the rotation group $SO(3)$ acts (on the right). The two actions commute.

Putting together the action of $SO(3)$ on the right and the action of the isometry group $\mathscr{G}$ on the left:

$$\mathscr{G} \times E \times SO(3) \longrightarrow E, \quad (\mathscr{I}, f, g) \longmapsto d\mathscr{I} \circ f \circ g.$$

We see a nine-dimensional group $\mathscr{G} \times SO(3)$ acting on E, so there are lots of chances to reduce equations using symmetry.

The vector fields $A_1, A_2, A_3, B_1, B_2, B_3$ satisfy the Lie bracket formulas:

$$[A_i, A_j] = \varepsilon_{ijk} A_k, \tag{13}$$

$$[A_i, B_j] = \varepsilon_{ijk} B_k, \tag{14}$$

$$[B_i, B_j] = \varepsilon_{ijk} G A_k, \tag{15}$$

where $\varepsilon_{ijk} = \pm 1$ depending on the sign of the permutation of $\{1, 2, 3\}$ and is 0 if two are equal. Formulas (13) and (14) hold on *any* 3 - manifold. Formula (15) holds for space forms of curvature G.

If V is a vector field on E, then the Hamiltonian $\mathscr{H}_V : T^*E \to \mathbb{R}$ is defined by $\mathscr{H}_V(p) = p(V)$, p any covector. This defines six *linear Hamiltonians* $\mathscr{A}_i, \mathscr{B}_i$ from A_i, B_i. The functions $\mathscr{A}_1, \mathscr{A}_2, \mathscr{A}_3, \mathscr{B}_1, \mathscr{B}_2, \mathscr{B}_3$ are the generators of the algebra of left-invariant functions $\mathscr{L}\mathscr{G}$.

To compute Poisson brackets in this algebra, it suffices to compute brackets of the generators. For this we use the formula: $\{\mathscr{H}_V, \mathscr{H}_W\} = -\mathscr{H}_{[V,W]}$

In particular:

$$\{\mathscr{A}_i, \mathscr{A}_j\} = -\varepsilon_{ijk} \mathscr{A}_k,$$

$$\{\mathscr{A}_i, \mathscr{B}_j\} = -\varepsilon_{ijk} \mathscr{B}_k,$$

$$\{\mathscr{B}_i, \mathscr{B}_j\} = -\varepsilon_{ijk} G \mathscr{A}_k.$$

An element of the algebra is a function $P(\mathscr{A}_1, \mathscr{A}_2, \mathscr{A}_3, \mathscr{B}_1, \mathscr{B}_2, \mathscr{B}_3)$ of six variables. 'Geometric' variational problems on curves give rise to left-invariant Hamiltonian systems.

The (generalized) Frenet equations for a framed curve are

$$\gamma'(t) = T \qquad \begin{aligned} T' &= & & k_2 U & +k_3 V, \\ U' &= & -k_2 T & & +k_1 V, \\ V' &= & -k_3 T & -k_1 U. \end{aligned}$$

The usual Frenet equations correspond to $k_1 = \tau, k_2 = k, k_3 = 0$. If instead $k_1 = 0$, then we have the natural or inertial frame.

If $f(t) = (\gamma(t); T, U, V)$ is a curve in E, then the Frenet equations become:

$$\frac{df}{dt} = B_1(f) + k_1 A_1(f) - k_3 A_2(f) + k_2 A_3(f). \tag{16}$$

This defines a control system: $k_i(t)$ are controls; given $f(0)$ we get a unique framed curve satisfying (16). Then we may seek controls satisfying the condition that the "cost"

$$c = \int \mathscr{L}(k_1, k_2, k_3) ds$$

is minimal.

Examples:

1. **Elastic curves** Using the standard Frenet frame ($k_3 = 0$), the cost functional is $\frac{1}{2} \int k_2^2 ds = \frac{1}{2} \int k^2 ds$. If instead we use the inertial frame ($k_1 = 0$), the cost functional is $\frac{1}{2} \int k_2^2 + k_3^2 ds = \frac{1}{2} \int k^2 ds$. This holds since $\nabla_T T = k_2 U + k_3 V = kN$.
2. **Kirkhoff rods** Using the general frame, the cost function is $\frac{1}{2} \int \alpha(k_3^2 + k_2^2) + \beta k_1^2 ds$
3. **τ-elastic rods** Consider the variational problem of minimizing total squared curvature for curves of fixed constant torsion $\tau = c$. Using the standard Frenet frame, the cost functional is $\frac{1}{2} \int k_2^2 ds = \frac{1}{2} \int k^2 ds$.

Given the control system and cost functional, one can apply the Pontrjagin Maximum Principle to produce a left-invariant Hamiltonian system on T^*E whose trajectories project to solutions of the optimal control problem, as follows:

1. Lift (16) to get a time-dependent Hamiltonian on T^*E (depending on control(s)) and subtract[4] the cost functional $\mathscr{L}$. This gives

$$\begin{aligned} \mathscr{H}(p; k_i) &= \mathscr{B}_1(f) + k_1 \mathscr{A}_1(f) - k_3 \mathscr{A}_2(f) \\ &\quad + k_2 \mathscr{A}_3(f) - \mathscr{L}(k_1, k_2, k_3). \end{aligned}$$

2. Maximize $\mathscr{H}$ with respect to choice of controls $k_i(t)$ (for each fixed t.) This is done by solving $\frac{\partial \mathscr{H}}{\partial k_i}$ for controls and eliminating them.
   The result is then a time-independent Hamiltonian.

We illustrate this with the example of the elastic rod.

$$\mathscr{H}(p; k_1, k_2, k_3) = \mathscr{B}_1 + k_1 \mathscr{A}_1 - k_3 \mathscr{A}_2 + k_2 \mathscr{A}_3 - \frac{\alpha}{2}(k_2^2 + k_3^2) - \frac{\beta}{2} k_1^2.$$

---

[4] This is a simplified description!

25

$$\frac{\partial \mathcal{H}}{\partial k_1} = 0 = \mathcal{A}_1 - \beta k_1,$$

$$\frac{\partial \mathcal{H}}{\partial k_2} = 0 = \mathcal{A}_3 - \alpha k_2,$$

$$\frac{\partial \mathcal{H}}{\partial k_3} = 0 = \mathcal{A}_2 - \alpha k_3.$$

Eliminating the controls gives the Hamiltonian of the Kirchhoff elastic rod:

$$\mathcal{H} = \mathcal{B}_1 + \frac{\mathcal{A}_2^2 + \mathcal{A}_3^2}{2\alpha} + \frac{\mathcal{A}_1^2}{2\beta}.$$

Now we can establish the Liouville-Arnol'd Integrability of the (symmetric) Kirchhoff rod problem. An important tool for this is the existence of two Casimirs for the algebra $\mathcal{LG}$.

The quadratic Hamiltonians

$$\mathcal{P} = \mathcal{A}_1 \mathcal{B}_1 + \mathcal{A}_2 \mathcal{B}_2 + \mathcal{A}_3 \mathcal{B}_3,$$

$$\mathcal{Q} = \mathcal{B}_1^2 + \mathcal{B}_2^2 + \mathcal{B}_3^2 + G(\mathcal{A}_1^2 + \mathcal{A}_2^2 + \mathcal{A}_3^2),$$

are in the center of $\mathcal{LG}$. That is, $\{\mathcal{P}, \mathcal{H}\} = 0$ and $\{\mathcal{Q}, \mathcal{H}\} = 0$ for all $\mathcal{H}$ in $\mathcal{LG}$. This can easily be verified: we only need check on the generators $\mathcal{A}_i$ and $\mathcal{B}_i$ because of the product rule.

Let $\mathcal{RG}$ be the algebra of *right*-invariant Hamiltonians; it is generated by the lifts of right-invariant vector fields. If $\mathcal{H} \in \mathcal{LG}$ and $\mathcal{K} \in \mathcal{RG}$, then $\{\mathcal{H}, \mathcal{K}\} = 0$. (This is because left and right actions commute, so the vector fields commute).

So if $\mathcal{H} \in \mathcal{LG}$, then by choosing $\mathcal{R}_1$ and $\mathcal{R}_2$ to be linear right-invariant Hamiltonians with $\{\mathcal{R}_1, \mathcal{R}_2\} = 0$ [which can be done in any of the three space-forms] we have five independent Hamiltonians in involution:

$$\mathcal{H}, \mathcal{P}, \mathcal{Q}, \mathcal{R}_1, \text{ and } \mathcal{R}_2.$$

To prove a given $\mathcal{H}$ is integrable, we need one more integral.

In our example, one can verify that $\mathcal{H} = \mathcal{B}_1 + \frac{\mathcal{A}_2^2 + \mathcal{A}_3^2}{2\alpha} + \frac{\mathcal{A}_1^2}{2\beta}$ commutes with $\mathcal{A}_1$. The other Hamiltonians also commute with it automatically. Thus the Kirchhoff elastic rod is Liouville integrable.

Next consider the elastic curve. The usual Frenet frame does not work so well here, because the optimal control problem is *singular*. The equations can not be solved to eliminate the controls. Instead, we use the inertial frame. The resulting Hamiltonian is

$$\mathcal{H} = \mathcal{B}_1 + \frac{\mathcal{A}_2^2 + \mathcal{A}_3^2}{2}.$$

Again, this commutes with $\mathcal{A}_1$, so the Euler elastica is integrable.

26

For the τ-elastic curve, there is only one control, so we can use the Frenet frame and eliminate the control. The resulting Hamiltonian is

$$\mathcal{H} = \mathcal{B}_1 + c\mathcal{A}_1 + \frac{\mathcal{A}_3^2}{2}.$$

In this example it can be checked that $\mathcal{C} = \mathcal{B}_3 - c\mathcal{A}_3$ is a constant of motion *when the curvature $G = 1$ and $c = \pm 1$.*

One more example worth noting is given by $\mathcal{L}(k,\tau) = k^2\tau$. This leads to a non-polynomial example:

$$\mathcal{H} = \mathcal{B}_1 + \mathcal{A}_3\sqrt{\mathcal{A}_1}.$$

Let

$$\mathcal{C} = \mathcal{A}_1^2 + \mathcal{A}_2^2 + \mathcal{A}_3^2 - 4\mathcal{A}_1\sqrt{\mathcal{B}_3} - 4G\mathcal{A}_1.$$

Then one can check that $\{\mathcal{C}, \mathcal{H}\} = 0$; so this defines an integrable system.

For further examples of integrable geometric variational problems, see [22]. One intriguing *non-example* is the functional $\mathcal{L} = k^2 + \tau^2$. This does not appear to yield an integrable system.

Another source of geometric variational problems is found by replacing the orthonormal frame bundle with some other frame bundle. An interesting example of this is the notion of *affine* and *subaffine elastica*. The basic idea here is to replace Euclidean space with an affine space. (Good references for this kind of geometry include [28] and [33].) Given a curve, one can define a preferred frame by the condition that the volume be 1 at each point. In two dimensions, this is just the condition that $\det(\gamma, \gamma') \equiv 1$. From this it follows that $\gamma'' + \kappa\gamma = 0$ for some function κ, which is the *affine curvature*. An affine elastica is a critical point of $\int \kappa^2 d\sigma$, where σ is the (affine) arclength parameter. This turns out to be integrable; the related notion of subaffine elastica in three dimensions is also integrable. For details, see ([10], [11], and [31].)

It is beyond the scope of these lectures to discuss soliton theory for infinite-dimensional Hamiltonian systems. We will only mention here that the integrability of the elastic rod is related to the integrability of the partial differential equation for an evolving space curve $\gamma(s,t)$:

$$\frac{\partial \gamma}{\partial t} = \frac{\partial \gamma}{\partial s} \times \frac{\partial^2 \gamma}{\partial s^2}.$$

This is known as the Betchov - Da Rios equation, also called the localized induction equation (LIE). A completely integrable Hamiltonian system, (LIE) possesses infinitely many conserved quantities, all of them integrals involving the curvature and torsion of the curve. The first four such quantities are the arclength, the total torsion, the total squared curvature, and the quantity $\mathcal{L}(k,\tau) = k^2\tau$. Note that the elastic rod is a critical point of a linear combination of the first three of these, while the fourth is constant along such a rod. For further details, see, e.g., [9].

5. CURVE STRAIGHTENING

Suppose a thin springy wire is fashioned into a smooth closed loop by joining the ends together, and the wire is held in some fixed configuration in $\mathbb{R}^3$. Then in the

Bernoulli-Euler model the bending energy is proportional to the total squared curvature. If we now release the rod and it moves in such a way as to reduce its energy as efficiently as possible, it will want to follow the "negative gradient" of the energy. (This is not physically realistic, however, since it assumes Aristotelian rather than Newtonian dynamics; also, we will ignore the fact that the wire can not physically pass through itself.) This is the idea behind the *Curve Straightening Flow*, which we will now be considering.

In order to actually define such a flow, we need several ingredients. First, we need the space of allowable curves, within which this flow will be defined. Next we need a way of defining a gradient, namely a Hilbert Space structure. Finally, we need to know that the flow is actually defined; that is, that the partial differential equation governing the flow has short-time and long-time solutions.

Assuming such a flow exists, we could then hope to describe the critical points of the functional (that is, the elastic curves) as limit points of trajectories of the flow. In fact, it turns out that under suitable hypotheses such a gradient flow is very well-behaved. (See [29] for the difficult technical details.) This allows us to rigorously determine existence and stability or instability of solutions in a wide range of situations.

First we formulate the appropriate definition of our space of curves. Let M be a Riemannian manifold. Let

$$\Omega = \{\gamma : [0,1] \longrightarrow M \,|\, \|\gamma'\| \equiv \ell \neq 0, \gamma'' \in L^2\},$$

and let

$$\Lambda = \{\gamma \in \Omega \,|\, \gamma(0) = \gamma(1), \gamma'(0) = \gamma'(1)\}.$$

Λ is a Hilbert manifold: If M is Euclidean space we can define the norm by, for instance,

$$\|\gamma\|_2^2 = \gamma(0)^2 + \gamma'(0)^2 + \int_0^1 \|\gamma''\|^2 \, ds.$$

For a general Riemannian manifold, the definition is rather more complicated. See [21] for the technical details.

$\mathscr{F}^\lambda : \Lambda \longrightarrow \mathbb{R}$ is defined by

$$\mathscr{F}^\lambda(\gamma) = \frac{1}{2} \int_\gamma k^2 + \lambda \, ds.$$

More generally, a critical point of $\mathscr{F}^0$ with constrained length is a critical point of $\mathscr{F}^\lambda$ for some λ (Lagrange multiplier). λ may be thought of as a *length penalty*. Thus if $\lambda > 0$ arbitrarily long curves will have high $\mathscr{F}^\lambda$ values. Very short curves have high $\int k^2 ds$. So when λ is positive, the functional is bounded from below.

$\mathscr{F}^\lambda$ is a smooth function on a Hilbert manifold, so it defines a flow via the negative gradient, called Curve-Straightening. The key analytic result ([20], [21]) is:

Theorem 9 *If $\lambda > 0$, then $\mathscr{F}^\lambda$ satisfies the Palais-Smale condition (C). Therefore, the trajectories of $-\nabla \mathscr{F}^\lambda$ converge to critical points (or at least have critical points as adherence points). Furthermore the minimax principle and Morse theory hold for $\mathscr{F}^\lambda$.*

The theorem applies for M compact, and with modification for space forms. In the latter case, one can use congruence classes of curves to prevent trajectories from "escaping to infinity". The theorem also applies to spaces of curves of a fixed length.

Example: In $\mathbb{R}^2$, the only closed elastic curves are coverings of circles and figure-eight curves. There is precisely one critical point for $\mathscr{F}^\lambda$ ($\lambda > 0$) of rotation index $n \neq 0$; curve-straightening takes any closed curve of rotation index n to the n - fold circle. This demonstrates the Whitney-Graustein theorem, which asserts that any two closed curves of the same rotation index are regularly homotopic. In rotation index 0, the n-fold coverings of the figure eight curve are all critical points. Note that in this case Morse theory predicts multiple critical points; the space of closed curves of rotation index zero has nontrivial homology.

There are no critical points for $\lambda = 0$, because dilation reduces total squared curvature. Curve-straightening cannot satisfy the Palais-Smale condition in this case, since dilation causes curves to reduce total squared curvature; thus they will expand to infinite length.

Anders Linner has shown ([24]) that the curve straightening flow does not preserve convexity and that embedded curves need not stay embedded (although ultimately they must approach circles). See also [25] and [26].

Note: Yingzhong Wen initiated the study of "L^2 - curve straightening" [34], which has some nice geometric features. This is not actually a gradient flow in the sense described above, but rather a fourth-order semilinear parabolic differential equation. See [8] for recent interesting numerical work on this flow.

Recall the classification theorem of elastic curves in $\mathbb{R}^3$: For each pair of relatively prime integers (m,n) with $m > 2n$ there is a unique closed elastic curve (up to congruence) which lies on an embedded torus of revolution and represents an (m,n) torus knot.

Using the Palais-Smale condition one can prove more:

Theorem 10 *Let p and q be a pair of relatively prime integers with $0 < p \leq q$. Let $\mathscr{G}$ be the group of rotations around the z-axis generated by rotation through angle $\theta = \frac{2\pi p}{p+q}$. Then there is a non-circular closed elastica $\gamma_{p+q,p}$ which is $\mathscr{G}$-symmetric and $\mathscr{G}$-regularly homotopic to the p-fold circular elastica. It is a minimax critical point of total squared curvature (and hence unstable). It is only planar when $p = q$, in which case it is the figure-eight elastica.*

5.1. The Hyperbolic Plane

In $\mathbb{H}^2$, the closed elastic curves are much more abundant than in the Euclidean plane. In particular, there are critical points for $\lambda = 0$; we call such curves free elastica(e).

If the curvature of the hyperbolic plane is G, then the circle C of radius $\frac{\sinh^{-1}(1)}{\sqrt{-G}}$ is a free elastica, called the 'equator' of $\mathbb{H}^2$. (The name is by analogy to the equator of the sphere, which is a free elastica by virtue of being a geodesic.)

Free elastic curves can be classified using the Killing field $J(s) = \frac{k^2}{2}T + k'N$. Although rotation fields, translation fields, and horocycle fields all arise as examples of J, only the rotation fields are compatible with closed solutions. The following results are proved in [17]:

Theorem 11 *Let γ be a free elastica in $\mathbb{H}^2$. Then either γ is C^m for some m, or γ is a member of the family of solutions $\{\sigma_{m,n}\}$ having the following description:*

if $m > 1$ and n are integers satisfying $\frac{1}{2} < \frac{m}{n} < \frac{\sqrt{2}}{2}$ there is (up to congruence) a unique curve $\{\sigma_{m,n}\}$ which closes up in n periods of its curvature $k = k_0 \operatorname{cn}(\frac{k_0 s}{2}, p)$ while making m orbits about the fixed point q of the rotation field J.

Theorem 12 *Let γ be a regular closed curve in $\mathbb{H}^2$, the hyperbolic plane with curvature G. Then*

$$\mathcal{F}(\gamma) = \int_\gamma k^2 \, ds \geq 4\pi\sqrt{-G},$$

with equality precisely for the equator C.

Among closed curves of rotation index 1 in the hyperbolic plane the equator is the only critical point for $\mathcal{F}$, and it is a global minimizer. One cannot conclude, however, that curve-straightening takes any closed curve to the equator, since the Palais-Smale condition fails to hold. This is an intriguing, and so far unresolved, issue. Work of Steinberg ([32]) gives supporting evidence for convergence of the flow in this case; however, see [26] for contrary evidence.

It is also interesting to note that among curves of rotation number $m \geq 3$, the multiply-wrapped equator is actually unstable, and there is no minimizer for $\mathcal{F}$ in such regular homotopy class.

Theorem 12 has an important application to a higher-dimensional geometric problem, namely that of *Willmore tori of revolution* in $\mathbb{R}^3$.

For the *Chen-Willmore problem*, one considers immersions $\Psi : M^2 \longrightarrow \mathbb{R}^3$ and the total squared mean curvature functional

$$\mathcal{H}(\Psi) = \int\int_M H^2 dA,$$

where H is mean curvature and dA is the area element. Willmore showed in 1965 ([35]) that $\int\int_M H^2 dA \geq 4\pi$ on any closed surface in $\mathbb{R}^3$, with equality only for the round sphere.

The Chen-Willmore conjecture is that $\mathcal{H}(\Psi) \geq 2\pi^2$ when M is a torus. Robert Bryant ([5]) and Ulrich Pinkall independently observed the following:

Theorem 13 *Let γ be a regular closed curve in the hyperbolic plane represented by the upper half plane above the $x - axis$. If Ψ is the torus obtained by revolving γ around the $x - axis$, then $\mathcal{H}(\Psi) = \frac{\pi}{2}\mathcal{F}(\gamma)$*

From the inequality we derive [18]:

Corollary 14 *The Willmore inequality holds for tori of revolution.*

There are other ways to relate elastic curves to Willmore manifolds (critical points for total squared mean curvature). Let γ be critical for $\mathscr{F}^1$ in $\mathbb{S}^2$. U. Pinkall observed ([30]) that the inverse image of γ under the Hopf map $\pi : \mathbb{S}^3 \longrightarrow \mathbb{S}^2$ is a Willmore torus; stereographic projection gives a Willmore torus in $\mathbb{R}^3$. If we apply Theorem 4, this gives an infinite family of embedded Willmore surfaces in $\mathbb{R}^3$. Variations on this theme can be found in, e.g., [2] and [3].

This construction also extends to higher dimensions. In [4], the Hopf map from the five-sphere $\mathbb{S}^5$ to complex projective space $\mathbb{C}P^2$ is used to pull back elastic curves to Willmore surfaces. In this case, there is no integrability result for elastic curves. However, special solutions can be explicitly obtained: among the elastic curves are a countable family of *closed* curves whose curvature, torsion, and third curvature are all constant: helices in $\mathbb{C}P^2$. These pull back to a family of Willmore surfaces with constant mean curvature in $\mathbb{S}^5$.

ACKNOWLEDGMENTS

The work described in these lectures is the result of the efforts of many people. A large amount of the author's work is joint research with Joel Langer. The author also thanks co-authors Manuel Barros, Oscar Garay, and Thomas Ivey, and former students Rongpei Huang, Anders Linner, and Daniel Steinberg, all of whom contributed key ideas. Finally, the author wishes to thank Velimir Jurdjevic, who generously shared his expertise in optimal control theory.

REFERENCES

1. S. Antman, "The special cosserat theory of rods," in *Nonlinear Problems of Elasticity,* Springer-Verlag, Berlin, 1995.
2. J. Arroyo and O. Garay, "Hopf vesicles in $\mathbb{S}^3(1)$," in *Global Differential Geometry: the mathematical legacy of A. Gray,* Contemp. Math. **288** (2001), 258–262.
3. M. Barros, *Math. Proc. Camb. Phil. Soc.* **121**, 321–324 (1997).
4. M. Barros, O. Garay and D. Singer, *Tokohu Math J. (2)* **51**, 177–192 (1999).
5. R. Bryant and P. Griffiths, *Amer. J. Math.* **108** , 525–570 (1986).
6. P. Byrd and M. Friedman, *Handbook of elliptic integrals for engineers and physicists*, Springer-Verlag, Berlin, 1954.
7. B. Y. Chen, *Boll. Un. Mat. Ital.* **10**, 380–385. (1974)
8. G. Dziuk, E. Kuwert and R. Schätzle, *SIAM J. Math. Anal.* **33**, 1228–1245 (2002).
9. H. Hasimoto, *J. Fluid Mech.* **51**, 477–486 (1972).
10. R. Huang, "Affine and Subaffine Elastic Curves in $\mathbb{R}^2$ and $\mathbb{R}^3$," thesis, Case Western Reserve University, 1999.
11. R. Huang and D. Singer, *Proc. Amer. Math. Soc.* **130**, 2725–2735 (2002).
12. T. Ivey and D. Singer, *Proc. London Math. Soc.* **79**, 429–450 (1999).
13. V. Jurdjevic, *Amer. J. Math.* **117**, 93–124 (1995).
14. ———,*Integrable Hamiltonian Systems on Complex Lie Groups,* Mem. Amer. Math. Soc. **178**, 2005.
15. S. Kobayashi and K. Nomizu, *Foundations of Differential Geometry,* vol. 1, Wiley-Interscience, 1963.
16. N. Koiso, *Osaka J. Math.* **29**, 539–543 (1992).
17. J. Langer and D. Singer, *J. Diff. Geom.* **20** , 1–22 (1984).
18. ———,*Bull. London. Math. Soc.* **16**, 531–534 (1984).
19. ———,*J. London Math. Soc.* **30** , 512–520 (1984).

20.   ——,*Topology* **24**, 75–88 (1985).
21.   ——,*Ann. Global Anal.Geom.* **5**, 133–150 (1987).
22.   ——,*Comment. Math. Helv.* **69**, 272–280 (1994).
23.   ——,*SIAM Review* **38**, 605–618 (1996).
24. A. Linner, *Trans. Amer. Math. Soc.* **314**, 605–617 (1989).
25.   ——,*Nonlinear Anal.* **21**, 575–593 (1993).
26.   ——,*Trans. Amer. Math. Soc.* **350**, 3743–3765 (1998).
27. A. E. H. Love, *A Treatise on the Mathematical Theory of Elasticity*, Cambridge University Press, Cambridge, England 1927.
28. K. Nomizu & T. Sasaki, *Affine differential geometry*, Cambridge University Press, 1994.
29. R. S. Palais, "Critical point theory and the minimax principle," in *Global Analysis,* Proc. Sympos. Pure Math. **15**, 185–212 (1970).
30. U. Pinkall, *Invent. Math.* **81**, 379–386 (1985).
31. U. Pinkall, *Results Math.* **27**, 328–332 (1995).
32. D. H. Steinberg, "Elastic curves in hyperbolic space," thesis, Case Western Reserve University (1995).
33. B. Su, *Affine differential geometry*, Beijing, 1983.
34. Y. Wen, *J. Differential Equations* **120**, 89–107 (1995).
35. T. J. Willmore, *An. Sti. Univ. "Al I. Cuza" Iasi Sect. I a Mat. (N.S.)* **11B**, 493–496 (1965).

Variational Problems which are Quadratic in the Surface Curvatures

Bennett Palmer

Department of Mathematics, Idaho State University, Pocatello, Idaho, 83209, U.S.A
email: palmbenn@isu.edu

Abstract. We study variational problems for surfaces in Euclidean space for functionals involving curvatures. Particularly, we study both the classical Willmore surfaces and their anisotropic analogue.

Keywords: plate, bending energy, Willmore functional , anisotropic
PACS: 87.16.Dg, 02.40.Hw

INTRODUCTION

This paper consists of notes prepared for a course given for the international graduate school entitled Curvature and Variational Modeling in Physics and Biophysics which took place at the Santiago de Compostela in September 2007. The overall theme of these notes can roughly be summarized as "plates, conformal geometry and the Willmore functional".

There are basically two origins to this subject. The first is the theory of elasticity that grew out of Sophie Germain's original work on the Chladni plates. The other is the subject of conformal differential geometry as developed by Wilhelm Blaschke and his students, particularly Gerd Thomsen.

In the early nineteenth century, the German physicist and amateur violinist Ernst Chladni presented a demonstration in which sand was sprinkled on metal plates. Chladni then played the edge of the plate with his violin bow and a variety of interesting shapes were observed to be formed by the sand. (The interested reader in encouraged to do a web search using "Chladni plates" to locate one of the numerous available videos of this experiment.) Today, we would recognize the patterns formed by the sand to be the nodal lines of non linear vibrations, but at the time such an explanation was not available. The Istitut de France offered a substantial prize for their explanation.

The problem excited the interest of the leading mathematicians of the day including James Bernoulli, Poisson and Lagrange who expressed the opinion that the solution would involve tools which had yet to be developed. After several attempts and much controversy, the French mathematician Sophie Germain was awarded the prize. Remarkably, Germain was a self-taught number theorist with no formal training in analysis. Her work was aided by Lagrange's encouragement and corrections of her manuscript. It is interesting to note that the plate energy was originally introduced in connection with surfaces with boundary, a subject which has been for the most part neglected by geometers. Stoker, [1], has pointed out that it is noteworthy that the non linear theory of plates

CP1002, *Curvature and Variational Modeling in Physics and Biophysics*
edited by O. J. Garay, E. García-Río, and R. Vázquez-Lorenzo
© 2008 American Institute of Physics 978-0-7354-0521-9/08/$23.00

appeared in essentially correct form before the linear theory.

Blaschke was highly influenced by the Erlangen Program of Felix Klein which interpreted geometry in terms of the invariants of group actions. For a group G acting on Euclidean space, Blaschke sought to determine the G invariants of an immersed surface. For example, when G is the group of rigid motions, one obtains the 'usual' invariants of a surface such as its mean and Gaussian curvature. One could also take other choices of G, for example the group of projective transformations, the special linear group or the group of conformal transformations. In the latter case, one obtains the conformal differential geometry.

After determining the invariant tensors and the integrability relations between them, special classes of surfaces would need to be selected for deeper study. Wisely, Blaschke turned his attention to those surfaces which were critica of the simplest conformally invariant variational problem which he called "conformal minimal surfaces". Today, these surfaces are known as Willmore surfaces, after the English mathematician Tom Willmore who reintroduced them in the nineteen sixties. Blaschke realized that the fact that these surfaces arose from a variational problem with a large symmetry group would manifest itself in special characteristics of their geometry. He and his collaborators obtained many interesting and deep results some of which were rediscovered in the nineteen eighties. The subject fell into obscurity for many years, probably because the the high order of the equations involved made it difficult to obtain an interesting variety of examples.

The notes begin with a discussion of conformal geometry using Blaschke's point of view. This is based on the use of the conformal Gauss map which involves using Lorentzian geometry to express the conformal geometry of surfaces in three dimensional space. The effect is to lower the order of the equation for a Willmore surface to second order. After discussing important results of Bryant, we discuss the our previous work on stability of Willmore surfaces and Willmore surfaces with boundary. The second half of these notes concerns anisotropic surface energies.

CONFORMAL GEOMETRY AND WILLMORE SURFACES

The space of spheres

From now on $\langle , \rangle$ will denote the usual inner product on $\mathbf{R}^n$. We denote the pseudometric on the five the five dimensional Minkowski space $\mathbf{R}_1^5$ with a dot. The signature of this pseudometric is $(+,+,+,+,-)$.

We denote the set of non zero null vectors in $\mathbf{R}_1^5$, (i.e. the light cone minus the origin) by $\mathscr{L}$. We let $M^3(c)$ be the unique simply connected space form with constant curvature c which we take to be -1, 0 or $+1$. Note that there are isometric, in particular conformal, embeddings

$$M^3(c) \hookrightarrow \mathscr{L},$$

given as follows. When $c = -1$, we have

$$M^3(-1) = \mathbf{H}^3 = \{X \in \mathbf{R}_1^4 \mid X \cdot X = -1\} \hookrightarrow \mathscr{L}$$
$$X \mapsto (1,X).$$

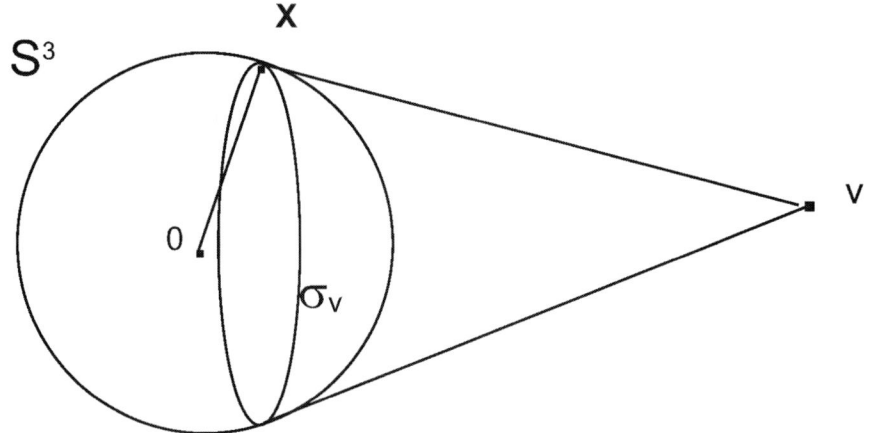

FIGURE 1. Bijection between the exterior of $\mathbf{S}^3$ and the non equatorial, unoriented 2-spheres.

When $c = 0$, we have

$$M^3(0) = \mathbf{R}^3 \hookrightarrow \mathscr{L}$$
$$X \mapsto (X, \frac{\langle X,X \rangle - 1}{2}, \frac{\langle X,X \rangle + 1}{2})$$

and when $c = 1$, we have

$$M^3(1) = \mathbf{S}^3 = \{X \in \mathbf{R}^4 \mid \langle X,X \rangle = 1\} \hookrightarrow \mathscr{L}$$
$$X \mapsto (X, 1).$$

Let σ be a non equatorial two dimensional sphere in $\mathbf{S}^3$. We associate to σ the point v in $\mathbf{R}^4$ which is the vertex of the cone tangent to $\mathbf{S}^3$ along σ. This, in fact, defines a bijection between the exterior of $\mathbf{S}^3$ and the non equatorial, unoriented 2-spheres which we write as $v \leftrightarrow \sigma_v$.

In order to include the equatorial spheres, we replace v first with $\underline{v} := (v, 1) \in \mathbf{R}_1^5$ and then consider the projective class $[\underline{v}] \in \mathbb{PR}_1^5$. Note that $\underline{v} \cdot \underline{v} > 0$ holds and so we have

$$[\underline{v}] \in \{[w] \in \mathbb{PR}_1^5 \mid w \cdot w > 0\}.$$

As $v \to \infty$ along a straight line, we have

$$[(v,1)] = [(\frac{v}{|v|}, \frac{1}{|v|})] \to [(\hat{v}, 0)]$$

with $\hat{v} \in \mathbf{S}^3$. In this way we include $\mathbf{S}^3$ itself as the space of unoriented equatorial 2-spheres and we have a bijection:

$$\{[w] \in \mathbb{PR}_1^5 \mid w \cdot w > 0\} \leftrightarrow \{\text{unoriented } 2 - \text{spheres in } \mathbf{S}^3\}.$$

If we want to consider oriented 2-spheres, we take the double cover of the set on the left by the deSitter space which is the unit sphere in $\mathbf{R}_1^5$.

$$\begin{array}{ccc} \mathbf{S}_1^4 & \longrightarrow & \{[w] \in \mathbb{P}\mathbf{R}_1^5 \mid w \cdot w > 0\} \\ Y & \longmapsto & [Y] \end{array}$$

The two lifts of a class $[w]$ represent the same sphere with its two orientations.

Now we consider a fixed $Y \in \mathbf{S}_1^4$ and we recover the sphere which it represents. In Figure 1, note that

$$\begin{aligned} X \in \sigma_v & \Leftrightarrow \langle v - X, X \rangle = 0 \\ & \Leftrightarrow \langle v, X \rangle - 1 = 0 \\ & \Leftrightarrow \underline{v} \cdot \underline{X} = 0, \end{aligned}$$

where $\underline{X} = (X, 1)$. This identifies the sphere represented by $Y \in \mathbf{S}_1^4$ with

$$Y^\perp \cap \mathscr{L} \cap \{x_5 = 1\}.$$

Next note that $N = (v - x)/|v - x|$ is the unit normal to σ_v in $\mathbf{S}^3$. Solving for v gives

$$v = N + \frac{X}{|v - X|} = N + \frac{X}{\sqrt{v^2 - 1}}.$$

Therefore

$$[\underline{v}] = [(N + \frac{X}{\sqrt{v^2-1}})] = [\pm(\sqrt{v^2-1}\,X + N, \sqrt{v^2-1})] = [\pm(hX + N, h)], \quad (1)$$

where $h = \sqrt{v^2 - 1}$ is the mean curvature of σ_v in $\mathbf{S}^3$. The representatives $\pm(hX + N, h)$ are exactly the ones which lie in $\mathbf{S}_1^4$.

The action of group the $O(4,1)$ on $\mathbf{R}_1^5$ preserves $\mathscr{L}$ and induces an action on $\mathbf{S}^3 \approx \mathscr{L}/\mathbf{R}^*$. It is well known, [2] that the induced action gives the conformal (Moebius) transformations, i.e. $\mathbb{P}O(4,1) = \mathrm{Conf}(\mathbf{S}^3)$. For g in $O(4,1)$ we denote the corresponding Moebius transformation by $[g]$ such that the following diagram commutes.

$$\begin{array}{ccc} \mathscr{L} & \xrightarrow{\;g\;} & \mathscr{L} \\ \downarrow & & \downarrow \\ \mathscr{L}/\mathbf{R}^* \approx \mathbf{S}^3 & \xrightarrow{\;[g]\;} & \mathscr{L}/\mathbf{R}^* \approx \mathbf{S}^3 \end{array}.$$

Spherical Congruences

We will now develop the basic extrinsic conformal geometry of an immersed surface. It is most convenient to do this by considering the surface to be immersed in the three dimensional sphere. In this case the calculations are a bit easier than they would be if the surface was in $\mathbf{R}^3$ and the three sphere has the advantage of being compact. We wish to emphasize however that all of the results are just as valid for surfaces in $\mathbf{R}^3$ since three dimensional Euclidean space is conformally equivalent to the three sphere minus one point.

Let Σ be a smooth surface. By an (oriented) *spherical congruence* we mean a smooth map

$$Z : \Sigma \longrightarrow \mathbf{S}_1^4 \, .$$

This is just a two parameter family of oriented spheres. Now consider an immersion

$$X : \Sigma \to \mathbf{S}^3 \, .$$

We will say that X *envelopes* Z if

$$Z \cdot \underline{X} \equiv 0 \tag{2}$$

and

$$Z \cdot d\underline{X} \equiv 0 \tag{3}$$

hold. By the discussion above, this means that the point X lies on the sphere represented by Z to first order.

Proposition 1 *Let $X : \Sigma \to \mathbf{S}^3$ be an immersed surface such that the set of umbilic points is nowhere dense. Then there exists a unique spherical congruence Y such that*

(i) X envelopes Y

and

(ii) $Y : (\Sigma, ds_X^2) \to \mathbf{S}_1^4$, is conformal in the sense that $ds_Y^2 = B ds_X^2$ where B is a non negative function.

Before proving this we will need to introduce some notation.

Let $X : \Sigma \to \mathbf{S}^3$ be a sufficiently smooth immersion of an oriented surface. The orientation will be determined by specifying a normal map $N : \Sigma \to \mathbf{S}^3$. The induced metric, ds_X^2 induces a conformal structure on Σ in a natural way and we can, locally near an arbitrary $p \in \Sigma$, introduce a complex coordinate z and write

$$ds_X^2 =: e^{\mu} \, |dz|^2 \, .$$

Similarly, we can express the second fundamental form as

$$-\langle dN(\cdot), dX(\cdot) \rangle =: \Re\{\phi dz^2 + h e^{\mu} dz d\bar{z}\} \, .$$

In the formula above, h is the mean curvature of the immersion and the quantity ϕdz^2 is an invariantly defined quadratic differential called the *Hopf differential*. The *Frenet equations* for the immersion are then given by

$$X_{zz} = \mu_z X_z + (\phi/2)N \tag{4}$$
$$X_{z\bar{z}} = h(e^{\mu}/2)N - (e^{\mu}/2)X \tag{5}$$
$$N_z = -hX_z - \phi e^{-\mu} X_{\bar{z}} \tag{6}$$

The integrability equations for this system are the equations of Gauss

$$|\phi|^2 e^{-2\mu} = h^2 + 1 - 2e^{-\mu}\partial_z\partial_{\bar{z}}\mu = h^2 + 1 - K, \qquad (7)$$

and Codazzi

$$\phi_{\bar{z}} = e^{\mu} h_z. \qquad (8)$$

Proof of Proposition 1. We define Y by

$$Y := h\underline{X} + (N, 0). \qquad (9)$$

Computations using the Frenet equations yield

$$Y_z = h_z\underline{X} - \frac{\phi e^{-\mu}}{2}(X_{\bar{z}}, 0) \qquad (10)$$

from the last two equations it follows that X envelopes Y. Note also from (10) that

$$\langle Y_z, Y_z \rangle = 0, \quad \langle Y_z, Y_{\bar{z}} \rangle = |\phi|^2 e^{-\mu}/2. \qquad (11)$$

The first equation means that Y is conformal and the second together with the Gauss equation implies

$$ds_Y^2 = (h^2 + 1 - K)ds_X^2 \qquad (12)$$

and so (ii) holds also.

We now prove uniqueness. The equations in (2), (3) can also be written $Z \cdot \underline{X} \equiv 0, \langle Z, d\underline{X} \rangle \equiv 0$. Since X is an immersion this forces Z to lie in the 2-dimensional subspace of $\mathbf{R}_1^5$ which contains $\underline{X}$ and Y. It therefore follows that the most general spherical congruence is of the form $Z = Y + \lambda\underline{X}$ for some function λ. A computation using (10) then gives $Z_z = (h_z + \lambda_z)\underline{X} - \phi e^{-\mu}/2(X_{\bar{z}}, 0) + \lambda(X_z, 0)$ so that $Z_z \cdot Z_z = -\phi\lambda$. Since the set of umbilics is assumed to be nowhere dense, we see that Z is conformal if and only if $\lambda \equiv 0$. **q.e.d.**.

From the expression (12) and the formula (1) we see that at each $p \in \Sigma$, $Y(p)$ represents the 2-sphere having the same normal and mean curvature as the immersion at p. In classical differential geometry this sphere was called the *central sphere*. The map Y was reintroduced by Bryant, [3], who called the *conformal Gauss map*. We will see that the map Y encodes the conformal invariants of the immersion X.

Proposition 2 *The conformal Gauss map is conformally invariant in the following sense. Suppose $X : \Sigma \to \mathbf{S}^3 \approx \mathcal{L}/\mathbf{R}^*$ is as above, $g \in O(4,1)$ and $[g]$ is the corresponding map in $\mathbb{P}O(4,1)$. Then if $X' := [g]X$, the conformal Gauss map Y' of X' is given by $Y' = gY$.*

Proof. This follows easily from the uniqueness statement of the previous proposition. If $X' := [g]X$ then gY satisfies the conditions (2), (3) for the immersion X'. **q.e.d.**

The proposition shows that any geometric invariant of Y as a surface in $\mathbf{S}_1^4$ is a conformal invariant of X in $\mathbf{S}^3$. In particular, since the metrics induced by X and Y satisfy (12), we have that

$$W[X] = \int_\Sigma h^2 + 1 - K \, d\Sigma_X = \int_\Sigma d\Sigma_Y = \text{Area}[Y]$$

is conformally invariant. The functional W is essentially the Willmore functional of the surface X. If we consider the immersion $X' := \pi X$ where π is the stereographic projection, then

$$\int_\Sigma h^2 + 1 - K \, d\Sigma_X = \int_\Sigma H^2 - K' d\Sigma_{X'},$$

where H and K' are respectively the mean and Gaussian curvature of X'. The integral of the curvature is, of course, a topological invariant if the surface is closed or if the boundary is kept fixed to first order.

The expression for the metric induced by Y given above shows that its singularities coincide exactly with the umbilics of the immersion X. Let U denote the set of umbilics for the immersion X. We will next give several important results of Bryant, [3].

Proposition 3 *The immersion X defines a Willmore surface if and only if $Y|_{\Sigma \setminus U}$ has zero mean curvature in $\mathbf{S}_1^4$.*

Proof. The sufficiency is clear. The necessity is slightly surprising since there are many more variations of the map Y than there are variations through conformal Gauss maps. A straightforward computation shows however that the mean curvature field of Y is given by

$$\mathcal{H} = \frac{1}{2}(\Delta_Y Y + 2Y) = \frac{1}{2}(\Delta_Y h + 2h)\underline{X}. \tag{13}$$

If X_ε is a variation of X supported in $\Sigma \setminus U$ and Y_ε are the corresponding conformal Gauss maps then

$$\partial_\varepsilon \text{Area}(Y_\varepsilon)_{\varepsilon=0} = -2\int_\Sigma \mathcal{H} \cdot \partial_\varepsilon (Y_\varepsilon)_{\varepsilon=0} d\Sigma_Y. \tag{14}$$

Letting "dot" denote $\partial_\varepsilon(\cdot)_{\varepsilon=0}$, we have $\dot{Y} = \dot{h}\underline{X} + h\underline{\dot{X}} + (\dot{N},0)$ and so

$$
\begin{aligned}
\partial_\varepsilon \text{Area}(Y_\varepsilon)_{\varepsilon=0} &= -\int_\Sigma (\Delta_Y h + 2h)\underline{X} \cdot (\dot{h}\underline{X} + h\underline{\dot{X}} + (tN,0)) d\Sigma \\
&= -\int_\Sigma (\Delta_Y h + 2h)X \cdot \dot{N} d\Sigma_Y \\
&= \int_\Sigma (\Delta_Y h + 2h)N \cdot \dot{X} d\Sigma_Y \\
&= \int_\Sigma (\Delta_X h + 2h(h^2 + 1 - K))N \cdot \dot{X} d\Sigma_X.
\end{aligned}
$$

The result follows since $\langle N, \dot{X} \rangle$ is a smooth arbitrary function on Σ.

39

Isothermicity

We will consider an immersed surface in $\mathbf{S}^3$. Away from the umbilic points, the conformal Gauss map is a space-like immersion and so the induced metric on the normal bundle is Lorentzian. Recall that $\underline{X}$ defines a null section of the normal bundle $\perp Y$ of Y. We can choose a second normal section Z with

$$\underline{X} \cdot Z \equiv 1 .$$

Let $D^{\perp}$ denote the connection in the normal bundle $\perp Y$ of Y in $\mathbf{S}_1^4$. Using (4), and working away from umbilics of the immersion X, one easily computes

$$D_z^{\perp} \underline{X} = (\phi_{\bar{z}}/\phi)\underline{X}, \qquad D_{\bar{z}}^{\perp} \underline{X} = (\bar{\phi}_{\bar{z}}/\bar{\phi})\underline{X}, \tag{15}$$

and

$$D_z^{\perp} Z = -(\phi_{\bar{z}}/\phi)Z, \qquad D_{\bar{z}}^{\perp} Z = -(\bar{\phi}_z/\bar{\phi})Z. \tag{16}$$

We introduce a Lorentz "orthonormal" frame for $\perp Y$,

$$e_3 := (1/\sqrt{2})(\underline{X} + Z), \qquad e_4 := (1/\sqrt{2})(\underline{X} - Z)$$

for which the connection 1-form and curvature are defined by

$$de_3 =: \omega_{34} e_4, \qquad d\omega_{34} =: K^{\perp} d\Sigma.$$

We have, using (15) and (16),

$$\omega_{34} = (\bar{\phi}_z/\bar{\phi})dz + (\phi_{\bar{z}}/\phi)d\bar{z}$$

from which we obtain

$$\begin{aligned} K^{\perp} d\Sigma : \quad &= \quad d\omega_{34} = ((\phi_{\bar{z}}/\phi)_z - (\bar{\phi}_z/\bar{\phi})_{\bar{z}} dz \wedge \bar{z} \\ &= \quad (\Delta_Y \arg \phi) d\Sigma. \end{aligned}$$

Recall that a surface in $\mathbf{S}^3$ is called *isothermic* if, away from its umbilic points, the lines of curvature are given by the zero sets of a pair of locally defined conjugate harmonic functions u and v. Using these harmonic functions as an isothermal coordinates on the surface, forming the complex coordinate $z = u + iv$ and recalling that the lines of curvature are given by

$$\Im(\phi dz^2) = 0$$

one sees that isothermicity is equivalent to the local existence of a complex coordinate so that the coefficient of the Hopf differential is real valued. This clearly is the same as the condition that $\arg \phi$ is harmonic and hence we obtain that the isothermic surfaces are those surfaces such that the curvature of the normal bundle of the conformal Gauss map vanishes identically.

A surface $X : \Sigma \to M^3(c)$ is a Willmore surface if and only if it satisfies $\Delta_X H + 2H(H^2 + c - K) = 0$. Here H is the mean curvature of X in $M^3(c)$. It is clear from this that any minimal surface ($H \equiv 0$) is a Willmore surface as is any conformal image of such a surface. It is well known that any minimal surface in a space form is isothermic. We state without proof the following theorem of G. Thomsen obtained in 1923.

Theorem 1 *(Thomsen[4]) Let $X : \Sigma \to \mathbf{S}^3$ be a Willmore immersion without umbilics. Then the surface is isothermic if and only if there exists a conformal map $\Psi : \mathbf{S}^3 \to M^3(c)$ such that $\Psi \circ X$ is minimal.*

Thomsen's original statement did not exclude umbilics and indeed a minimal surface with umbilics is conformal to a Willmore surface in $\mathbf{S}^3$. However for the necessity part of the theorem they must be excluded as the following example of K. Voss shows. Consider an elastic curve γ in the plane which is not a line and whose curvature vanishes at some point p. The cylinder $C := \gamma \times \mathbf{R}$ is a Willmore surface in $\mathbf{R}^3$ which is isothermic and the line $\{p\} \times \mathbf{R}$ consists entirely of umbilic points. Since the set of umbilics is invariant under ambient conformal maps and since umbilics in a minimal surface in $M^3(c)$ are isolated, we see that C cannot be extrinsically conformal to a minimal surface.

We now turn out attention to an important class of isothermic Willmore surface, those which are conformal to minimal surfaces in Euclidean space. We follow the work of Bryant, [3].

Proposition 4 *Assume that $X : \Sigma \to \mathbf{S}^3$ is a Willmore surface. Then the quartic form defined locally by*

$$Q = (Y_{zz} \cdot Y_{zz})dz^4 =: Q^{(4,0)}dz^4 \,,$$

is holomorphic.

Proof. The assertion is that for any local coordinate $\partial_{\bar{z}}Q^{(4,0)} = 0$. Note that, using (13), $\mathscr{H} \equiv 0$ can be expressed

$$Y_{z\bar{z}} + (Y_z \cdot Y_{\bar{z}})Y = 0 \,.$$

Using this to replace $Y_{z\bar{z}}$ below, we obtain

$$
\begin{aligned}
\partial_{\bar{z}}(Y_{zz} \cdot Y_{zz}) &= 2Y_{z\bar{z}z} \cdot Y_{zz} \\
&= -2(Y_z \cdot Y_{\bar{z}})_z Y \cdot Y_{zz} - 2(Y_z \cdot Y_{\bar{z}})Y_z \cdot Y_{zz} \\
&= 2(Y_z \cdot Y_{\bar{z}})_z Y_z \cdot Y_z - 2(Y_z \cdot Y_{\bar{z}})Y_z \cdot Y_{zz} \\
&= 0 \,,
\end{aligned}
$$

since $Y_z \cdot Y_z \equiv 0$ since Y is conformal and $2Y_z \cdot Y_{zz} = \partial_z(Y_z \cdot Y_z)$. **q.e.d.**

Lemma 1 *Let $X : \Sigma \to M^3(c)$ be a class C^4 immersion of a Willmore surface. Then the surface is real analytic.*

Proof. This is a consequence of elliptic regularity. Since the result is local, we can assume, without loss of generality, that the surface is given as a graph of a C^4 function u over a domain in the plane. The equation $\Delta H + 2H(H^2 - K) = 0$ can be expressed as a fourth order quasilinear, elliptic equation $E(Du, D^2u, .., D^4u) = 0$. The operator E depends real analytically on all the derivatives $D^k u$, $k = 1..4$. Standard results, [5], then imply that the solution is real analytic. **q.e.d.**

Proposition 5 *If $X : \Sigma \rightarrow \mathbf{S}^3$ is an immersion of a Willmore surface on which $Q \equiv 0$ holds, then the surface is either part of a round sphere or there exists a conformal transformation $f : \mathbf{S}^3 \rightarrow \mathbf{R}^3$ such that $f \circ X$ is a minimal immersion.*

Sketch of the Proof. Assume that the surface is not totally umbilic. By the lemma, there is an open, dense set of non umbilic points and we restrict our attention to this set.

We introduce a frame for the normal bundle $\{\underline{X}, Z\}$ as above. The superscript $\cdot^{\perp}$ will mean that the quantity is projected onto the normal bundle $\perp Y$.

We can write $Y_{zz}^{\perp} =: \alpha X + \phi Z$, and $Q = 2\alpha\phi \equiv 0$ implies that $\alpha \equiv 0$, (since $\phi = 0$ characterizes umbilics). Thus, $Y_{zz}^{\perp} =: \phi Z$. This gives

$$D_z^T Z \cdot Y_z = -Z \cdot Y_{zz} = 0,$$

and

$$D_z^T Z \cdot Y_{\bar{z}} = -Z \cdot Y_{z\bar{z}} = 0.$$

The second equation follows from the harmonicity of Y. These equations tell us that $D^T Z \equiv 0$. If $Z_1 := e^a Z$ for a smooth function a, then clearly $D^T Z_1 \equiv 0$ also.

A local calculation which will be omitted, shows that the intrinsic norm of the quartic form Q with respect to the metric ds_Y^2 is given by

$$\|Q\|^2 = 16 e^{-4\rho} |Q^{(4,0)}|^2 = (1 - K_Y)^2 + (K^{\perp})^2. \tag{17}$$

Therefore $Q \equiv 0$ implies that $K^{\perp} \equiv 0$ holds, and thus it is possible to find a new frame $\{X_1 := e^{-a}\underline{X}, Z_1 := e^a Z\}$ which is parallel in the normal bundle of Y. In particular $D^{\perp} Z_1 \equiv 0$. Since $D^T Z_1 \equiv 0$, it follows that there exists a constant null vector in $C \in \mathbf{R}_1^5$ with $C \equiv Z_1$ along Y. This can be seen by differentiating $E \cdot Z_1$ for any constant vector E.

There exists $g \in O(4, 1)$ such that $gC = (0, 0, 0, 1, 1)$. Then, by Proposition 2, $[g]\bar{X} =: \bar{X}_1$ is an immersion whose conformal Gauss may Y_1 satisfies $Y_1 \cdot gC \equiv 0$.

Theorem 2 *If $X : \Sigma \rightarrow \mathbf{S}^3$ be an immersion of a closed Willmore surface of genus zero then the conclusion of Proposition 5 holds.*

Proof. By the Uniformization Theorem, Σ is conformal to the complex plane $\mathbf{C}$ with a point adjoined at infinity. Using the usual complex coordinate in the plane, we write $Q = q(z)dz^4$ with q holomorphic in $\mathbf{C}$.

The behavior of Q at ∞ is determined by making the change of coordinate $\zeta = 1/z$ and studying the behavior of

$$Q = q(1/\zeta)[d(1/\zeta)]^4 = \frac{q(1/\zeta)}{\zeta^8} d\zeta^4,$$

at $\zeta = 0$. Since Q is assumed to be holomorphic at ∞ also, $q(1/\zeta)\zeta^{-8}$ must be holomorphic at $\zeta = 0$. Thus $q(1/\zeta) = q(z) \rightarrow 0$ as $z \rightarrow \infty$. In particular, q is bounded in $\mathbf{C}$ and so, by Liouville's Theorem, must reduce to a constant. The constant must be zero because of the limit discussed above.

This shows that any holomorphic quartic form on a genus zero Riemann surface vanishes identically and the result follows from the previous proposition. **q.e.d.**

The previous result also follows easily from the Riemann Roch Theorem.

Stability

It is explained above that the first variation of area for the conformal Gauss map vanishes exactly when the surface is a Willmore surface. What happens for the second variation?

The Clifford torus $T = \mathbf{S}^1 \times \mathbf{S}^1 \to \mathbf{S}^3$ is the conjectured minimum of W among all genus one surfaces. It has been shown by Weiner, [8], that the Clifford torus is stable as a Willmore surface. The conformal Gauss map of this surface is an isometric, minimal embedding $Y : T \to \mathbf{S}^3 \subset \mathbf{S}^4_1$. It is well known that all closed minimal surfaces in $\mathbf{S}^3$ are unstable for the area functional and so the second variation of area for the conformal Gauss map of this surface is negative for some variation. This example shows that in order to use the conformal Gauss map to study the second variation of the Willmore functional, it is necessary to identify which variations of this map arise as variations through conformal Gauss maps. The material above is based on [6].

Let

$$Y : \Sigma \to \mathbf{S}^4_1$$

be an immersion of a torus. We assume that the induced metric is spacelike and that the mean curvature vector of the immersion is identically zero. Since the immersion is spacelike, the induced metric induces a conformal structure on Σ and we let z be a complex coordinate for this structure. It is well known that when the mean curvature vanishes, the quartic differential defined by

$$Q := Y_{zz} \cdot Y_{zz} dz^4 =: Q^{(4,0)} dz^4,$$

is holomorphic. By lifting the coefficient $Q^{(4,0)}$ to the complex plane and applying Liouville's theorem, one sees that $Q^{(4,0,)}$ must be a constant. We will assume that this constant is not zero. By the discussion in the previous section, this means that the torus is not the conformal image of a minimal surface in $\mathbf{R}^3$. By a simple change of the variable z, we can assume that

$$Q^{(4,0)} \equiv 1 .$$

Since the immersion is spacelike, locally we can introduce a frame A, B for the normal bundle of Y, $\perp Y$, such that

$$A \cdot A = 0 = B \cdot B, \qquad A \cdot B = 1. \tag{18}$$

The Frenet equations for Y can then be expressed:

$$Y_{zz} = \rho_z Y_z + \frac{e^{-i\psi}}{\sqrt{2}} A + \frac{e^{i\psi}}{\sqrt{2}} B$$

43

$$Y_{z\bar{z}} = \frac{-e^{\rho}}{2}Y$$

$$A_z = -\sqrt{2}e^{-\rho}e^{i\psi}Y_{\bar{z}} - i\psi_z A$$

$$B_z = -\sqrt{2}e^{-\rho}e^{-i\psi}Y_{\bar{z}} + i\psi_z B$$

The integrability conditions for these equations are, respectively, the Gauss and Ricci equations:

$$\Delta_0\rho - 8e^{-\rho}\cos 2\psi + 2e^{\rho} = 0 \qquad (19)$$

$$\Delta_0\psi + 4e^{-\rho}\sin 2\psi = 0, \qquad (20)$$

where $\Delta_0 = 4\partial_z\partial_{\bar{z}}$. This means that the curvatures of the tangent and normal bundles are given by,

$$1 - K_Y = 4e^{-2\rho}\cos 2\psi,$$

$$K^{\perp} = -4e^{-2\rho}\sin 2\psi.$$

We now introduce the Jacobi operator of the immersion Y. Since the mean curvature is zero, the second variation of area for a variation $Y_{\varepsilon} = Y + \varepsilon\xi + O(\varepsilon^2)$, $\xi \in \Gamma(\perp Y)$ can be expressed

$$\delta_{\xi}^2\text{Area}(Y) = -\int_{\Sigma}\xi \cdot J[\xi]d\Sigma. \qquad (21)$$

Here it is assumed that ξ has compact support. The operator J is given by, [7],

$$J[\xi] = \Delta^{\perp}\xi + 2\xi + \mathscr{B}\xi,$$

where

$$\Delta^{\perp}\xi = \sum_i(D_i^{\perp}D_i^{\perp} - D_{\nabla_{i}e_i}^{\perp})\xi,$$

is the rough Laplacian in the normal bundle and $\mathscr{B}$ is an endomorphism of the normal bundle defined for sections by,

$$(\mathscr{B}\xi) \cdot \eta = (D^T\xi) \cdot (D^T\eta), \qquad \xi, \eta \in \Gamma(\perp Y).$$

If we express the normal section ξ in terms of our frame,

$$\xi = \alpha A + \beta B,$$

then,

$$\Delta^{\perp}A = |\nabla\psi|^2 A, \qquad \Delta^{\perp}B = |\nabla\psi|^2 B,$$

$$\mathscr{B}A = 4e^{-2\rho}((\cos 2\psi)A + B)$$

$$\mathcal{B}B = 4e^{-2\rho}(A + (\cos 2\psi)B)$$

and the Jacobi operator is given by,

$$
\begin{aligned}
J[\xi] &= [\Delta\alpha + \alpha(|\nabla\psi|^2 + 2 + \mathcal{B}A \cdot B) + 2(D^\perp_{\nabla\alpha}A \cdot B) + \beta(\mathcal{B}B \cdot B)]A \\
&\quad + [\Delta\beta + \beta(|\nabla\psi|^2 + 2 + \mathcal{B}A \cdot B) + 2(D^\perp_{\nabla\beta}B \cdot A) + \alpha(\mathcal{B}A \cdot A)]B \\[2mm]
&= [\Delta\alpha + \alpha(|\nabla\psi|^2 + 2 + 4e^{-2\rho}\cos 2\psi) + 4e^{-\rho}i(\alpha_z\psi_{\bar z} - \alpha_{\bar z}\psi_z) + \beta 4e^{-2\rho}]A \\
&\quad + [\Delta\beta + \beta(|\nabla\psi|^2 + 2 + 4e^{-2\rho}\cos 2\psi) + 4e^{-\rho}i(\psi_z\beta_{\bar z} - \beta_z\psi_{\bar z}) + \alpha 4e^{-2\rho}]B.
\end{aligned}
$$

Define self-adjoint and skew adjoint operators acting on smooth functions by

$$L[u] := \Delta u + u(|\nabla\psi|^2 + 2 + \mathcal{B}A \cdot B) = \Delta u + u(|\nabla\psi|^2 + 2 + 4e^{-2\rho}\cos 2\psi).$$

$$\Lambda[u] := 2(D^\perp_{\nabla u}B \cdot A) = 4e^{-\rho}i(\psi_z u_{\bar z} - u_z\psi_{\bar z}) = 2\langle \mathcal{J}(\nabla u), \nabla\psi\rangle,$$

where $\mathcal{J}$ denotes the almost complex structure, i.e. rotation by 90° in the tangent space. Then

$$J[\xi] = [(L - \Lambda)[\alpha] + \beta(\mathcal{B}B \cdot B)]A + [(L + \Lambda)[\beta] + \alpha(\mathcal{B}A \cdot A)]B.$$

Proposition 6 *Let $X : \Sigma \to \mathbf{S}^3$ be a Willmore torus with conformal Gauss map Y. Let $\xi \in \Gamma(\perp Y)$. Then there exists a variation $X_\varepsilon : (-c,c) \times \Sigma \to \mathbf{S}^3$, with $X_0 = X$ whose conformal Gauss maps Y_ε satisfy*

$$\partial_\varepsilon(Y_\varepsilon)^\perp_{\varepsilon=0} = \xi,$$

if and only if

$$J(\xi) \cdot \underline{X} \equiv 0, \tag{22}$$

holds.

Proof. Assume that $\mathcal{H} \equiv 0$, i.e. that X is a Willmore surface. Consider a one parameter variation X_ε of X and let Y_ε be the corresponding conformal Gauss maps. By (13), we have

$$2\delta\mathcal{H} = \delta\big((\Delta_Y h + 2h)\underline{X}\big) = [\delta(\Delta_Y h + 2h)]\underline{X}. \tag{23}$$

Using (21), one can easily deduce that

$$J[\xi] = 2\delta\mathcal{H},$$

where $\xi := (\delta Y)^\perp$. From this and (23), the sufficiency follows.

Now let ξ be a smooth section of the normal bundle of Y satisfying (22). We can locally choose a framing $\{A, B\}$ for $\perp Y$ satisfying (18) with $A =: e^\tau\underline{X}$ for a smooth function τ. Write $\xi = \alpha A + \beta B$. Then previous calculations given above, $J[\xi] \cdot \underline{X} = J[\xi] \cdot A = 0$ implies

$$(L + \Lambda)[\beta] + \alpha(\mathcal{B}A \cdot A) = 0.$$

45

Note that $(\mathscr{B}A \cdot A) = 4e^{-2\rho}$ so this means that α *is determined by* β if (22) holds.

Consider the variation of X given by $X_\varepsilon = \cos(\varepsilon(-e^{-\tau}\beta))X + \sin(\varepsilon(-e^{-\tau}\beta))N$. Note

$$-e^{-\tau}\beta = \langle \dot{X}, N \rangle = -\langle X, \dot{N} \rangle = -\underline{X} \cdot (\dot{h}\underline{X} + h\underline{\dot{X}} + \dot{N}) = -\underline{X} \cdot \dot{Y} = -e^{-\tau}A \cdot \dot{Y}.$$

By the first part of the proof, the section of $\perp Y$ given by $\dot{Y}^{\perp}$ must satisfy $J[\dot{Y}^{\perp}] \cdot \underline{X} \equiv 0$. Therefore, $\dot{Y}^{\perp} \equiv \xi$, i.e. ξ arises as from a variation through conformal Gauss maps. **q.e.d.**

Theorem 3 *Let* $X : \Sigma \to \mathbf{S}^3$ *be as above. Then* X *is a stable Willmore immersion if and only if*

$$-\int_\Sigma \xi \cdot J(\xi) \, d\Sigma_Y$$

holds for all $\xi \in \Gamma(\perp Y)$ *satisfying*

$$J(\xi) \cdot \underline{X} \equiv 0.$$

In a general Lorentzian 4-manifold, spacelike zero mean curvature surfaces are neither local maxima nor minima for the area functional under compactly supported variations of the surface. Let $\mathcal{N}$ be a 4-dimensional Lorentzian manifold which satisfies the *null convergence condition*

$$\mathrm{Ricci}_{\mathcal{N}}(Z,Z) \geq 0, \quad \forall Z \in T\mathcal{N} \; s.t. \, Z \cdot Z = 0.$$

In [9], it was shown that if $\Sigma \to \mathcal{N}$ is a spacelike, zero mean curvature immersion, then Σ is locally stable with respect to compactly supported variations through surfaces with null mean curvature. It is somewhat surprising that surfaces with null mean curvature arise in relativity theory where they are called *marginally trapped surfaces*.

Proposition 7 *Suppose the eigenvalue problem*

$$(L+\Lambda)f + \lambda f(\mathscr{B}B \cdot B) = 0$$

has an eigenvalue satisfying $|\lambda| < 1$. *(Note that the eigenvalues of* $L + \Lambda$ *are, in general, complex). Then, the surface is not stable.*

If $J\xi \cdot A \equiv 0$, then $(L+\Lambda)[\beta] = -\alpha(\mathscr{B}A \cdot A)$, and so

$$
\begin{aligned}
-(\xi, J\xi) &= -\int (\xi \cdot A)(J\xi \cdot B) d\Sigma \\
&= -\int \beta((L-\Lambda)[\alpha] + \beta(\mathscr{B}B \cdot B) d\Sigma \\
&= -\int \alpha(L+\Lambda)[\beta] + \beta^2(\mathscr{B}B \cdot B) d\Sigma \\
&= \int (\alpha^2 - \beta^2)(\mathscr{B}B \cdot B) d\Sigma.
\end{aligned}
$$

Here we have used that $(\mathscr{B}A \cdot A) = (\mathscr{B}B \cdot B)$ and that $L+\Lambda$ is the adjoint of $L-\Lambda$. If f is as above, then we have

$$(L+\Lambda)\Re(f) = -\Re(\lambda f)(\mathscr{B}B \cdot B)$$

$$(L+\Lambda)\Im(f) = -\Im(\lambda f)(\mathscr{B}B \cdot B).$$

If we apply the above, with $\beta_1 = \Re(f), \beta_2 = \Im(f)$ and sum the results, we obtain

$$\int [(\alpha_1^2 - \beta_1^2) + (\alpha_2^2 - \beta_2^2)](\mathscr{B}B \cdot B)d\Sigma = \int (|\lambda|^2 - 1)|f|^2(\mathscr{B}B \cdot B)d\Sigma.$$

If $|\lambda| < 1$ holds then second variation for at least one of the fields determined by β_1 and β_2 must be negative.**q.e.d.**

Remark When $X : \Sigma \to \mathbf{S}^3$ is a minimal surface, then $K^\perp \equiv 0$ and $\Lambda = 0$. The eigenvalue problem in the proposition becomes,

$$\Delta_X f + (3 - 2K_X)f = -\lambda f,$$

where the subscript X indicates the quantities are those computed using the metric induced by X. The condition for instability $-1 < \lambda < 1$ can be interpreted by saying that the stability operator for the area functional $\Delta_X + (4 - 2K_X)$, has an eigenvalue in the interval $(-2,0)$. This condition was found by Weiner in [8].

Theorem 4 *If the norm of the quartic differential Q satisfies*

$$||Q||^2 \geq 1 , \tag{24}$$

with strict inequality at at least one point, then the surface is unstable.

Proof. We take $\beta := \sin \psi$, to obtain using (20). Then, $J[\xi] \cdot A \equiv 0$ becomes,

$$-4\alpha e^{-2\rho} = -4e^{-2\rho}(\sin \psi \cos 2\psi - \sin 2\psi \cos \psi) + 2\sin \psi.$$

Using a double angle formula, this gives

$$\alpha = \sin \psi (1 - \frac{1}{2}e^{2\rho}).$$

Therefore $\alpha^2 - \beta^2 = \sin^2 \psi[(1 - \frac{1}{2}e^{2\rho})^2 - 1]$ which is non positive if and only if $e^{2\rho} < 4$ holds. However with the normalization used above, we have $||Q|| = 4e^{-2\rho}$ and so the result follows. **q.e.d.**

Using the variation given above, the condition that the second variation is negative can be written

$$0 > \int_\Sigma (1 - \frac{1-K}{||Q||})(\frac{1}{||Q||-1})d\Sigma.$$

This is an open condition which is considerably more general than the condition (24).

The first examples of non isothermic Willmore tori were given by Pinkall [10]. These tori are the lifts of elastic curves $\gamma : \mathbf{S}^1 \to \mathbf{S}^2$ under the Hopf fibration $\pi : \mathbf{S}^3 \to \mathbf{S}^2$. Such surfaces are flat and we can introduce coordinates on T so that the metric has the form $dS_X^2 = ds^2 + dt^2$ and the mean curvature $h = h(t)$ satisfies the Euler-Lagrange equation

$$h'' + 2h(h^2 + 1) = 0. \tag{25}$$

Therefore, we have

$$(h')^2 + (h^2 + 1)^2 \equiv \text{constant} =: c + 1, \quad c \ge 0, \tag{26}$$

with $c = 0$ only in the case when $h \equiv 0$. The only time when this occurs is when $\gamma(S^1)$ is an equator and T is the Clifford torus.

The Hopf differentials of these tori are given by $\phi = h - i$. From this it follows that, except for the Clifford torus, the surfaces are not isothermic and, in particular, they are not conformally related to a minimal surface in any $M^3(c)$. A lengthy but straightforward calculation using (25), (26) and (17) shows that $Q = (-1/4)(c+1)d(t + is)^4$ and the norm of Q with respect to dS_γ^2 is

$$\|Q\|^2 = \frac{(c+1)^2}{(h^2+1)^4} = \left(\frac{(h')^2}{(h^2+1)^2} + 1 \right)^2,$$

using (26). So $\|Q\| \ge 1$ holds with equality only at the critical points of h and it follows from the previous theorem that, except for the Clifford torus, all these tori are unstable. The stability of the Clifford torus was shown by Weiner [8]. We wish to point out that the instability of these surfaces was first shown by Langer and Singer, [11], by showing the instability of the elastic curve γ. However, our proof only uses the S^1 symmetry of the surface to verify the inequality $\|Q\| \ge 1$.

BOUNDARY VALUE PROBLEMS

In this section we will treat uniqueness theorems for several boundary value problems for Willmore surfaces using geometric methods. Although the Willmore functional, or more generally, the energy functional of an elastic plate, was introduced in connection with a bordered surface, this subject appears not to have been widely treated in the mathematics literature. Some exceptions are the papers of Nitsche, [12], [13].

We will first treat the simplest such problem we can formulate, that of a Willmore surface bounded by a round circle. In order to obtain some intuition about what to expect, we first consider the linearized version of this problem which is the problem, [24], of finding a biharmonic function on, say, the unit disc D which vanishes on the circle. This problem is clearly not well posed. For any harmonic function h on D, we can solve the Dirichlet problem $\Delta u = h$, in D with $u \equiv 0$ on ∂D and thus obtain infinitely many biharmonic graphs over D with circular boundaries. It is not reasonable to expect to do better in the non linear case so we need to specify a second boundary condition which will make the problem well-posed. A natural one to chose is to require that the surface intersect the plane of the boundary in a constant angle. Then, both the condition that the surface be an extremal and the two boundary conditions will be in some sense conformally invariant.

Theorem 5 *Let D denote the unit disc and let $X : (D, \partial D) \to (\mathbf{R}^3, S^1)$ be a $C^4(\bar{D})$ immersion of a Willmore surface. Assume that the surface intersects the plane $\{x_3 = 0\}$ in a constant angle along ∂D. Then the image $X(D)$ is a spherical cap or a flat disc.*

Let $X : (D, \partial D) \to (\mathbf{R}^3, S^1)$ be any $C^1(\bar{D})$ immersion. We orient ∂D in the usual way (counterclockwise) and let t and n respectively denote the the tangent and outward pointing conormal defined along the boundary. We let v the unit normal field on D such that $n \times t = v$ holds on ∂D.

Let E_3 denote the vertical unit vector in $\mathbf{R}^3$. If we define an angle γ by $\cos \gamma := \langle X, v \rangle$ then it is easy to check that

$$\langle X, v \rangle = \cos \gamma = -\langle n, E_3 \rangle \tag{27}$$

$$\langle X, n \rangle = \sin \gamma = \langle v, E_3 \rangle \tag{28}$$

hold.

We will use the conformal Gauss map of the surface to derive important integral identities for the surface. We recall that in the case of a surface in $\mathbf{R}^3$, this map is expressed,

$$Y = H(X, \frac{X^2 - 1}{2}, \frac{X^2 + 1}{2}) + (v, \langle X, v \rangle, \langle X, v \rangle) \tag{29}$$

Away from umbilics in Σ, Y defines a conformal spacelike immersion from (Σ, ds_X^2) into $\mathbf{S}_1^4$. Recall that a surface is a Willmore surface if and only if Y defines a zero mean curvature immersion on Σ minus the umbilic set. In this case we have that Y satisfies

$$\Delta_X Y + 2(H^2 - K)Y = 0 \tag{30}$$

on Σ. As a consequence we have the following lemma.

Lemma 2 *(Flux formula) Let $X : \Sigma \to \mathbf{R}^3$ be a Willmore immersion and let $Y = (Y_1, ..., Y_5)$ be its conformal Gauss map. Then for $1 \le \alpha < \beta \le 5$, the forms defined by*

$$\omega_{\alpha\beta} := Y_\alpha * dY_\beta - Y_\beta * dY_\alpha$$

are closed.

Proof. Compute
$$d\omega_{\alpha\beta} = (Y_\alpha \Delta Y_\beta - Y_\beta \Delta Y_\alpha) * 1 = 0$$

by (30). **q.e.d.**

We next specialize the flux formula to a Willmore surface with circular boundary. We will let k_n denote the normal curvature of the boundary curve.

Lemma 3 *Let Σ be a compact, oriented surface with boundary homeomorphic to a circle and let $X : (\Sigma, \partial \Sigma) \to (\mathbf{R}^3, S^1)$ be a $C^4(\bar{\Sigma})$ Willmore immersion. Define a (not necessarily constant) angle γ by (27). Then the following hold:*

49

$$\int_{\partial \Sigma} k_n - H \, ds = 0 \tag{31}$$

$$\int_{\partial \Sigma} \cos(\gamma) \partial_n H + \sin(\gamma) H (H - k_n) \, ds = 0 \tag{32}$$

$$\int_{\partial \Sigma} \sin(\gamma) \partial_n H - \cos(\gamma) H (H - k_n) \, ds = 0 \tag{33}$$

Proof. We will use the following

$$\partial_n v_3 = \langle d v(n), E_3 \rangle = \langle n, E_3 \rangle \langle n, d v(n) \rangle = \langle n, E_3 \rangle (k_n - 2H) \tag{34}$$

and

$$\partial_n \langle X, v \rangle = \langle X, d v(n) \rangle = \langle X, n \rangle \langle n, d v(n) \rangle = \langle X, n \rangle (k_n - 2H) \tag{35}$$

Note that $Y|_{\partial \Sigma} = H(x_1, x_2, 0, 0, 1) + (v_1, v_2, \sin \gamma, \cos \gamma, \cos \gamma)$. Using this we compute on ∂D

$$
\begin{aligned}
\omega_{34} &= \sin \gamma * d(H(\frac{X^2 - 1}{2})) - \cos \gamma * d(H x_3 + v_3) \\
&= [\sin \gamma (H \langle X, n \rangle + \langle X, d v(n) \rangle - \cos \gamma (H \langle E_3, n \rangle + \langle E_3, d v(n) \rangle)] ds \\
&= [\sin \gamma (H \sin \gamma - \sin \gamma (2H - k_n)) + \cos \gamma (H (\cos \gamma) + \cos \gamma (k_n - 2H))] ds \\
&= [k_n - H] ds.
\end{aligned}
$$

Next note that $y_5 = y_4 + H$ holds and so

$$
\begin{aligned}
\omega_{45} &= y_4 * dH - H * dy_4 \\
&= \cos \gamma * dH - H * d(H \frac{X^2 - 1}{2} + \langle X, v \rangle) \\
&= [\cos \gamma \, \partial_n H - H(H \langle X, n \rangle + \partial_n \langle X, v \rangle)] ds \\
&= [\cos \gamma \, \partial_n H - H(H \langle X, n \rangle + \langle X, n \rangle (k_n - 2H))] ds \\
&= [\cos \gamma \, \partial_n H - H \sin \gamma (H + (k_n - 2H))] ds \\
&= [\cos \gamma \, \partial_n H + H \sin \gamma (H - k_n)] ds
\end{aligned}
$$

Again using $y_5 = y_4 + H$ we have

$$
\begin{aligned}
\omega_{35} &= \sin \gamma * dH - H * dy_3 + \omega_{34} \\
&= \sin \gamma * dH - H * (H x_3 + v_3) + \omega_{34} \\
&= [\sin \gamma \, \partial_n H - H(H \langle n, E_3 \rangle + \partial_n v_3)] ds + \omega_{34} \\
&= [\sin \gamma \, \partial_n H + H \cos \gamma (H + (k_n - 2H))] ds + \omega_{34} \\
&= [\sin \gamma \, \partial_n H - H \cos \gamma (H - k_n)] ds + \omega_{34}.
\end{aligned}
$$

The result then follows by applying Lemma 3.**q.e.d.**

Proof of Theorem 5. If the surface intersects the plane $\{x_3 = 0\}$ in a constant angle, then $\gamma \equiv$ const. and we easily obtain

$$\int_{\partial \Sigma} \partial_n H \, ds = 0 \tag{36}$$

and

$$\int_{\partial \Sigma} H(H - k_n) \, ds = 0 \tag{37}$$

from (32) and (33).

Using a classical theorem of Joachimsthal we have that since the angle between the surface and the plane is constant along the boundary, the boundary is a line of curvature. Letting "prime" denote differentiation with respect to arc length on ∂D we have

$$\text{const.} = \langle X, v \rangle = -\langle X'', v \rangle = \langle X', v' \rangle = -k_n$$

and so k_n is constant on ∂D.

Let $k_1 := k_n$ and $k_2 := 2H - k_n$ denote the principal curvatures on ∂D. Then (31) and (37) yield

$$\int_{\partial D} k_1 - k_2 \, ds = 0 \tag{38}$$

and

$$\int_{\partial D} k_1^2 - k_2^2 \, ds = 0 \tag{39}$$

By Hölder's inequality, we have for $j = 1, 2$

$$\left| \int_{\partial D} k_j \, ds \right| \le \int_{\partial D} |k_j| \, ds \le \left(\int_{\partial D} k_j^2 \, ds \right)^{1/2} (2\pi)^{1/2} \tag{40}$$

However, since $k_1 =$ const. we have equality in (40) for $j = 1$ and hence also for $j = 2$ by (38) and (39). We can then conclude from the necessary condition for equality in Hölder's inequality, that $k_2 \equiv$ const. also. In particular $h \equiv$ const. holds on the boundary and by (38), every boundary point is an umbilic.

Let II_Y denote the second fundamental form of the map Y. As shown by Bryant [3], the form

$$Q := II_Y^{(4,0)} \tag{41}$$

defines a holomorphic quartic differential on any Willmore surface. A calculation shows that Q is given locally in terms of a complex coordinate on D by $Q =: Q^{(4,0)} \, dz^4$, where

$$Q^{(4,0)} = \begin{cases} (\phi^2/4)(H^2 + \Delta \log \phi), & \text{if } \phi \ne 0; \\ -\phi_z H_z, & \text{if } \phi = 0. \end{cases}$$

Here ϕdz^2 is the Hopf differential which is the $(2,0)$ part of the second fundamental form of X. We recall that the Codazzi equations on D take the form

$$\phi_{\bar{z}} = e^\mu H_z \tag{42}$$

51

where $e^{\mu}|dz|^2$ is the local expression of the metric.

We will need only the second part of the formula for $Q^{(4,0)}$ which can be obtained as follows. Write (30) in the form $Y = HX + (v, q, q)$, where $q = \langle X, v \rangle$. Differentiation gives $Y_z = HX_z - \phi e^{-\mu} X_{\bar{z}}$. At a point where $\phi = 0$, differentiating again gives $Y_{zz} = H_{zz}X + H_z X_z - \phi_z e^{\mu} X_{\bar{z}}$. Therefore, at the point in question $Q^{(4,0)} = Y_{zz} \cdot Y_{zz} = -\phi_z H_z$ as claimed.

Since the zeros of ϕ correspond to umbilics, we have $\phi \equiv 0$ on ∂D. Letting $z = re^{i\theta}$ denote the usual coordinate on D, we have on ∂D

$$0 = \partial_\theta \phi = i(z\phi_z - \bar{z}\phi_{\bar{z}}) \tag{43}$$

and

$$0 = \partial_\theta H = i(zH_z - \bar{z}H_{\bar{z}}). \tag{44}$$

Therefore on ∂D we have, using (42), (43), and (44)

$$-q = \phi_z H_z = \bar{z}z\phi_z H_z = (\bar{z})^2 \phi_{\bar{z}} H_z = (\bar{z})^2 e^{\mu} H_z H_z = (\bar{z})^3 e^{\mu} z H_z H_z = (\bar{z})^4 e^{\mu} H_z H_{\bar{z}}.$$

It then follows that the function $z^4 q$ is holomorphic on the disc and is real valued on the boundary. By the maximum principle it follows that $z^4 q \equiv a$ for some real constant a. Since $q = a/z^4$ is holomorphic in D it then follows that $a = 0$ and so $q \equiv 0$ holds in D. It then follows from results of [3] that either $X(D)$ is, after a conformal transformation, a minimal immersion or $X(D)$ is part of a sphere. Since the set of umbilics is a conformal invariant, we would in the first case obtain a minimal surface in $\mathbf{R}^3$ having a boundary component made up entirely of umbilics. It follows then that the surface is a flat disc since its Hopf differential vanishes identically. The only remaining possibility is that the surface is a spherical cap. **q.e.d.**

A example of Babich and Bobenko, [14], see Figure 2, has the property that it cuts a horizontal plane in a constant angle along a circle of *umbilics* of the surface. This surface is a topological disc but contains one singularity. In fact, the surface is self-similar near the point which corresponds to the origin in the figure.

By using the Implicit Function Theorem, it is possible to produce infinitely many dic-type Willmore surfaces which are bounded by a circle, [15].

FREE BOUNDARY

We will consider here a more general bending energy functional

$$E = \int_\Sigma ak_1^2 + 2bk_1 k_2 + ak_2^2 \, d\Sigma, \tag{45}$$

where $x, y \mapsto ax^2 + 2bxy + ay^2$ is a positive semi- definite quadratic form with constant coefficients. The energy E can easily be expressed:

$$E = \int_\Sigma \alpha H^2 + \beta K \, d\Sigma,$$

52

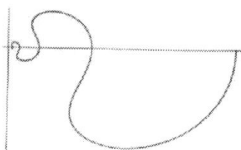

FIGURE 2. Generating curve of the Babich-Bobenko example.

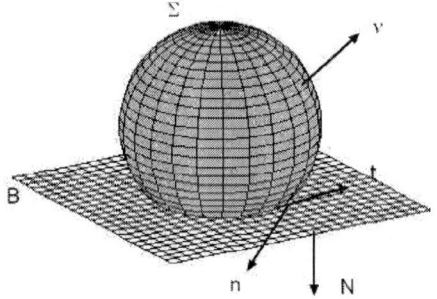

FIGURE 3. Different normal vector fields on the immersion.

with $a = \alpha/4$, $2b = (\alpha/2) + \beta$. The functional E is variationally equivalent to the Willmore functional if $\partial\Sigma = \emptyset$ or if the boundary is kept fixed to first order. For the free boundary problem we consider now, the problems for various choices of the constants are inequivalent. Their critica will, in general, satisfy different boundary conditions although they will satisfy the same equation in the interior of the surface.

We consider a domain $\Omega \subset \mathbf{R}^3$ with smooth boundary S. It is not required that $\Omega \cup S$ be compact. We consider the critical points of E among all immersions of compact surfaces $X : \Sigma \to \Omega$ such that $X(\partial\Sigma) \subset S$. Let t be the unit tangent vector to $\partial\Sigma$, let n be the unit conormal to $\partial\Sigma$, let v be the unit normal to Σ such that $v = n \times t$ on the boundary and let N denote the unit normal to S which points out of Ω. Let (σ_{ij}) be the components of the second fundamental form of Σ with respect to a frame which agrees with $\{n,t\}$ on $\partial\Sigma$. (See Figure 3.)

We express a variation of X as $X_\varepsilon = X + \varepsilon \dot{X} + \mathcal{O}(\varepsilon^2)$ where $\dot{X} =: \xi + \eta v$ with ξ tangent to Σ. A calculation then shows

$$
\begin{aligned}
\delta E &= \int_\Sigma \eta \left(\Delta H + 2H(H^2 - K) \right) d\Sigma \\
&+ \oint_{\partial\Sigma} (\alpha H + \beta \sigma_{22}) \eta_n - (\alpha H_n - \beta(\sigma_{12})_t) \eta + (\alpha H^2 + \beta K)\langle \xi, n \rangle \, ds .
\end{aligned}
\tag{46}
$$

The only constraint on the variations in that $\langle \dot{X}, N \rangle \equiv 0$ on $\partial\Sigma$.

We find that the Euler-Lagrange equations for the free boundary problem are

$$
\Delta H + 2H(H^2 - K) = 0, \qquad \text{in} \quad \Sigma .
\tag{47}
$$

$$
\alpha H + \beta \sigma_{22} = 0, \qquad \text{on} \quad \partial\Sigma .
\tag{48}
$$

$$
(\alpha H^2 + \beta K)\langle v, N \rangle + (\alpha H_n - \beta(\sigma_{12})_t)\langle n, N \rangle = 0 \quad \text{on} \quad \partial\Sigma .
\tag{49}
$$

We now specialize to the case where Ω is the half space $x_3 > 0$ and S is the plane $x_3 = 0$. Let Σ be a solution of (47)-(49). We assume that the surface is in equilibrium and compute the first variation with $\dot{X} = E_3 = v_3 + E_3^T$. Even though this variation does not preserve the constraint that $\partial\Sigma$ stays on S, the first variation is nevertheless zero since the variation is a symmetry of the problem. Using (47) and (48), we obtain from (46),

$$
0 = \oint_{\partial\Sigma} -(\alpha H_n - \beta(\sigma_{12})_t)v_3 + (\alpha H^2 + \beta K)\langle E_3, n \rangle \, ds .
\tag{50}
$$

Note that in (49) we have $(\alpha H^2 + \beta K) \geq 0$ and $\langle n, N \rangle = \langle n, -E_3 \rangle \geq 0$ since the surface is contained the upper half-plane. It then follows from (49) that $(\alpha H_n - \beta(\sigma_{12})_t)v_3 \geq 0$ must hold. But this means that both the terms in the integrand of the last integral are non positive, so they both must vanish,

$$
(\alpha H_n - \beta(\sigma_{12})_t)v_3 \equiv 0 ,
\tag{51}
$$

and

$$
(\alpha H^2 + \beta K)\langle E_3, n \rangle \equiv 0 ,
\tag{52}
$$

on $\partial\Sigma$. Since $\langle n, E_3 \rangle^2 + v_3^2 \equiv 1$ on $\partial\Sigma$, $\langle n, E_3 \rangle$ and v_3 cannot simultaneously vanish. At boundary points where $v_3 \neq 0$, we can multiply (49) by v_3 and use (51) to obtain $\alpha H^2 + \beta K = 0$. At points where $v_3 = 0$, we get the same conclusion from (52) and the remark above. We conclude that $(\alpha H^2 + \beta K) \equiv 0$ holds and by the assumptions on the functional, both principal curvatures vanish identically on the boundary.

We first consider the case $\alpha > 0 = \beta$. From the discussion above, we find that $H \equiv 0 \equiv H_n$ on the boundary. However H satisfies the second order elliptic equation (47) and the only solution of such an equation vanishing to first order on the boundary is $H \equiv 0$, so the surface is a minimal surface. If we next note that the boundary of the surface is contained in the plane, we find by the Maximum Principle that the surface is itself part of the plane.

Another case we will discuss is $\alpha = 1$, $\beta = -1$. This is the conformally invariant functional

$$
E = \int_\Sigma H^2 - K \, d\Sigma .
$$

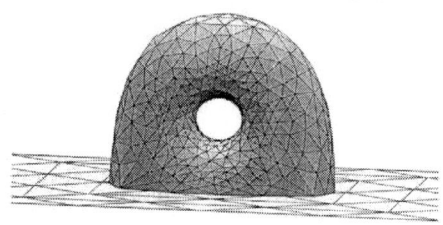

FIGURE 4.   A genus one Willmore surface with free boundary on a plane.

In this case we obtain from the argument above that $H^2 - K \equiv 0$ on the boundary of the surface, i.e. the boundary consists entirely of umbilics. If the topology of the surface is restricted to be that of a disc, then the following result holds, [15].

Theorem 6 *Let D denote the unit disc and let $X : (D, \partial D) \to (\mathbf{R}^3_+, \mathbf{R}^2)$ be a $C^4(\bar{D})$ immersion which is a solution of (47)-(49) with $\alpha = 1$, $\beta = -1$. Then $X(D)$ is either a spherical cap or flat disc meeting the plane in a constant angle.*

Figure 4 was produced using the Surface Evolver program and indicates that for higher topology there may be other Willmore surfaces with free boundary on a plane.

ANISOTROPIC SURFACE ENERGY

An anisotropic surface energy assigns to a surface an energy whose density at each point depends on the direction of the surface. We will first consider the simplest type of such an energy called a *parametric functional*.

Let $F : \mathbf{S}^2 \to \mathbf{R}^3$ be a sufficiently smooth, non negative function. To a smooth, oriented surface $X : \Sigma \to \mathbf{R}^3$ with normal map ν, we assign the value

$$\mathscr{F}[X] := \int_{\Sigma} F(\nu)\, d\Sigma.$$

(We will assume that Σ is at least relatively compact so that the integration makes sense.) This type of functional was introduced by the crystallographer G. Wulff [16] in order to model the shape of a small crystal. It also occurs in connection with nematic liquid crystals, [17], [18]. For example, Virga, [17], [19], has shown that such an energy determines the interface of small nematic liquid crystal drop in an isotropic environment. When the drop is small, the director field will tend to a constant and the surface energy will be of the form given above.

We briefly recall the *Wulff Construction*. Consider the radial plot of F on $\mathbf{S}^2$, that is the map $\mathbf{S}^2 \to \mathbf{R}^3$, given by $\nu \mapsto F(\nu) \cdot \nu$. For each $\nu \in \mathbf{S}^2$ draw the plane Π perpendicular to ν through the point $F(\nu) \cdot \nu$ and throw away the component of $\mathbf{R}^3 \setminus \Pi$ which does not contain $\mathbf{S}^2$. When this is done at each point of the sphere, what remains is a convex

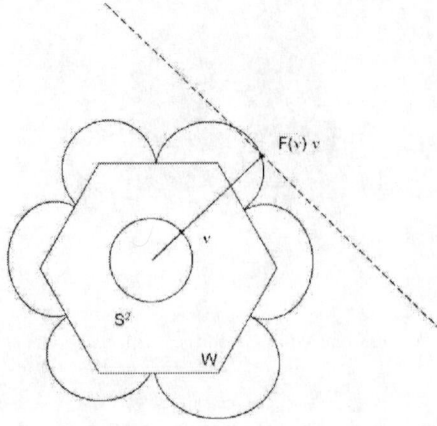

FIGURE 5. The Wulff construction.

region whose boundary will be called the *Wulff shape* or *Wulff crystal*. It wil be denoted by W. (In some references the region itself is given this name.) A result loosely known as *Wulff's Theorem* states that

Theorem 7 W *minimizes* $\mathscr{F}$ *among all closed surfaces enclosing the same volume.*

Thus, W solves the isoperimetric problem for the functional $\mathscr{F}$ and plays the role that the two dimensional sphere plays for the area functional.

We now assume that F is sufficiently smooth. We consider a variation of the surface X given by

$$X_\varepsilon = X + \varepsilon(\xi + \eta\nu) + \mathcal{O}(\varepsilon^2), \tag{53}$$

where ξ is tangent to Σ. A simple calculation shows that $\delta\nu = -\nabla\eta + d\nu(\xi)$. Let DF denote the gradient of F on $\mathbf{S}^2$. Note that $DF(\nu)$ can also be considered as a tangent vector field on Σ.

The first variation of $\mathscr{F}$ is

$$
\begin{aligned}
\delta\mathscr{F}[X] &= \int_\Sigma \langle DF, -\nabla\eta + d\nu(\xi)\rangle + F(\operatorname{div}\xi - 2H\eta)\,d\Sigma \\
&= \int_\Sigma \eta(\operatorname{div}DF - 2HF\eta) + \langle DF, d\nu(\xi)\rangle + F\operatorname{div}\xi\,d\Sigma \\
&= \int_\Sigma \eta(\operatorname{div}DF - 2HF\eta) + \langle d\nu(DF),\xi\rangle + F\operatorname{div}\xi\,d\Sigma \\
&= \int_\Sigma \eta(\operatorname{div}DF - 2HF\eta)\,d\Sigma + \oint_{\partial\Sigma} \langle DF,n\rangle + F\langle\xi,n\rangle\,ds,
\end{aligned}
$$

using that $d\nu(DF) = \nabla F$ and that $\operatorname{div}(F\xi) = \langle\nabla F,\xi\rangle + F\operatorname{div}\xi$.

If we restrict to compactly supported variations, we see that the the the first variation vanishes exactly when

$$0 \equiv \Lambda := -(\operatorname{div}DF - 2HF). \tag{54}$$

56

The function Λ is called the *anisotropic mean curvature*. Note that when $F \equiv 1$, Λ is twice the usual mean curvature. Critical points of $\mathscr{F}$ are characterized by the Euler-Lagrange equation $\Lambda = 0$ and more generally, critical points of $\mathscr{F}$ subject to a volume constraint are characterized by the equation $\Lambda \equiv$ constant.

We now will assume that W is a smooth, convex surface. Note that this is much stronger than requiring F to be smooth. If we let μ_j, $j = 1,2$ denote the principle curvatures of W with respect to the inward pointing normal, then the convexity implies $\mu_j < 0$ for $j = 1,2$. In some of our examples, we relax this and only require the non strict inequality.

In order to produce an F so that W is smooth, one can start with any smooth convex surface W. Since W is convex, the normal map N is a diffeomorphism onto $\mathbf{S}^2$. Let Q be the support function of W and define $F := Q \circ N^{-1}$. Then it is not hard to see that W will be the Wulff shape for the functional defined by this choice of F. In what follows, we will use the fact that the normal map is a diffeomorphism to consider all geometric quantities defined on W to also be defined on the sphere $\mathbf{S}^2$.

In this case, we will call $\mathscr{F}$ an *elliptic parametric functional*. There is a more general version in which F may depend on both v and the position vector X which is useful for inhomogeneous materials but we will not consider this case here. In the literature, the case $F = F(v)$ is referred to as a *constant coefficient* elliptic parametric functional.

When W is smooth and uniformly convex, the equation for prescribed anisotropic mean curvature is *absolutely elliptic* in the sense of Hopf, [20] . This means that the linearization of an equation for prescribed Λ is elliptic at any surface, not just a solution surface. It follows that a Maximum Principle analogous to the well known one for constant mean curvature surfaces holds. For example, if two surfaces with the same constant anisotropic mean curvature are in oriented contact at a point p and one of the surfaces lies on one side of the other near p, then the two surfaces must coincide on a neighborhood of p.

For closed surfaces with constant anisotropic mean curvature, we have the following uniqueness result, [21].

Theorem 8 *If $X : \Sigma \to \mathbf{R}^3$ is a closed surface with constant anisotropic mean curvature, then the surface is stable if and only if it is the Wulff shape modulo homotheties and those rigid motions of $\mathbf{R}^3$ which are symmetries of W.*

Let D^2F denote the Hessian of F on $\mathbf{S}^2$ and define $A := D^2F + FI$ where I is the identity on each tangent space to $\mathbf{S}^2$. By parallel translation in $\mathbf{R}^3$, $A_{v(p)}$ defines a linear transformation on each tangent space $T_p\Sigma$. Although Adv is not, in general, self-adjoint, it nevertheless has real eigenvalues which we denote by λ_i, $i = 1,2$. They will be called *anisotropic principal curvatures*. They satisfy:

$$\Lambda = \lambda_1 + \lambda_2 = -\operatorname{trace} A \circ dv , \tag{55}$$

and

$$\lambda_1 \lambda_2(p) = K_\Sigma(p)/K_W(v(p)) \tag{56}$$

Here K_W denotes the curvature of the Wulff shape.

The surfaces of revolution with constant anisotropic mean curvature are called *anisotropic Delaunay surfaces*, [22]. It is to be expected that these surfaces exist only

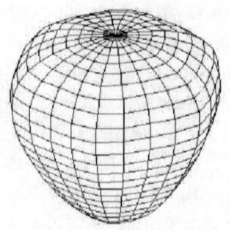

FIGURE 6. A Wulff shape of the form (57)

when the free energy is rotationally invariant, $F(v_3)$, i.e. when the Wulff shape is a surface of revolution. There is a generalization to the case when W has a product structure, [23]. We assume that W can be parameterized

$$\chi(\sigma, \tau) = (\sigma, \tau) \mapsto (u(\sigma)\alpha(\tau), u(\sigma)\beta(\tau), v(\sigma)), \qquad (57)$$

where (u, v) and (α, β) are convex curves. We then look for a surface given by

$$(s, \tau) \mapsto (x(s)\alpha(\tau), x(s)\beta(\tau), z(s)),$$

having constant anisotropic mean curvature for the functional defined for the given Wulff shape. It can be shown that this is the case provided the following hold.

$$2ux + \Lambda x^2 = c, \qquad (58)$$

$$z = \int_{v_0} x_u \, dv. \qquad (59)$$

In (58), Λ and c are constant. This equation must be solved to obtain $x = x(u)$ whose derivative appears in (59). The orientation of an anisotropic Delaunay surface may be chosen so that $\Lambda \leq 0$ holds and then the anisotropic Delaunay surfaces fall into six cases as follows:

- (I-1) $\Lambda = 0$ and $c = 0$: *horizontal plane.*
- (I-2) $\Lambda = 0$ and $c \neq 0$: *anisotropic catenoid.*
- (II-1) $\Lambda < 0$ and $c = 0$: *Wulff shape (up to vertical translation and homothety).*
- (II-2) $\Lambda < 0$ and $c = ((\mu_2|_{v_3=0})^2|\Lambda|)^{-1}$: *cylinder of radius* $(\mu_2|_{v_3=0}|\Lambda|)^{-1}$.
- (II-3) $\Lambda < 0$ and $((\mu_2|_{v_3=0})^2|\Lambda|)^{-1} > c > 0$: *anisotropic unduloid.*
- (II-4) $\Lambda < 0$ and $c < 0$: *anisotropic nodoid.*

Here, μ_2 is the principal curvature of the Wulff shape along a circle of latitude. The surfaces in each case above are complete and they have properties similar to those of the corresponding constant mean curvature surfaces. For example, the generating curve of

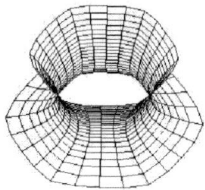

FIGURE 7. Anisotropic catenoid for the energy having Wulff shape given in Figure 6.

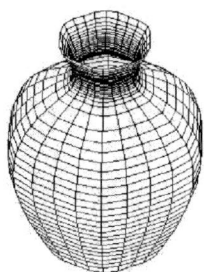

FIGURE 8. An anisotropic unduloid for the energy having Wulff shape given in in Figure 6.

an anisotropic unduloid is an embedded periodic curve, while the generating curve of an anisotropic nodoid is a periodic curve with self-intersections.

By analogy with the bending energy (45), we define an *anisotropic bending energy* by:

$$E_{a,b} := \int_{\Sigma} a(\lambda_1^2 + \lambda_2^2) + 2b\lambda_1\lambda_2 \, d\Sigma, \quad a > 0. \tag{60}$$

Here a and b are constants. When $F \equiv 1$ holds, the λ_i's are just the usual principal curvatures and the definition of E corresponds to the usual bending energy, [24]. These functional are scale invariant but are not, in general, conformally invariant. If the plate is closed ($\partial\Sigma = \emptyset$) or if the boundary values are kept fixed to first order under deformations, then $E_{a,b}$ is variationally equivalent to the anisotropic Willmore functional

$$\Xi[\Sigma] := \int_{\Sigma} \Lambda^2 \, d\Sigma.$$

In particular, $\delta E_{a,b} = 0$ if and only if $\delta\Xi = 0$.

59

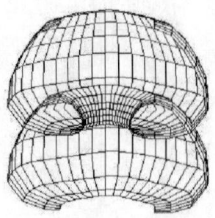

FIGURE 9. An anisotropic nodoid for the energy having Wulff shape given in in Figure 6.

We state the following simple result from [25] and [22].

Theorem 9 *Let $\Sigma \to \mathbf{R}^3$ be any immersed topological sphere. Then,*

$$\Xi(\Sigma) \geq \Xi(W) = 4\text{Area}(W)$$

holds with equality in the inequality if and only if Σ is either a homothety of W or Σ differs from W by a rigid motion which is a symmetry of W.

Proof. The definitions in (55) and (56) give that

$$0 \leq (\lambda_1 - \lambda_2)^2 = \Lambda^2 - 4K_\Sigma/K_W \ .$$

It is known that $\lambda_1 \equiv \lambda_2$ only for surfaces which differ from W by a homothety or a rigid motion which is a symmetry of W, [26]. Let $d\Omega$ denote the usual area form on S^2.

$$\begin{aligned}
\Xi(\Sigma) = \int_\Sigma \Lambda^2 \, d\Sigma \ &\geq \ \int_\Sigma 4K_\Sigma/K_W \, d\Sigma \\
&= \ \int_{S^2} 4/K_W \, d\Omega \\
&= \ 4\text{Area}(W) \\
&= \ \int_W \Lambda^2 \, dW.
\end{aligned}$$

Here we have used the change of variables theorem to equate the first and second lines together with the fact that the Gauss map of a genus zero surface has degree one. The last equality is that the anisotropic mean curvature of W with the outwards orientation is -2. **q.e.d**

The three theorems already stated in this section give a nice picture of how certain results for isotropic functionals can be extended to the anisotropic case.

We consider now the first variation of the functional Ξ. We include the boundary terms for use later. The pointwise variation of the anisotropic mean curvature for the variation (53) is given by

$$\delta\Lambda = L[\eta] + \langle \nabla\Lambda, \xi \rangle \ ,$$

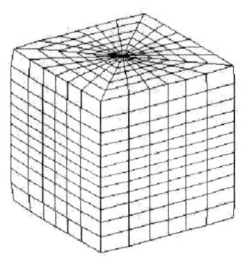

FIGURE 10. A Wulff shape of the form (57).

where L is the self adjoint operator

$$L[\eta] = \mathrm{div} A \nabla \eta + \langle A \cdot dv, dv \rangle \eta .$$

$$
\begin{aligned}
\delta \Xi &= \int_{\Sigma} 2\Lambda \delta\Lambda + \Lambda^2 (\mathrm{div}\xi - 2H\eta)\, d\Sigma \\
&= \int_{\Sigma} 2\Lambda (L[\eta] + \langle \nabla\Lambda, \xi \rangle) + \Lambda^2 (\mathrm{div}\xi - 2H\eta)\, d\Sigma \\
&= \int_{\Sigma} 2\eta (L - H\Lambda)[\Lambda]\, d\Sigma + \oint_{\partial\Sigma} 2\Lambda \langle A\nabla\eta, n \rangle - 2\eta \langle A\nabla\eta, n \rangle + \Lambda^2 \langle \xi, n \rangle\, ds.
\end{aligned}
$$

By restricting to compactly supported variations, we obtain the Euler Lagrange equation

$$(L - H\Lambda)[\Lambda] = 0 . \tag{61}$$

This is a fourth order non linear equation for the surface. Any C^4 surfaces which satisfies this equation will be called an *anisotropic Willmore surface*.

We will now show how to obtain some simple examples of anisotropic Willmore surfaces.

- **Rescalings of W.** It is clear from Theorem 9 that any such surface must solve the equation (61).
- **Anisotropic catenoids** [22]. This is a special case of the construction of anisotropic Delaunay surfaces given above. The Wulff shape is assumed to be given by (57). The anisotropic catenoid is a surface with anisotropic mean curvature $\Lambda \equiv 0$ parameterized as follows. Its profile curve $(x(\sigma), z(\sigma))$ is given by:

$$x = \frac{c}{2u}, \quad z = \frac{-c}{2} \int^{\sigma} \frac{dv(\sigma)}{u^2} .$$

The catenoid can then be parameterized:

$$X(\sigma, t) = (x(\sigma)\alpha(t), x(\sigma)\beta(t), z(\sigma)), \tag{62}$$

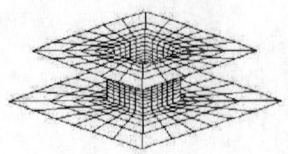

FIGURE 11.   Anisotropic catenoid for the energy having Wulff shape given in Figure 10.

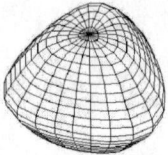

FIGURE 12.   A Wulff shape of the form (57).

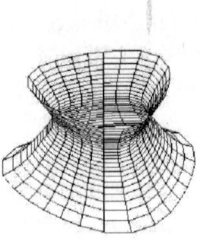

FIGURE 13.   Anisotropic catenoid for the energy having Wulff shape given in Figure 12.

where $c \neq 0$ is an arbitrary constant. The catenoid has the property that *sufficiently small pieces of it minimize the anisotropic energy defined by W among all surfaces having the same boundary.*

- **Helicoids.** Let

$$\mathscr{F} = \int F(v_3) \, d\Sigma$$

be a rotationally invariant elliptic functional. We will consider surfaces representable as graphs

$$x, y \mapsto (x, y, z(x, y)).$$

Let $d^2x = dxdy$ and note that the induced measure is given by $d\Sigma = v_3^{-1} d^2x$, so

$$\mathscr{F} = \int F(v_3) v_3^{-1} \, d^2x.$$

If we consider a variation of the function $z : z \to z + \varepsilon \dot{z} + \dots$, the corresponding change in v_3 is

$$\dot{v}_3 = -v_3^3 \langle Dz, D\dot{z} \rangle,$$

and so (by a simple calculation)

$$\delta(F \, d\Sigma) = \frac{v_3}{\mu_2} \langle Dz, D\dot{z} \rangle \, d^2x.$$

Integrating this last expression by parts yields the Euler-Lagrange equation $\Lambda = 0$ is equivalent to

$$\mathrm{Div}(\frac{v_3}{\mu_2} Dz) = 0.$$

Let (u, v) be the profile curve of the Wulff shape W. The axial coordinate of the Wulff shape is given by,

$$u = \frac{\sqrt{1 - v_3^2}}{\mu_2}.$$

Using this and the formula $v_3 = (1 + |Dz|^2)^{-1/2}$, allows us to rewrite the Euler Lagrange equation as

$$\mathrm{Div}(u \frac{Dz}{|Dz|}) = 0. \tag{63}$$

Here u is the function of v_3 given above re-expressed as a function of $|Dz|$.

Proposition 8 *For any $a \in \mathbf{R}^+$, the (multi-valued) function $z = a\theta := a \arctan(y/x)$ solves (63). In particular, for any rotationally invariant functional, the usual helicoids*

$$x, y \mapsto (x, y, a\theta)$$

have zero anisotropic mean curvature.

Proof. Set $\vec{r} = (x,y)$, $J\vec{r} = (-y,x)$ and $r = \sqrt{x^2+y^2}$. Then

$$D(a\theta) = a\frac{J\vec{r}}{r^2}, \quad \frac{D(a\theta)}{|D(a\theta)|} = \frac{J\vec{r}}{r}.$$

We then have $v_2 = (1+|Da\theta|^2)^{-1/2} = (1+a^2/r^2)^{-1/2}$ so that $u(v_3) = u((1+a^2/r^2)^{-1/2}) =: U(r)$. Combining these facts, we find that the $\Lambda \equiv 0$ is equivalent to

$$\text{Div}(\frac{U(r)}{r}J\vec{r}) = 0.$$

Note that

$$\text{Div}(\frac{U(r)}{r}J\vec{r}) = \partial_r(\frac{U(r)}{r})\langle Dr, J\vec{r}\rangle + \frac{U(r)}{r}\text{Div}J\vec{r}.$$

However $Dr = \vec{r}/r \perp J\vec{r}$ and $\text{Div}(J\vec{r}) = 0$ so the result follows. **q.e.d**

For more general functionals it is sometimes possible to construct a surface with $\Lambda \equiv 0$ which is a periodic multi-valued graph over the punctured plane.

- **Anisotropic Willmore surfaces of revolution.** We assume that the Wulff shape W is a surface of revolution (with vertical axis) and look for another surface of revolution Σ which satisfies the Euler-Lagrange equation (61). For this we will use the boundary terms in the first variation formula. We take the the variation field δX to be a symmetry of the Lagrangian, Then if Σ is a part of a surface of revolution satisfying (61), we have

$$0 = \delta\Xi = \oint_{\partial\Sigma} 2\Lambda\langle A\nabla\eta, n\rangle - 2\eta\langle A\nabla\Lambda, n\rangle + \Lambda^2\langle \xi, n\rangle \, ds. \tag{64}$$

Assuming that the boundary consists of two horizontal circles C_{Top} and C_{Bottom}. We take first, $\delta X = E_3 = E_3^T + v_3 v$ and then take $\delta X = X = X^T + qv$ where q is the support function. These variations fields correspond to vertical translation and homothety which are symmetries of Ξ. For both these variational fields, the integrands in (64) are constant on each of the boundary circles. The integrals are just a constant times the arc length $2\pi x$ and the results are the two equations given below:

$$x_s\frac{2\Lambda_s}{\mu_1} + z_s\Lambda(\frac{k_2}{\mu_2} - \frac{k_1}{\mu_1}) = \frac{C_1}{x}, \tag{65}$$

$$-q\frac{2\Lambda_s}{\mu_1} + \langle X, X_s\rangle\Lambda(\frac{k_2}{\mu_2} - \frac{k_1}{\mu_1}) = \frac{C_2}{x}. \tag{66}$$

Computing the determinant $x_s\langle X, X_s\rangle + qz_s = x$ and inverting this system, yields

$$\frac{2\Lambda_s}{\mu_1} = \frac{1}{x^2}(-C_2 z_S + C_1\langle X, X_s\rangle) \tag{67}$$

$$\Lambda(\frac{k_2}{\mu_2} - \frac{k_1}{\mu_1}) = \frac{1}{x^2}(C_2 x_s + C_1 q) \tag{68}$$

if we make a vertical translation of the generating curve $(x,z) \rightarrow (x, z+c)$ changes the support function q according to $q \rightarrow q - cx_s =: q_1$. Thus choosing $cc_1 = -c_2$ changes (68) to the simpler

$$\Lambda(\lambda_2 - \lambda_1) = \frac{cq}{x^2}, \tag{69}$$

where we have renamed q_1 as q. In the isotropic case $F \equiv 1$, this equation occurs in [27].

Remarks. (i) In the case of a surface of revolution , the equation (61) becomes,

$$\frac{1}{x}(\frac{x\Lambda_s}{\mu_1})_s + (\frac{k_1^2}{\mu_1} + \frac{k_2^2}{\mu_2} - \Lambda H)\Lambda = 0. \tag{70}$$

It can be shown that, except for the case when the surface is a round cylinder, equations (69) implies equation (70). For this one needs the "Codazzi equation "

$$x^2\Lambda_s = (x^2(\lambda_1 - \lambda_2))_s,$$

which is easily derived from (72) and (73) below.

(ii) An anisotropic Willmore surface of revolution with zero or one boundary component is, up to rescaling, the Wulf shape W or a flat disc. To see this, note that in either of these cases we can let one of the boundary circles shrink to a point and we obtain in (67) and (68) that the C_i's are both zero. Thus from (69) $\Lambda(\lambda_2 - \lambda_1) = 0$ holds. If $\Lambda \equiv 0$, is is easy to see that the surface must be a flat disc since the Gaussian curvature is non positive everywhere. If $\lambda_1 \equiv \lambda_2$, then the surface is known, [26], to be homothetic to a part of W.

In order to make these conclusions, it is essential that the surface be sufficiently regular, i.e. of class C^4. For example, one can invert the catenoid to get a closed surface but this surface fails to have the required differentiability.

Denote by $(u(\sigma), v(\sigma))$, $(x(s), z(s))$ the arc length parameterizations of the generating curve of the Wulff shape and of a second surface of revolution Σ respectively. At points where the normals to these two curves agree, we have

$$u_\sigma = x_s, \qquad v_\sigma = z_s. \tag{71}$$

We denote the principal curvatures of W (with respect to the outward pointing normal) by μ_i, $i = 1, 2$ and we denote the principal curvatures of Σ by k_i, $i = 1, 2$. Then $z_{ss} = -k_1 x_s$, $v_{\sigma\sigma} = \mu_1 u_\sigma$. It follows that from (71) that,

$$-\lambda_1 = \frac{z_{ss}}{v_{\sigma\sigma}} = \frac{v_{\sigma s}}{v_{\sigma\sigma}} = \frac{d\sigma}{ds}.$$

Then using that $v_z = v_\sigma \sigma_s s_z$, we obtain

$$\lambda_1 := k_1/\mu_1 = -v_z. \tag{72}$$

Again, u and x are related by the equations (71). We also have that

$$k_2 = -z_s/x, \qquad \mu_2 = v_\sigma/u,$$

so by (71)

$$\lambda_2 = -u/x. \qquad (73)$$

If we now think of the generating curve of Σ as being given as a graph $x = x(z)$, we can then express the equation (69) as

$$u^2 - x^2 v_z^2 = \frac{c(x - zx_z)}{\sqrt{1 + x_z^2}}. \qquad (74)$$

We can easily obtain from (71) that

$$x_z = u_v. \qquad (75)$$

Because W is convex, we have that $u_{vv} < 0$ holds globally and thus it is possible to solve $V = V(u_v)$. Using (75), we can write the generating curve of W in the form $u = u(v)$ and replace u in (74) by $u(V(u_v)) =: U(x_z)$. Likewise, we can write

$$v_z = \frac{\partial V}{\partial u_v}(u_v) \cdot \frac{\partial u_v}{\partial z} = \frac{\partial V}{\partial u_v}(u_v) \cdot x_{zz}.$$

Using this to replace v_z in (74) and solving algebraically for the highest order derivative, gives the quasilinear *second order* ODE:

$$x_{zz} = \pm \frac{1}{xV_{u_v}(x_z)} \left[U^2(x_z) - \frac{c(x - zx_z)}{\sqrt{1 + x_z^2}} \right]^{1/2} \qquad (76)$$

for the generating curve $z = z(x)$. For a single example, it is possible that there is a sign change in (76), that is for different parts of the surface, a different choice of the sign must be chosen.

In Figures 14 through 16 we show pieces of several examples. In all cases the Wulff shape W_p is derived from a generating curve given by $u^p + v^p = 1$. We use the notation (p, c, x_i, x_{zi}, s) to denote the solution of (76) for the Wulff shape W_p having initial conditions $x(0) = x_i$, $x_z(0) = x_{zi}$ and where $s = \pm 1$ is used to denote the choice of sign.

• **Cylinders over anisotropic elastica.**
We consider a 1-dimensional, smooth Wulff shape Ω and the corresponding 1-dimensional anisotropic energy

$$\mathscr{F}[C] = \int_C F(v_2)\, ds.$$

Here C is a C^2 curve parameterized by arc length and $n := (n_1, n_2) = (z', -x')$ is its normal. Also F is the support function of W considered as a function on the circle via the inverse of the normal map. If $C + \varepsilon \eta(s)n$ is a deformation of C, then

$$\delta \mathscr{F}[C] = -\int_C \lambda \eta\, ds,$$

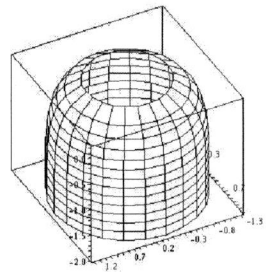

FIGURE 14. A piece of an anisotropic Willmore surface, $(4,.5,1,-1,-1)$.

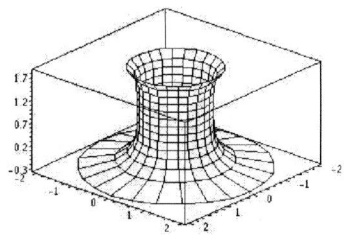

FIGURE 15. A piece of an anisotropic Willmore surface, $(4,.5,1,-1,1)$.

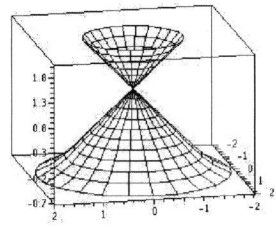

FIGURE 16. A piece of an anisotropic Willmore surface, $(4,1,1,-1,1)$.

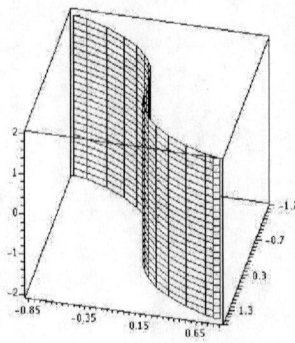

FIGURE 17. An anisotropic Willmore surface which is a cylinder over an anisotropic elastic curve. The Wulff shape is the surface of revolution generated by $u^4 + v^4 = 1$.

where $\lambda = k/\mu$. Here k denotes the curvature of C and

$$1/\mu = (1 - n_2^2)F''(n_2) - n_2 F'(n_2) + F(n_2),$$

is negative the reciprocal of the curvature of W. We introduce another energy for sufficiently smooth curves by

$$\mathscr{E}[C] := \int_C \lambda^2 - \tau \, ds,$$

where τ is a real constant. A lengthy calculation gives the Euler Lagrange equation

$$(\frac{\lambda_s}{\mu_1})_s + (\frac{k^2}{2\mu})\lambda + \frac{\tau k}{2} = 0. \tag{77}$$

Solutions will be called *generalized anisotropic elastica*. When $\tau = 0$ we will call the curves simply generalized elastic. If now the curve Ω is considered as the generating curve of a surface of revolution W, then any cylinder over an anisotropic elastic curve will be an anisotropic Willmore surface for the energy functional whose Wulff shape is W.

A free boundary problem

We will consider briefly a boundary value problem for the anisotropic bending energy (45). We assume that the Wulff shape is rotationally symmetric ($F = F(v_3)$) and uniformly convex. This means that the principal curvatures of W, μ_j, $j = 1, 2$ are everywhere positive. We will express (45) as

$$E_{a,b} = \int_\Sigma \alpha \Lambda^2 + \beta K_\Sigma / K_W \, d\Sigma.$$

We have discussed above the first variation of the integral of Λ^2 and we use

$$\delta \int_\Sigma K_\Sigma / K_W \, d\Sigma = \oint_{\partial\Sigma} \langle \delta\chi, \nu \times d\chi \rangle \, ds,$$

where $\delta\chi = A\delta\nu = A(-\nabla\eta + d\nu(\xi))$.

As in the section above where free boundary problems were considered in the isotropic case, we consider a surface in the half plane $x_3 > 0$ having free boundary on the plane $x_3 = 0$. Using the notation of that section, we obtain the first variation formula

$$\delta E_{a,b} \;=\; 2\alpha \int_\Sigma 2\eta (L - H\Lambda)[\Lambda] \, d\Sigma +$$

$$\oint_{\partial\Sigma} (\frac{2\alpha\Lambda + \beta\frac{\sigma_{22}}{\mu_2}}{\mu_1}) \eta_n - (\frac{2\alpha\Lambda_n}{\mu_1} - \beta(\frac{\sigma_{12}}{\mu_1\mu_2})_t) \eta + (\alpha\Lambda^2 + \beta\frac{K_\Sigma}{K_W}) \langle \xi, n \rangle \, ds.$$

Following the same steps as in the isotropic case, we arrive at the equations for equilibrium: We find that the Euler-Lagrange equations for the free boundary problem are

$$(L - H\Lambda)[\Lambda] = 0, \text{ in } \Sigma. \tag{78}$$

$$2\alpha\Lambda + \beta\frac{\sigma_{22}}{\mu_2} = 0, \qquad \text{on} \quad \partial\Sigma. \tag{79}$$

$$(\alpha\Lambda^2 + \beta\frac{K_\Sigma}{K_W}) \langle \nu, E_3 \rangle + (\frac{2\alpha\Lambda_n}{\mu_1} - \beta(\frac{\sigma_{12}}{\mu_1\mu_2})_t) \langle n, E_3 \rangle = 0 \quad \text{on} \quad \partial\Sigma. \tag{80}$$

The discrepancy between the constants here and in the isotropic case is because $\Lambda = 2H$ in the isotropic case. We will make the rather strong assumption that the quadratic form $\mathcal{Q}(\lambda_1, \lambda_2) := a(\lambda_1^2 + \lambda_2^2) + 2b\lambda_1\lambda_2$ which appears in the definition of $E_{a,b}$ is positive definite. Following the same steps as in the isotropic case, we arrive at the conclusion that $\alpha\Lambda^2 + \beta K_\Sigma / K_W \equiv 0$ on the boundary. By using that $\mathcal{Q}$ is positive definite, we conclude that the tensor field $A \cdot d\nu$ vanishes identically on the boundary since both of its eigenvalues vanish there. For the chosen frame on the boundary, we have,

$$A \cdot d\nu = \begin{pmatrix} -\sigma_{11}/\mu_1 & -\sigma_{12}/\mu_1 \\ -\sigma_{12}/\mu_2 & -\sigma_{22}/\mu_2 \end{pmatrix}$$

so we see that the vanishing of $A \cdot d\nu$ implies that the geodesic torsion σ_{12} of the boundary is identically zero.

From (78) and (80), we have that Λ satisfies $(L - H\Lambda)[\Lambda] = 0$ in Σ with $\Lambda \equiv 0 \equiv \Lambda_n$ on $\partial\Sigma$. Since $u \to (L - H\Lambda)[u]$ is a second order elliptic operator, we can conclude that $\Lambda \equiv 0$ in Σ, i.e. the surface is an anisotropic minimal surface. A Maximum Principle similar to the one for minimal surfaces holds also in the anisotropic case, an anisotropic minimal surface lies in the convex hull of its boundary. (Alternatively one can use that a surface with $\Lambda \equiv 0$, has non positive Gaussian curvature.) Therefore, if there is only one boundary component which is a planer curve, the surface must be flat.

ACKNOWLEDGMENTS

The author would like to thank Professors Miyuki Koiso and Udo Hertrich -Jeromin for helpful conversations in the preparation of this paper.

REFERENCES

1. J. J. Stoker, *Bull. Amer. Math. Soc.* **48**, 247–261 (1942).
2. N. H. Kuiper, *Ann. Math.*, **50**, 916 (1949))
3. R. Bryant, *J. Differential Geom.*, **20**, 23-53 (1984).
4. G. Thomsen, *Abh. Math. Sem. Univ. Hamburg*, 31-56 (1923).
5. C. Morrey, *Multiple Integrals in the Calculus of Variations*, Grundlehren Series Volume 130, Springer-Verlag, Berlin,(1966).
6. B. Palmer, *Ann. Global Analysis Geom.*, **9**, no.3, 305-317(1991).
7. D. Brill, and F. Flaherty, *Comm. Math. Phys.*, **50** , no. 2, 157–165(1976).
8. J. L. Weiner, *Indiana Univ. Math. J.*, **27**, no. 1, 19–35 (1978).
9. L. J. Alías and B. Palmer, *Classical Quantum Gravity*, **14** , no. 8, 2107–2111(1997).
10. U. Pinkall, *Invent. Math.* , **81**, no. 2, 379–386(1985).
11. J. Langer and D. Singer, *Ann. Global Anal. Geom.*, **5**, no. 2, 133–150 (1987).
12. J. C. C. Nitsche, *Statistical Thermodynamics and Differential Geometry of Microstructured Materials*, IMA Vol. Math. Appl. **51**, 69-98 , Springer, New York(1993).
13. J. C. C. Nitsche, *Quart. of Appl. Math.*, **LI**, 363-387(1993).
14. M. Babich and A. Bobenko, *Duke Math. J.*, **72**, 1141-185 (1992).
15. B. Palmer, *Indiana Univ. Math. J.*, **49**, no. 4, 1581–1601(2000).
16. G. Wulff Zeitschrift f. Krystallogr., textbf 34, 449-530 (1901)
17. E. G. Virga, *Arch. Rational Mech. Anal.*,**107**, no. 4, 371–390 (1989).
18. A. Rey, *Soft Matter*, 3, 1349 -1368(2007).
19. E. G. Virga, *Variational Theories for Liquid Crystals*, Chapman and Hall, London,(1994).
20. H. Hopf, *Differential geometry in the large*, Second edition, Lecture Notes in Mathematics, 1000. Springer-Verlag, Berlin(1989).
21. B. Palmer, *Proc. Amer. Math. Soc.*, **126** , no. 12, 3661–3667(1998).
22. M. Koiso and B. Palmer *Indiana Univ. Math. J.* , **54**, no. 6, 1817–1852(2005).
23. M. Koiso and B. Palmer *To appear in: Indiana Univ. Math. J.*
24. R. Courant and D. Hilbert, *Methods of Mathematical Physics*, Volume I, John Wiley and Sons, New York, 1989.
25. U. Clarenz, *Interfaces Free Bound.*, **6** , no. 3, 351–359(2004).
26. R. C. Reilly, *Duke Math. J.*, **43** , no. 4, 705–721(1976).
27. P. Castro-Villarreal and J. Guven, *J. Phys. A: Math. Theor.*, **40**, 4273-4283, (2007).

Simple Geometrical Models with Applications in Physics

Manuel Barros

Departamento de Geometría y Topología, Universidad de Granada, 18071 Granada, Spain.
E-mail:mbarros@ugr.es

Abstract. The main idea governing this article is to assume that the dynamics associated with a physical model is encoded in the geometry of their so called extended structures. According to this consideration, we collect some geometrical variational models constructed with dynamical variables being curves and surfaces. Consequently, these approaches can be viewed as field theories governed by Lagrangians whose densities are functions of the geometrical extrinsic invariants of their elementary fields. The geometrical models that we exhibit apply to different physical phenomena going from relativistic particles to bosonic string theories. Even to different physical contexts apparently unrelated as sigmamodels and membranes theories, magnetic flows and elastic rods, etc.

Keywords: particle, magnetic field, elastica, Hopf map, Willmore surface
PACS: 02.40.Hw; 02.40.Ma; 03.50.De; 11.10.Ef; 11.30.Er

INTRODUCTION

Roughly speaking, the reasons and motivations that stimulated me to prepare the material of this paper could be inserted in a philosophical context that may be described as follows. Most of experimental sciences study nature. In particular, Physics studies physical objects which provide the physical systems. These are modelled inside the spaces by means of geometrical structures, so we need Geometry. More precisely, extrinsic Geometry. However, this is not enough. Physical systems evolve, move and change, in the space due to forces and interactions producing physical phenomena. To model the motion, we need Calculus. Consequently, to obtain models describing physical phenomena, we need Differential Geometry and more precisely extrinsic Differential Geometry. In other words, Geometry of Curves, Surfaces and Submanifolds. The requirements of Poincaré and reparametrization invariance give special relevance to the extrinsic geometric invariants. These are encoded in two main structures. On one hand, the second fundamental form: mean curvature and length of the second fundamental form. On the other hand, the normal connection and the corresponding normal curvature. For example

- The second fundamental form of any curve, viewed as a one dimensional submanifold, in a space no matter dimension, is nothing but its curvature function, the first curvature function, also called its proper acceleration. Notice that the intrinsic geometry of a curve is trivial.
- The normal curvature of a curve in a three dimensional background is encoded in its torsion or second curvature function.

CP1002, *Curvature and Variational Modeling in Physics and Biophysics*
edited by O. J. Garay, E. García-Río, and R. Vázquez-Lorenzo
© 2008 American Institute of Physics 978-0-7354-0521-9/08/$23.00

- The extrinsic geometry of a surface in a three dimensional space is mainly encoded in its mean curvature function.
- More generally, the extrinsic geometry of a hypersurface is encoded exclusively in its second fundamental form, or shape operator, and mainly in its mean curvature, because its normal connection is trivial.

In this framework, the philosophy around which this mini-course was constructed can be explained as follows. Many physical phenomena can be modelled in the context of the theory of submanifolds. Essentially, in a physical terminology, they are field theories. In a geometrical context, they are geometrical variational problems. The elementary fields or dynamical variables in both contexts are mappings, immersions or submanifolds, from a source space into a target one. In both terminologies, the densities of the Lagrangians, governing either the field theory or the variational approach, involve the extrinsic geometry of submanifolds with a fixed topology, because we admit that fluctuations do not change the topology of the evolved elements. So the beauty of this approach and its aesthetically attractive point is to assume that the quantum states, quantum numbers, are encoded in the geometry of the extended structures, submanifolds.

Throughout this paper, we will restrict ourselves to the case where the source space is either one dimensional, a curve, or two dimensional, a surface. Therefore, we will construct geometrical models to describe relativistic particles both massive and massless so as other with applications in different contexts, a priori, unrelated. We will see that the bosonic string theory, *a la Polyakov*, is mainly a kind of Willmore theory. Also the nonlinear two-dimensional $O(3)$ sigmamodel, when the source space is compact, is nothing but the variational Willmore problem. This variational approach can be also applied to theories of membranes and vesicles.

In the first section, we discuss a typical and very popular example to illustrate the historical relationship, plenty of meetings and missing affairs, between Mathematics, in particular Geometry, and Physics. In 1930's P. Dirac [24, 25] created the idea of magnetic monopole, and just simultaneously H. Hopf [34] introduced its famous map to prove the non triviality of the third homotopy group of the two sphere. Nowadays, we know that the magnetic monopoles and circle bundles over the two sphere both look alike. We also evoke the Dirac quantization principle according to which, if there exist a magnetic monopole somewhere, the electric charge everywhere becomes quantized, it comes only in integer multiples of some basic quantity of charge. This is the first evidence of the existence of magnetic charges.

The Landau-Hall problem for Gauss magnetic fields on surfaces was studied in [8]. In the second section, we exhibit these magnetic fields appearing naturally on any surface under the presence of a magnetic monopole. The existence of a magnetic charge, with strength g somewhere in the universe, induces a Gaussian magnetic field on each immersed surface. The Landau-Hall problem associated with the dynamics of a test electric charge q moving through the surface under the influence of such Gauss magnetic field can be variationally approached according to a Lagrangian, acting on trajectories γ, with density governed by the proper acceleration κ of the test charge, that is, the curvature of trajectories. More precisely

$$\mathscr{F}(\gamma) = \int_{\gamma} \left((\kappa(s) + \frac{1}{g} \right) ds.$$

This provides, under the point of view assumed in this paper, a new theoretical evidence of the existence of magnetic charges.

In the next section, we explain a new philosophy to search for Lagrangians describing spinning particles. They have several advantages regarding those obtained from the conventional approach. For instance, the new models are intrinsic, they are established in the space where particles evolve, without any extra degrees of freedom. We collect some already well studied examples. Then, in the next section, we explore, in detail, the order one rigidity model. The whole dynamics of this model, including closed worldlines, is completely exhibited in two dimensional systems as well as backgrounds with constant curvature. It is important to say that in the latter case, where the target space has the highest rigidity, particles evolve on three dimensional totally geodesic submanifolds. Therefore, we will consider three dimensional backgrounds. Even more, it is enough to study, up to topology, the three sphere in the Riemannian context and the anti de Sitter three space in the Lorentzian one. In this cases, the Hopf mappings, once more, are the master pieces in solving the dynamics.

The second part of this paper is dedicated to field theories where the elementary fields are surfaces. The key model is based on the Willmore program. It applies to several apparently unrelated physical phenomena, including bosonic string theories, nonlinear sigma models and theories of membranes and vesicles. Then, we explain a method to get examples of configurations in these theories. The main idea is to get a geometric integration of the equations that govern the dynamics of these phenomena. To do it, we use two chief ingredients. First the conformal invariance of the Willmore program according which it is an approach that actually is stated in the conformal structure associated with the metric of the target space. Second the principle of symmetric criticality. It has been widely used in Mathematics and Physics, several times without being explicitly noticed, even in cases where the principle does not work. It is based in the search of symmetries and their exploitation, when they exist, in problem solving. We will use a nice version due to Palais, [40], that can be suggestively stated as *any critical symmetric object is a symmetric critical one*. Certainly, the principle is not valid in this general form, as we will see later. However, there exist nice, simple sufficient conditions to guarantee its validity. We will use it to get the moduli space of all Willmore revolution tori as that of Willmore Hopf tori in an ample class of three dimensional conformal structures.

CIRCLE BUNDLES OVER A TWO SPHERE

Sometimes, Mathematics and Physics strive for, at least a priori, quite different ideas and agendas. Maybe, these divorces are less known than romances and reconciliations have often been uneasy. However, the confluence of ideas, one from Physics and the other from Mathematics, produces a remarkable sea of new ideas. The story we have to tell contains these characteristics and we wish to exhibit it as an illustrative example of this confluence of ideas. The story started in the thirties and it has two main stars: H. Hopf and P. Dirac.

In 1931, H. Hopf, motivated by his interest in the *higher homotopy groups* of spheres, constructed its popular map

$$\Pi : \mathbb{S}^3 \longrightarrow \mathbb{S}^2.$$

In particular, Hopf used this map to show that the third homotopy group of a two sphere is not trivial

$$\pi_3(\mathbb{S}^2) \neq 0,$$

being the first example of a continuous map between spheres $\mathbb{S}^m \to \mathbb{S}^n$, with $m > n$, that is not *nulhomotopic*. This fact was quite surprising in the 1930's.

The Hopf map is very important not only in Mathematics but also in Physics [49]. It has a powerful and rich structure which will be very useful and so it will be exploited along this mini-course. In particular, it is a circle bundle, *something locally similar to the projection*

$$P : \mathbb{S}^2 \times \mathbb{S}^1 \longrightarrow \mathbb{S}^2.$$

Both, the Hopf map and the above mentioned projection provide two *different* circle bundles over $\mathbb{S}^2$. An interesting problem is to get the complete space, *moduli space*, of circle bundles over $\mathbb{S}^2$.

Let $\Gamma : \mathbf{P} \to \mathbb{S}^n$ be a $\mathbf{G}$-bundle over a sphere, $\mathbf{G}$ being a pathwise connected Lie group. Set $\mathbf{V}_1 = \mathbb{S}^n \backslash \{p\}$ and $\mathbf{V}_2 = \mathbb{S}^n \backslash \{-p\}$. Then $\mathbb{S}^n = \mathbf{V}_1 \bigcup \mathbf{V}_2$ and $\mathbf{V}_1 \bigcap \mathbf{V}_2$ is an open strip containing the equator $\mathbb{S}^{n-1}$. Certainly $\{\mathbf{V}_1, \mathbf{V}_2\}$ provides a trivializing cover. That is, for $1 \leq i \leq 2$ we have homeomorphisms

$$\psi_i : \Gamma^{-1} \to \mathbf{V}_i \times \mathbf{G}, \qquad \psi_i(u) = (\Gamma(u), \varphi_i(u)), \qquad \varphi_i(u \cdot g) = \varphi_i(u) \cdot g.$$

Therefore, we only need a transition function

$$G_{12} : \mathbf{V}_1 \bigcap \mathbf{V}_2 \longrightarrow \mathbf{G},$$

to determine the bundle, whose restriction to the equator gives the so called **characteristic map** of the $\mathbf{G}$-bundle

$$\mathbf{T} : \mathbb{S}^{n-1} \longrightarrow \mathbf{G}, \quad \mathbf{T} = G_{12}/\mathbb{S}^{n-1}.$$

Now the following two facts provide the complete classification of $\mathbf{G}$-bundles over $\mathbb{S}^n$.

1. The correspondence that associates to each $\mathbf{G}$-bundle $\mathbf{P}(\mathbb{S}^n, \mathbf{G})$ its characteristic function $\mathbf{T} : \mathbb{S}^{n-1} \longrightarrow \mathbf{G}$ is one-to-one.
2. Two $\mathbf{G}$-bundles, $\mathbf{P}_1(\mathbb{S}^n, \mathbf{G})$ and $\mathbf{P}_2(\mathbb{S}^n, \mathbf{G})$, are isomorphic if and only if their characteristic mappings $\mathbf{T}_1, \mathbf{T}_2 : \mathbb{S}^{n-1} \longrightarrow \mathbf{G}$ are homotopic.

Consequently, the moduli space of $\mathbf{G}$-bundles over $\mathbb{S}^n$ is identified with $\pi_{n-1}(\mathbf{G})$. In particular, the space of circle bundles over $\mathbb{S}^2$ is identified with $\pi_1(\mathbb{S}^1) \equiv \mathbb{Z}$.

MAGNETIC MONOPOLES

The Dirac magnetic monopole is an example in Physics which is crucial to understand the topological structure of the space, a manifold, where we are working. To give a brief description of what a magnetic monopole is, or could be, it is necessary to know a little bit about electric and magnetic fields. Roughly speaking, electric fields arise from the presence of point charges, like electrons and protons, which exert forces on other point charges. However, magnetic fields arise from the motion of the same electric point charges and corresponding forces only affect to moving charges. This is a powerful reason to think that magnetism is a purely relativistic effect; it is ultimately caused by the motion of the reference frame of the particle on which it is exerted. Thus there are electric and magnetic fields, but only electric charges. There are no magnetic charges, as far as we know. However, there is nothing in the universe which bars the possibility of the existence of such a charge, and as we shall see, it leads to some nice symmetries, and finally to some incredible conclusions.

Most of essential physics come from Maxwell's equations for electromagnetism, which, indeed, represent one of the most elegant ways to state the fundamentals of electricity and magnetism

$$\operatorname{div}(\vec{E}) = \rho, \qquad \operatorname{curl}(\vec{E}) = -\frac{d\vec{B}}{dt},$$

$$\operatorname{div}(\vec{B}) = 0, \qquad \operatorname{curl}(\vec{B}) = \vec{J} + \frac{d\vec{E}}{dt},$$

where ρ and $\vec{J}$ are the electric charge and current distributions, respectively. When Paul Dirac looked at these equations, he noticed an incredible amount of potential symmetry. In particular, if we set $\rho = \vec{J} = 0$, then the transformation $(\vec{E}, \vec{B}) \rightarrow (-\vec{B}, \vec{E})$, for example, is a symmetry of the theory. The electric and magnetic fields suddenly seem very interchangeable. Fortunate or unfortunately, ρ and $\vec{J}$ do not always vanish and this lack of symmetry constitutes one of the oldest puzzles in Physics. However, this potential symmetry was enough to convince Dirac to go further. So, Dirac studied Maxwell's equations augmented to include a magnetic charge and current distribution

$$\operatorname{div}(\vec{E}) = \rho_e, \qquad \operatorname{curl}(\vec{E}) = \vec{J}_m - \frac{d\vec{B}}{dt},$$

$$\operatorname{div}(\vec{B}) = \rho_m, \qquad \operatorname{curl}(\vec{B}) = \vec{J}_e + \frac{d\vec{E}}{dt}.$$

Now the symmetry is present in our theory. However, does it describe a real world?

In 1931, P. Dirac published its first paper on magnetic monopoles, [24]. Although the *magnetic analogue* of an electric charge has never been observed in nature, P. Dirac felt that such an object is worth thinking about anyway. He linked the existence of the magnetic monopole with the quantization of electric charge, another puzzle that is still not fully understood. Dirac showed that **if there exists a magnetic monopole then the electric charge is quantized** (*Dirac quantization principle*). Magnetic monopoles have been also predicted by some theories which seek for unify the electro-weak and strong interactions. Very recently, a physicist team observed an anomalous Hall effect in a ferromagnetic crystal which can be explained by the existence of magnetic monopoles (see [26, 55]).

Let us start by considering an electric charge q, which at the origin of $\mathbb{R}^3$ determines an electric field $\vec{E}$. It is described by the Law of Coulomb

$$\vec{E} = \left(\frac{q}{r^2}\right)\vec{\partial}_r, \quad \mathbb{R}^3 - \{O\}.$$

In a mimetic way with the Coulomb law, if we choose a magnetic charge density $\rho_m = 4\pi g\delta(r)$, where δ stands for the Dirac delta function, then the magnetic field produced is given by

$$\vec{B} = \left(\frac{g}{r^2}\right)\vec{\partial}_r, \quad \mathbb{R}^3 - \{O\},$$

g being a constant, the strength of the magnetic monopole.

As $\mathbb{R}^3 - \{O\}$ is simply connected, the magnetic field is provided by a scalar potential $h \in C^\infty(\mathbb{R}^3 - \{O\})$, such that

$$\vec{B} = \nabla h, \quad h = -g/r.$$

The existence of a scalar potential has a considerable significance. However, for reasons directly related with the quantum mechanical description of charged particles, which we hope to make clear later, in the case of the magnetic field is the existence of a vector potential what would be desirable,

Problem: Do exist a $\vec{V}$ such that $\mathrm{curl}(\vec{V}) = \vec{B}$ in $\mathbb{R}^3 - \{O\}$?

The answer to this problem is *no*, because

$$\pi_2(\mathbb{R}^3 - \{O\}) \neq 0.$$

Otherwise, take a sphere $\mathbb{S}^2$ with radius R surrounding the monopole, for example, centered at the origin. Let C be the equator and choose the usual orientation to apply the Stokes theorem. Then we have

$$\int_{\mathbb{S}^2} \langle \mathrm{curl}(\vec{V}), d\vec{S}\rangle = \oint_C \langle \vec{V}, d\vec{r}\rangle + \oint_{-C} \langle \vec{V}, d\vec{r}\rangle = 0.$$

On the other hand, the magnetic flux on $\mathbb{S}^2$ is

$$\int_{\mathbb{S}^2} \langle \vec{B}, d\vec{S}\rangle = \int_{\mathbb{S}^2} \left(\frac{g}{r^2}\right) dS = \frac{g}{R^2}(4\pi R^2) = 4\pi g,$$

which provides a contradiction.

To avoid the above argument, Dirac used a strategy based on the, so known in the physics literature, Dirac strings. He proceed as follows. First, one can integrate the equation

$$\mathrm{curl}(\vec{V}) = \vec{B} = \frac{g}{r^2}\vec{\partial}_r,$$

to obtain, up to an integration constant a, the following solution

$$\vec{V}(r,\phi,\theta) = \frac{g}{r\sin\phi}(a - \cos\phi)\vec{\partial}_r.$$

This equation blows up at $\phi = 0$ and $\phi = \pi$. So the vector potential obtained, for general values of a, is singular along the z-axis. However, Dirac used special values of a to do somewhat better. Thus, if one chooses $a = 1$ then the vector potential is well-behaved in $\phi = 0$, because the numerator vanishes faster than the denominator. This provides a singular vector potential for a magnetic monopole in just the negative z-axis. Now, one chooses $a = -1$ to get that the vector potential is singular in the positive z-axis. Therefore, Dirac considered two pieces with trivial second homotopy group as follows

$$Z_- = \{(0,0,z) \in \mathbb{R}^3 : z \leq 0\} \quad \text{and} \quad U_+ = \mathbb{R}^3 - Z_-,$$

$$Z_+ = \{(0,0,z) \in \mathbb{R}^3 : z \geq 0\} \quad \text{and} \quad U_- = \mathbb{R}^3 - Z_+.$$

Certainly, $\pi_2(U_+) = \pi_2(U_-) = 0$ and this guarantees, as we saw above, the existence of vector potentials for a magnetic field. Namely, the vector fields

$$\vec{V}_+(r,\phi,\theta) = \frac{g}{r\sin\phi}(1 - \cos\phi)\vec{\partial}_r \qquad \text{in} \quad U_+$$

and

$$\vec{V}_-(r,\phi,\theta) = -\frac{g}{r\sin\phi}(1 + \cos\phi)\vec{\partial}_r \qquad \text{in} \quad U_-$$

satisfy

$$\mathrm{curl}(\vec{V}_+) = \vec{B} \qquad \text{in} \quad U_+$$

and

$$\mathrm{curl}(\vec{V}_-) = \vec{B} \qquad \text{in} \quad U_-$$

New Problem: What can we say in $U_+ \cap U_-$?

The magnetic field $\vec{B}$ in $U_+ \cap U_-$ admits a couple or vector potentials, $\vec{V}_+$ and $\vec{V}_-$, which, obviously, do not coincide. However, vector potentials, even when they exist, are not unique. Actually,

$$\text{If} \quad \mathrm{curl}(\vec{V}) = \vec{B} \quad \text{in} \quad \Omega \subset \mathbb{R}^3, \quad \text{then}$$

$$\mathrm{curl}(\vec{V} + \nabla h) = \vec{B}, \qquad \forall h \in C^\infty(\Omega).$$

Now, a simple computation shows that

$$\vec{V}_+ - \vec{V}_- = \nabla(2g\theta) \quad \text{in} \quad U_+ \cap U_-.$$

Looking for a Vector Potential

Let $(M^n, g = \langle , \rangle)$ be an n-dimensional Riemannian space with Levi-Civita connection ∇. A magnetic field in this background is just a closed two-form

$$F \in \Lambda_2(M), \quad dF = 0.$$

Associated with a magnetic field one has the Lorentz force, a skew symmetric operator given by

$$\langle \phi(X), Y \rangle = F(X, Y).$$

The main equation governing the dynamics of an electric charge inside a magnetic field is the Lorentz force equation

$$\nabla_{\gamma'} \gamma' = \phi(\gamma'),$$

whose solutions constitute the so called magnetic flow. These curves are the trajectories of a test electric charge inside the magnetic field.

To explain why we need a vector potential, we have to consider somewhat more complicated system than an isolated monopole. Just a test electric charge moving around the magnetic monopole.

Three dimensional systems are quite singular. In this kind of backgrounds, vector fields and two-forms can be canonically identified. To do that we use musical isomorphisms and Hodge operator:

$$\vec{B} \quad \mapsto \quad (\vec{B})^\sharp \quad \mapsto \quad \star\left((\vec{B})^\sharp \right),$$

the converse being

$$F \quad \mapsto \quad \star(F) \quad \mapsto \quad (\star(F))^\flat.$$

In this one-to-one correspondence, closed two-forms correspond with vector fields with divergence zero. Therefore, in dimension three, magnetic fields are nothing but free divergence vector fields. Furthermore, the Lorentz force equation can be written, using the cross product, as follows

$$\nabla_{\gamma'} \gamma' = \phi(\gamma') = \vec{B} \wedge \gamma',$$

which is the classical way to describe how the test electric charge responds to $\vec{B}$, i. e., the Hall effect. The basic physical principle underlying the Hall effect is the Lorentz force. When an electron moves along a direction perpendicular to an applied magnetic field, it experiences a force acting perpendicularly to both directions and moves in response to this. Even more, the existence of a vector potential for a magnetic field $\vec{B}$ is reflected in the fact that its dual two-form is exact

$$\mathrm{curl}(\vec{V}) = \vec{B} \quad \Leftrightarrow \quad F = \star\left((\vec{B})^\sharp \right) = d\,\omega$$

and the one-form ω is just the image of the vector potential under the musical isomorphism

$$\omega = (\vec{V})^{\sharp}.$$

It should be observed that the existence of a vector potential allows one to see the magnetic trajectories, of the test electric charge, as critical points of a certain action which involves in an essential way the vector potential. In other words, the Lorentz force equation can be viewed as the field equation in a variational approach. In fact, let Σ be the space of curves in M^3 connecting a pair of fixed points $p, q \in M$. Choose the Lagrangian $\mathscr{L} : \Sigma \to \mathbb{R}$ given by

$$\mathscr{L}(\gamma) = \frac{1}{2} \int \langle \gamma', \gamma' \rangle \, dt + \int_{\gamma} \omega(\gamma'(t)) \, dt.$$

A standard argument, involving some integrations by parts, allows us to compute $\delta \mathscr{L}_{\gamma} : T_{\gamma}(\Sigma) \to \mathbb{R}$. Therefore, for any $Z \in T_{\gamma}(\Sigma)$ we have

$$\delta \mathscr{L}_{\gamma}(Z) = \int_{\gamma} \langle \nabla_{\gamma'} \gamma' - \phi(\gamma'), Z \rangle \, dt.$$

Now, we can compute, locally, the dual potential associated with the magnetic field of a monopole. Remember that

$$\vec{B} = \left(\frac{g}{r^2} \right) \partial_r \qquad \text{in} \qquad \mathbb{R}^3 - \{O\}.$$

By identifying each point $p = (x, y, z) \in \mathbb{R}^3 - \{O\}$ with its position vector relative to the fixed origin, we have

$$\vec{B}(p) = \left(\frac{g}{r^3} \right) p, \qquad r = (x^2 + y^2 + z^2)^{1/2}, \qquad \text{in} \qquad \mathbb{R}^3 - \{O\}.$$

An easy computation shows that the corresponding two-form $F = \star\left((\vec{B})^{\sharp} \right)$ is given by

$$F = \left(\frac{g}{r^3} \right) (z \, dx \wedge dy + y \, dz \wedge dx + x \, dy \wedge dz).$$

This two-form is not exact on the whole $\mathbb{R}^3 - \{O\}$. However, we can locally get potentials in U_+ and U_-, namely

$$\omega_+ = \frac{g}{r(z+r)} (x \, dy - y \, dx), \qquad d\omega_+ = F \qquad \text{in} \qquad U_+$$

and

$$\omega_- = \frac{g}{r(z-r)} (x \, dy - y \, dx), \qquad d\omega_- = F \qquad \text{in} \qquad U_-$$

In spherical coordinates these potentials write down as

$$\omega_+ = g(1 - \cos \phi) \, d\theta \qquad \text{in} \qquad U_+$$

and

$$\omega_- = -g(1 + \cos \phi) \, d\theta \qquad \text{in} \qquad U_-,$$

which are independent of r and so they can be regarded as 1-forms on open sets in the $\{\phi\,\theta\}$-plane in $\mathbb{R}^3$. Now, spherical coordinates identify these open sets in the plane with the once punctured unit sphere, $\mathbb{S}^2\backslash\{(0,0,-1)\}$, for ω_+ and the once punctured unit sphere $\mathbb{S}^2\backslash\{(0,0,1)\}$, for ω_-. Moreover, the two form F can be regarded as one on the unit sphere, namely

$$F = g\,\Omega_2, \qquad \Omega_2 \quad \text{being the unit area element.}$$

Non-uniqueness of Vector Potential

The context we are considering changed dramatically with the advent of quantum theories. Therefore, the classical setting is described by a test electric charge q moving into a vicinity, say $\Omega \subset \mathbb{R}^3$, of the monopole whose own electromagnetic field has negligible effect on the monopole. However, in the new and current view the charge is not thought of as a point charge at all. Now, a charge is a quantum mechanical object which is described by its wavefunction. This is a complex valued function of the space and time

$$\psi : \Omega \times \mathbb{R} \longrightarrow \mathbb{C}, \qquad (p,t) \mapsto \psi(p,t),$$

which is believed to encode all of the physically measurable information about the charge. For example, the probability of finding the charge in some region **R**, at some instant t, is computed by

$$\int_{\mathbf{R}} |\psi|^2 \, dV, \qquad |\psi|^2 = \psi\bar{\psi}.$$

The wavefunction ψ of the charge q is obtained by solving the Schroedinger equation for the monopole-charge system

$$\frac{1}{2}\left(\nabla - \frac{q}{c^2}\vec{V}\right)\psi = i\hbar\frac{\partial \psi}{\partial t},$$

which involves, in a fundamental way, the vector potential for the magnetic monopole.

- **What about the wavefunction by replacing $\vec{V}$ by $\vec{V} + \nabla h$?**
  It can be shown that, replacing the vector potential, the wave function change as follows

$$\psi \to e^{iqh}\,\psi.$$

- **The wavefunction, when replacing the vector potential, changes only the phase but not the modulus (amplitude)**
  The phase changes in the wavefunction do not have, a priori, physical significance since all of physically measurable quantities associated with a charge q depend only on $|\psi|^2$. However, in 1959 Aharonov and Bohn, [1], suggested that, while the phase of a single charge is unmeasurable, the *relative phase* of two charges that interact should have observable consequences. This proposal was, in fact, experimented by Chambers in 1960, [23], with results that confirmed the expectations of Aharonov

80

and Bohm. Therefore, we have to search for the mathematical device for doing the vector potential's job, keeping track of the phase changes.

- **Exploring some rather exotic topological and geometrical territory. Constructing a circle bundle.**
    - First, note that, at each point on its trajectory, the phase of a charged particle is represented by an element $e^{i\omega}$ in $\mathbb{S}^1$, when the unit circle is observed as the subset of $\mathbb{C}$ made up of modulus one complex numbers.
    - As the trajectory lives in $\mathbb{R}^3 - \{O\}$, we imagine a copy of $\mathbb{S}^1$ *setting above* each point of $\mathbb{R}^3 - \{O\}$, acting as a sort of a notebook in order to record the phase of a charge whose trajectory goes through the point.
    - In this way, we have a kind of bundle $\mathbf{P}$, with fiber $\mathbb{S}^1$, which is called the space of phases. Certainly, we also have a natural projection, $\Pi : \mathbf{P} \rightarrow \mathbb{R}^3 - \{O\}$.
    - A charge moving in $\mathbb{R}^3 - \{O\}$ has, at each point p of its trajectory, a phase which is represented by a point, of the space of phases, in the fiber $\Pi^{-1}(p)$ on p.
    - Now, the changes of phase are explained by means of an action of $\mathbb{S}^1$ on itself fiber by fiber.
    - Furthermore, we have to solve a lifting problem. Given the trajectory $\alpha(r)$ of a charge in $\mathbb{R}^3 - \{O\}$ and the value $\psi_o \in \mathbf{P}$ of the phase, at some point $p_o = \alpha(r_o)$, the problem is to get a curve $\psi(r)$ (lifting), in the space of phases, with $\psi(r_o) = \psi_o$ and providing the evolution of the phase at all other points of the trajectory of the test charge under the influence of the monopole.
    - Finally, we will see later that we only need to know how the phase varies on the unit sphere $\mathbb{S}^2 \subset \mathbb{R}^3 - \{O\}$. So we will henceforth look for a **circle bundle** on $\mathbb{S}^2$, rather than on $\mathbb{R}^3 - \{O\}$.
- **This mathematical machinery was unknown to Dirac, but, even though he did not know, it was being developed simultaneously by H. Hopf.**
  To finish this introductory part, we will get the Dirac quantization principle.
  We have just obtained the couple of local vector potential for a monopole

$$\vec{V}_+(r,\phi,\theta) = \frac{g}{r\sin\phi}(1-\cos\phi)\,\vec{\partial}_r \qquad \text{in} \quad U_+,$$

and

$$\vec{V}_-(r,\phi,\theta) = -\frac{g}{r\sin\phi}(1+\cos\phi)\,\vec{\partial}_r \qquad \text{in} \quad U_-.$$

Let ψ_+ and ψ_- be the wavefunctions determined, via the Schroedinger equation, by $\vec{V}_+$ and $\vec{V}_-$. In $U_+ \cap U_-$, where both vector potential coexist, we know that

$$\vec{V}_+ = \vec{V}_- + \nabla(2g\theta) \quad \text{in} \quad U_+ \cap U_-.$$

Then, the corresponding wavefunctions are nicely related by

$$\psi_+ = e^{2iqg\theta}\,\psi_-.$$

Now, $U_+ \cap U_-$ contains the circle

$$\{(r, \phi, \theta) \in \mathbb{R}^3 - \{O\} : r = 1, \phi = \pi/2\}.$$

However, ψ_+ and ψ_- must assign exactly a complex value to each point of $U_+ \cap U_-$. As a consequence, for any fixed t, the mapping

$$\theta \quad \longmapsto \quad \theta + 2\pi$$

keep both ψ_+ and ψ_- unchanged. Therefore,

$$e^{2iqg(\theta+2\pi)} = e^{2iqg\theta}$$

and so

$$e^{2iqg\theta} \, e^{4iqg\pi} = e^{2iqg\theta},$$

which implies that

$$e^{4iqg\pi} = 1 \quad \Rightarrow \quad 4qg\pi = 2\pi n \quad \Rightarrow \quad qg = \frac{1}{2}n \quad n \in \mathbb{Z}.$$

If there exists a single magnetic monopole, of strength g, somewhere in the universe, then the electric charge everywhere becomes quantized, i.e., it comes only in integer multiples of some basic quantity of charge

$$q = \frac{n}{2g}, \quad n \in \mathbb{Z}.$$

GAUSS MAGNETIC FIELDS ARISING FROM A MONOPOLE

The two-form naturally associated with the magnetic field generated by a monopole can be viewed over the unit two-sphere as

$$F = g\Omega_2, \qquad \Omega_2 \quad \text{being the unit area element of} \quad \mathbb{S}^2.$$

Let **S** be an oriented surface immersed in $\mathbb{R}^3$ with Gauss map

$$N : \mathbf{S} \rightarrow \mathbb{S}^2.$$

The existence of a single magnetic monopole, with strength g, somewhere in the universe, induces a Gaussian magnetic field on the given surface by

$$N^*(g\Omega_2) = (g\,\mathbf{G})\,dA,$$

where **G** is the Gaussian curvature of the surface and dA its area element. The Lorentz force of a Gaussian magnetic field is given by

$$\phi(X) = (g\,\mathbf{G})\,\mathbf{J}, \quad \mathbf{J} \text{ being the complex structure on } \mathbf{S}.$$

It should be noticed that this magnetic field has an intrinsic nature and it does not depend on how the Riemannian surface is immersed in the Euclidean space.

Now, we are going to exhibit the magnetic flow of a Gaussian magnetic field arising from a variational approach, obviously, described intrinsically on the surface.

Suppose you have a Riemannian space $(M^n; g = \langle, \rangle)$ with dimension n, Levi-Civita connection ∇ and Riemannian operator $\mathbf{R}$. Choose a space $\mathbf{C}$ of suitable non-null curves living in the target space. We will precise later the term *suitable*. For any real number $m \in \mathbb{R}$ we consider the functional

$$\mathscr{F}_m : \mathbf{C} \to \mathbb{R} \qquad \mathscr{F}_m(\gamma) = \int_\gamma (\kappa(s) + m) \, ds,$$

where $\kappa =$ is the curvature or proper acceleration of curves in $\mathbf{C}$. To characterize the dynamics of this variational problem we first compute the field equations, whose solutions are the trajectories of the model. For $\gamma \in \mathbf{C}$, the tangent space $\mathbf{T}_\gamma \mathbf{C}$ of $\mathbf{C}$ in γ is identified with the space of vector fields along γ. If $W \in \mathbf{T}_\gamma \mathbf{C}$, then one can define a curve in $\mathbf{C}$ through γ in the direction of W, say for instance,

$$\Gamma : (-\varepsilon, \varepsilon) \to \mathbf{C} \qquad \Gamma(t) = \exp_{\gamma(s)} t W(s).$$

Then we have $\delta \mathscr{F}_m(\gamma) : \mathbf{T}_\gamma \mathbf{C} \to \mathbb{R}$ defined by

$$\delta \mathscr{F}_m(\gamma)[W] = \left[\frac{\partial}{\partial t} (\mathscr{F}_m(\Gamma(t))) \right]_{t=0}.$$

The first variation formula can be computed, as usual, from a standard argument which involves some integration by parts, to get

$$\delta \mathscr{F}_m(\gamma)[W] = \int_\gamma < \Omega(\gamma), W > ds + [\mathscr{B}(\gamma, W)]_0^L,$$

where $\Omega(\gamma)$ and $\mathscr{B}(\gamma, W)$ stand for the so called Euler-Lagrange and boundary operators, respectively. They are given by

$$\Omega(\gamma) = \nabla_T (\nabla_T N + (\kappa - m)T) + \mathbf{R}(N, T)T$$

and

$$\mathscr{B}(\gamma, W) = < \nabla_T W, N > + m < W, T > + \tau < W, B >.$$

It is not difficult to see that when $\mathbf{C}$ is made up of either closed or clamped curves then

$$\mathscr{B}(\gamma, W) = 0, \quad \forall W \in \mathbf{T}\gamma\mathbf{C}.$$

In this setting we have that

$$\gamma = \text{trajectory} \quad \Leftrightarrow \quad \delta \mathscr{F}_m(\gamma)[W] = 0, \quad \forall W \in \mathbf{T}\gamma\mathbf{C}.$$

Therefore, the trajectories are the solutions of the following Euler-Lagrange equation, or field equation,

$$\Omega(\gamma) = 0.$$

If the background is a surface with Gaussian curvature **G**, then the Euler-Lagrange equation providing the critical points, via the Frenet equation, turns out to be

$$m \kappa = \mathbf{G}.$$

When $m \neq 0$, this equation is nothing but the Lorentz force equation associated with the Gaussian magnetic field arising from a monopole with strength $g = 1/m$, that is,

$$N^*(g\,\Omega_2) = (g\,\mathbf{G})\,dA, \qquad g = \frac{1}{m}.$$

Then we have got

If there exists a single magnetic monopole, with strength g, somewhere in the universe, then it induces a Gaussian magnetic field on each immersed surface S. Hence, people living in such a surface perceive the Gaussian magnetic flow through trajectories which are critical points of $\mathscr{F}_{1/g}$.

PARTICLE MODELS ARISING FROM GEOMETRY

The searching for Lagrangians describing spinning particles, both massive and massless, has a long story. The conventional approach is based on an extension of the space-time, where particles are living, by auxiliary variables which, after quantization, provide the extra degrees of freedom required for integer and half-integer spin. The conventional approach is not satisfactory from a mathematical point of view. It is not sufficiently motivated and sometimes one does not know what the new extended space means. Even from the physical point of view, the conventional approach presents, from its own origin, certain problems. Though the details need not concern us, we will mention some of them.

- It gives up to a very old idea of describing particles from an intrinsic point of view, i.e., in the space-time where they live.
- It also drops to obtain models to describe, consistently, particles with proper maximal acceleration.
- Sometimes, the quantization of the models obtained in a conventional approach present difficulties to be quantized.

Conclusion. Since the conventional approach presents serious difficulties, we have to produce new ideas in looking for Lagrangians providing models to describe particles.

An interesting, and for now well explored, new approach started with a large series of articles, see for instance [3, 4, 7, 9, 27, 28, 29, 42, 43] and references therein. At the origin of this new approach, there exist important differences regarding the conventional one. Let us point out some of them.

- The new models are intrinsic. They are well stated and describe the particles inside the original space-time where the system is evolving. It is not necessary to add any new degree of freedom.

- The conventional Lagrangian, which imitates the fall free particle in a strange and extended space, with the intention of unifying gravity with other interactions, is replaced by a new Lagrangian, formulated in the original space-time, which is more complicated and subtle.
- To get an intrinsic formulation of Lagrangians, we have to pay something for it. The cost of the bill is given in terms of the inclusion of higher order derivatives in the Lagrangian densities. This make the new Lagrangians more complicated than in the conventional philosophy, where the Lagrangian density only depend on first derivatives (the length).
- The subtlety of this process emerges from the following considerations:
  – The Lagrangian density can not be arbitrarily constructed in terms of higher order derivatives.
  – Lagrangian densities must be invariant under motions, Poincaré and reparametrization invariance.
  – Thus, Lagrangian densities must be functions of the curvatures of the particle trajectories in the original space-time, which are the geometrical invariants.

The beauty of these models comes from an aesthetically new point which establish that the old added spinning degrees of freedom are nicely encoded in the geometry of the particle trajectories.

According to the above explained philosophy, to define a model of relativistic particles, in his classical status, we need

- First, a space-time where the dynamics of particles happens. That is, a Riemannian or Lorentzian space, say $(M^n; g = \langle,\rangle)$.
- In this n-dimensional space, a suitable (nicely regular) curve γ has $n-1$ curvature functions, $\kappa_1, \kappa_2, \cdots, \kappa_{n-1}$. The length function of the curve, as well as these curvatures, are invariant under the group of motions in $(M^n; g = \langle,\rangle)$, so they constitute the set of geometrical invariants associated with the curve. Sometimes, when the space is sufficiently rigid, those invariants are able to uniquely determine the curve, up to motions in the space.
- Now the Lagrangian densities in the new approach are suitable functions of the above stated curvatures, that is,

$$\Omega(\kappa_1, \kappa_2, \cdots, \kappa_{n-1}).$$

- Therefore, given a Lagrangian density, we consider a suitable space of curves **C** in $(M^n; g = \langle,\rangle)$. Then the variational problem associated with the action is

$$\mathscr{F} : \mathbf{C} \to \mathbb{R}, \quad \mathscr{F}(\alpha) = \int_\alpha \Omega(\kappa_1, \kappa_2, \cdots, \kappa_{n-1})(s)\, ds.$$

- This defines a model describing particles whose trajectories are the critical points of that variational problem.

Some Examples

Let us exhibit some examples of particle models according to the established philosophy. Certainly, the simplest choice for a Lagrangian density in the next context is

$$\Omega(\kappa_1, \kappa_2, \cdots, \kappa_{n-1}) = c, \quad \text{constant.}$$

This model describes free fall particles, i. e., particles evolving solely under the influence of gravity in $(M^n; g = \langle, \rangle)$. Therefore, the trajectories of these particles are the geodesics of $(M^n; g = \langle, \rangle)$.

In 1990, M. S. Plyushchay [43] proposed a beautiful Lagrangian to describe massless particles with helicity. That Lagrangian to model massless bosons is, up to a coupling constant (helicity), the proper acceleration of particles. That is, the curvature of trajectories

$$\Omega(\kappa_1, \kappa_2, \cdots, \kappa_{n-1}) = c\,\kappa_1 = c\,\kappa.$$

This is, up helicity, the total curvature of trajectories

$$\mathscr{F} : \mathbf{C} \to \mathbb{R}, \quad \mathscr{F}(\alpha) = c\int_\alpha \kappa(s)\,ds,$$

c being the helicity of particles, that is, the projection of vector spin onto the direction of linear momentum. As the spin, according to quantum mechanics, must be quantized, then helicity is discrete too and so quantized.

The above model was extended, by means of a Lagrange multiplier which is related with the mass of particles, to study massive particles. So the corresponding Lagrangian density is given by

$$\Omega(\kappa_1, \kappa_2, \cdots, \kappa_{n-1}) = c\,\kappa_1 + m.$$

The Lagrangian governing the so called extended (or constrained) Plyushchay model for massive bosons is

$$\mathscr{F} : \mathbf{C} \to \mathbb{R}, \quad \mathscr{F}(\alpha) = \int_\alpha (c\,\kappa(s) + m)\,ds.$$

The Lagrangian density given by

$$\Omega(\kappa_1, \kappa_2, \cdots, \kappa_{n-1}) = a\,\kappa_2 + b,$$

provides a model depending on the torsion of trajectories

$$\mathscr{F} : \mathbf{C} \to \mathbb{R}, \quad \mathscr{F}(\alpha) = \int_\alpha (a\,\kappa_2(s) + b)\,ds.$$

This model appeared in the Bose-Fermi transmutation mechanism and it has been widely studied, see for example [38, 44, 45].

The Lagrangian density being a linear function on both curvature and torsion of trajectories

$$\Omega_{mnp}(\kappa_1, \kappa_2, \cdots, \kappa_{n-1}) = m + n\,\kappa_1 + p\,\kappa_2,$$

provides a model depending on the torsion of trajectories

$$\mathscr{F}_{mnp} : \mathbf{C} \to \mathbb{R}, \qquad \mathscr{F}_{mnp}(\alpha) = \int_\alpha (m + n\,\kappa_1 + p\,\kappa_2)\,ds.$$

This Lagrangian has been widely studied to construct tachyonless models of relativistic particles (see [9] and references therein).

A break to explain what Tachyon means

Tachyon is a particle that moves faster than the speed of light. When one studies relativistic particles in a space-time, one has the linear momentum $\vec{P}$ and the so called squared mass, $M^2 = -\langle \vec{P}, \vec{P} \rangle$, of particles. According to the sign of the squared mass, i.e., minus the causal character of momentum, one has the so called mass spectra of particles

- **Tachyonic Sector**: $M^2 < 0$ $\quad \Rightarrow \vec{P}$ is space-like
- **Massless States**: $M^2 = 0$ $\quad \Rightarrow \vec{P}$ is light-like
- **Massive Sector**: $M^2 > 0$ $\quad \Rightarrow \vec{P}$ is time-like

Relativistic particles with Rigidity

Lagrangian densities depending on, a priori, an arbitrary function of the first curvature

$$\Omega(\kappa_1, \kappa_2, \cdots, \kappa_{n-1}) = F(\kappa_1)$$

have been also widely studied.

The Lagrangian governing the so called higher rigidity order model

$$\mathscr{F} : \mathbf{C} \to \mathbb{R}, \quad \mathscr{F}(\alpha) = \int_\alpha F(\kappa_1(s))\,ds.$$

In particular, they were considered in connection with the study of helical proteins (see [29]). The order two rigidity model, where $F(\kappa) = \kappa^2 + \lambda$ and λ is a constant, is the master piece in the classical theory for elastica due to Daniel Bernoulli. It has been recently, widely considered in a large series of papers, [5, 31, 35, 36, 37].

THE ORDER ONE RIGIDITY MODEL

Throughout the next sections, we will study, with some details, the dynamics associated with the constrained Plyushchay model. The framework or target space is an n-dimensional semi-Riemannian space, $(M^n; g = \langle, \rangle)$, with Levi-Civita connection ∇ and Riemannian operator $\mathbf{R}$. The space of elementary fields is a space of suitable non-null curves **C** living in the target space. For any real number $m \in \mathbb{R}$, we consider the functional

$$\mathscr{F}_m : \mathbf{C} \to \mathbb{R}, \qquad \mathscr{F}_m(\gamma) = \int_\gamma (\kappa(s) + m) \, ds,$$

where we have chosen, sending an apology to physicists, helicity one for particles. Of course, $\kappa = \kappa_1$ is the curvature or proper acceleration.

In this framework, we ask for dynamics of this model. To determine the trajectories, we first need to compute the field equations and then try to solve them. However, we have just computed this equation already, when **C** is made up of either closed or clamped curves. Then the Euler-Lagrange equation turns out to be

$$\Omega(\gamma) = \nabla_T \left(\nabla_T N + (\kappa - m)T \right) + \mathbf{R}(N, T)T = 0,$$

which, via the Frenet equations, can be written as

$$\Omega(\gamma) = (\varepsilon_2 \varepsilon_3 \tau^2 + \varepsilon_1 \varepsilon_2 m \kappa)N - \varepsilon_3 \tau_s B - \varepsilon_3 \tau \eta - \mathbf{R}(N, T)T = 0,$$

η being always orthogonal to $\{T, N, B\}$ along the curve. Let $\{\varepsilon_1, \varepsilon_2, \varepsilon_3\}$ be the causal characters of tangent T, unit normal N and unit binormal B, respectively.

It should be noted that the influence of the target space on this equation is exerted through the sectional curvature of the osculating plane of trajectories. So, besides 2-dimensional case, which has been slightly considered already, we can obtain the whole dynamics on spaces with constant curvature. Therefore, we will consider both cases separately.

Two Dimensional Systems

Let us start giving some information on particle systems. In particular, we will try to explain that dimension two, in the theory of particles, is very special.

- The quantum state of a system is the set of numbers, quantum numbers, that fully describe a quantum system.
- Quantum numbers describe values of conserved quantities in the dynamics of the system: energy, linear momentum, angular momentum, spin, mass, helicity, etc.
- Since any quantum system can have one or more quantum numbers, it is a futile job to provide a list of all possible quantum numbers.
- In dimension three or larger, particle system are strongly constrained to quantum numbers. In particular, particles are bosons, from Satyendra Bose (integer spin, radiation particles) or fermions, from Enrico Fermi (half integer spin, matter particles).
- In dimension two, the quantum states of observed particles range continuously between bosons and fermions, taking any quantum value between them. In 1982, F. Wilczek, [52], coined the term anyon to name these particles.

However, anyon is a mathematical term and one can find mathematical reasons to explain the special behavior in dimension two. We use the following terminology

$SO(n,1)$ means the (n+1)-dimensional Lorentz group

and

$\mathbf{P}(n,1)$ means the (n+1)-dimensional Poincaré group,

so that

$$\mathbf{P}(n,1) = SO(n,1) \bigoplus \mathbb{R}^{n+1}.$$

Geometry. There exist projective representations of the Lorentz transformation group $SO(2,1)$ which do not arise from linear representations. The same happens for its double covering space $Spin(2,1)$. These representations are called **anyons**. However this is impossible for $SO(n,1)$, $n > 2$.

Topology. The fundamental group of $SO(2,1)$, as well as for $\mathbf{P}(2,1)$, is $\mathbb{Z}$, which is an obstruction to be $Spin(2,1)$ simply connected. Now, for $n > 2$, the fundamental group of $SO(n,1)$, and $\mathbf{P}(n,1)$, is $\mathbb{Z}_2$, which implies that $Spin(n,1)$ is simply connected, so that it is the universal cover of $SO(n,1)$.

As the models have their own interest in mathematics, despite their interest in physics, we will consider two cases according to the causal character of the Riemannian surface $(\mathbf{S}, g = \langle , \rangle)$. However, both formally behave similar.

(A) Riemannian Surfaces. Let $(\mathbf{S}, g = \langle , \rangle)$ be a Riemannian surface with Gaussian curvature $\mathbf{G}$. Then the trajectories of particles are the solutions of the following equation

$$m\,\kappa = \mathbf{G},$$

where Gaussian curvature appears restricted to the curves.

(B) Lorentzian Surfaces. Let $(\mathbf{S}, g = \langle , \rangle)$ be a Lorentzian surface with Gaussian curvature $\mathbf{G}$. Then the dynamics of the anyons in the rigidity order one model are solutions of the following field equation

$$m\,\kappa = \varepsilon_2 \mathbf{G}.$$

We have a funny consequence in both settings: trajectories of the free model, i.e., the massless model $m = 0$, correspond with those curves made up of parabolic points. In particular, the Lorentzian case can be paraphrased as follows: anyons in the massless, rigidity order one, model evolve along curves with parabolic points.

Certainly, these equations have obvious consequences, some of them can be obtained as exercises. In particular, one can construct models with large classes of non trivial trajectories. We will restrict ourselves to exhibit the following one which, in some sense, evokes the popular model of Beem and Bussemann.

Anyons Trajectories in the Beem-Bussemann Model

For the sake of simplicity we will pay attention to solutions with time-like worldlines. However the argument holds also for space-like ones. We consider the Lorentzian, warped-product, surface

$$M = I \times_f (-\mathbb{S}^1), \qquad I \subset \mathbb{R}, \qquad f : I \to \mathbb{R} \quad \text{positive}$$

and

$$g = dt^2 - f^2 d\theta^2 \qquad \text{being the warped metric.}$$

It is clear that the vector field ∂_t defines a geodesic flow in (M,g) which is the unit normal flow to the time-like foliation whose leaves are the slices $\{\{t\} \times \mathbb{S}^1 : t \in \mathbb{R}\}$.

On the other hand, the curvature of this slices is $\kappa = \frac{f'}{f}$, showing that slices are circles in (M,g). Then we compute the curvature operator and get

$$R(\partial_t, T)T = -m\kappa\partial_t.$$

So, the slices in this model which are anyon worldlines are characterized by

$$m\partial_t(f) = \partial_t\partial_t(f).$$

Then we find that the curved space-time $(M = I \times \mathbb{S}^1, g = dt^2 - f^2 d\theta^2)$, with $f : I \to \mathbb{R}$ given by $f(t) = e^{mt}$, admits a foliation by circles which are anyon worldlines.

It should be noticed that the above construction can be reproduced by changing I into a Riemann space, say (N,h), and f being a positive smooth function satisfying the following property. Let Σ be the set of critical points of f, that is,

$$\Sigma = \{p \in N : \nabla f = 0\}, \qquad \nabla f = \text{gradient.}$$

Then $U = \frac{\nabla f}{|\nabla f|}$ defines a unit-speed geodesic flow on $N - \Sigma$ and the field equation holds along this flow. In the case of space-like slices, this situation is equivalent to the existence of a geodesic and irrotational unit vector field in the direction of ∇f. Furthermore, if this is time-like, then it is (at least locally) a proper time synchronizable observer field.

The Dynamics on Constant Curvature Backgrounds

Assume now that $(M^n; g = \langle,\rangle)$ has constant curvature c. Then

$$\mathbf{R}(N,T)T = \varepsilon_1 c.$$

The field equations turn out to be written as

$$\begin{aligned}
\varepsilon_2\varepsilon_3\tau^2 + \varepsilon_1\varepsilon_2 m\kappa &= \varepsilon_1 c, \\
\tau_s &= 0, \\
\tau\eta &= 0.
\end{aligned}$$

These equations will have soon important consequences. The trajectories of particles have constant torsion. The dynamics of particles actually happens in dimension three. The worldlines of particles, of the constrained non free model, are helices, whose curvature and torsion are both constant and related by

$$\kappa = \frac{\varepsilon_1 c - \varepsilon_2\varepsilon_3\tau^2}{\varepsilon_1\varepsilon_2 m},$$

which constitutes a real one-parameter class of helices.

The free total curvature, or Plyushchay, model is consistent, i.e., it has a nontrivial dynamics if and only if the target space is, up to topology, either a three sphere or an anti-de-Sitter three space according to it is Riemannian or Lorentzian, respectively, because

$$m = 0 \quad \Rightarrow \quad \varepsilon_2\varepsilon_3\tau^2 = \varepsilon_1 c.$$

So, the dynamics of the free total curvature model in spaces of constant curvature is reduced, up to scaling the metric, to the following geometric problem

Find all curves in either the unit three sphere $\mathbb{S}^3$ or the unit three anti-de-Sitter space AdS$_3$, whose torsion satisfies $\tau^2 = 1$.

At this point, it is clear that dynamics in $\mathbb{S}^3$ and **AdS$_3$** are specially interesting. We will solve, separately, both problems using geometry.

THE DYNAMICS IN THE THREE SPHERE

In the unit three sphere we have to solve the following problem to obtain the whole dynamics of both models:

- **Constrained model.** Determine the one-parameter class of helices

$$\{(\kappa, \tau) \in \mathbb{R}^2 : m\kappa + \tau^2 = 1\}.$$

- **Free model.** Determine the class of curves with torsion τ satisfying $\tau^2 = 1$.

We are going to solve both problem using geometrical tools involved in the Hopf map. To fix the notation, let us write down the 3-sphere $\mathbb{S}^3 = \{z = (z_1, z_2) \in \mathbb{C}^2 : |z_1|^2 + |z_2|^2 = 1\}$, the 1-sphere $\mathbb{S}^1 = \{e^{it} : t \in \mathbb{R}\}$ and the classical left action given by

$$\mathbb{S}^1 \times \mathbb{S}^3 \rightarrow \mathbb{S}^3, \qquad e^{it}.(z_1, z_2) = (e^{it}.z_1, e^{it}.z_2).$$

The orbits generate a Killing vector field, with constant length, the Hopf vector field, $V(z) = i.z$, in the unit three sphere. The space of orbits is identified with a two sphere and the natural projection is a submersion, which becomes a Riemannian one when we consider the two sphere of radius $1/2$

$$\Pi : \mathbb{S}^3 \rightarrow \mathbb{S}^2(1/2), \quad \text{Riemannain submersion.}$$

Riemannian submersions constitute a kind of construction very useful in Geometry as well as in Physics (see [18, 39] for details on this topic). Certainly, we can used the paraphernalia associated with Riemannian submersions. In particular, we can talk about horizontal and vertical spaces. As for the Hopf map, we have

$$T_z\mathbb{S}^3 = \mathscr{H}_z \oplus \langle V_z \rangle.$$

Also, we have the notion of horizontal lifts of curves and vector fields on the basis. Therefore, we will denote with over-bars the horizontal lifts of corresponding objects.

For instance, if β is a curve in the two sphere, then $\bar{\beta}$ will denote a horizontal lift of β in the three sphere. If X is a vector field in the two sphere, then $\bar{X}$ will denote a horizontal lift of X in $\mathbb{S}^3$, and so on. We also have the fundamental equations of the submersion, which are also known as O'Neill's equations. In the case of the Hopf map, these equations turn out to be

$$\overline{\nabla}_{\bar{X}}\bar{Y} = \overline{\nabla_X Y} - (\langle JX, Y \rangle \circ \Pi)V, \overline{\nabla}_{\bar{X}}V = \overline{\nabla}_V \bar{X} = i\bar{X},$$
$$\overline{\nabla}_V V = 0.$$

The key point to understand the dynamics of this model is the idea of Hopf tube. For any curve $\beta(s)$ in $\mathbb{S}^2$, we have its complete lift

$$\mathbf{T}_\beta = \Pi^{-1}(\beta) = \{e^{it} . \bar{\beta}(s) : (s,t) \in \mathbb{R}^2\}.$$

Note that it is the union of all horizontal lifts of the given curve. The solution of our problem is nicely encoded in the geometry of these sets. To get it, we list some of their more important properties.

- Complete lifts are surfaces which can be nicely parametrized by horizontal lifts (t =constant) and orbits (s =constant).
- They are flat surfaces with the induced metric from that on the unit three sphere. Then $\mathbf{T}_\beta$ will be called the Hopf tube with section the curve β.
- A Hopf tube is embedded in $\mathbb{S}^3$ if its section is a simple (with no self-intersections) curve in $\mathbb{S}^2$.
- If β is a closed curve, then the Hopf tube $\mathbf{T}_\beta$ is a flat torus, whose isometry type will be computed later.
- Most of the geometry of Hopf tubes is reflected in the geometry of their corresponding sections. For example, the whole extrinsic geometry of a Hopf tube in $\mathbb{S}^3$ is governed by the curvature function of its section in $\mathbb{S}^2$.

Since Hopf tubes are flat surfaces, it is obvious that a geodesic α in a Hopf tube $\Pi^{-1}(\beta)$ is completely determined by a pair of real moduli: the curvature function ρ of the section β in $\mathbb{S}^2$ and the slope g of the geodesic, α in the Hopf tube $\Pi^{-1}(\beta)$. The slope is measured regarding orbits. Now, assume that β has constant curvature $\rho = $ constant in $\mathbb{S}^2$, that is, it is a geodesic circle. Then the corresponding Hopf tube is a circular one. It is not difficult to see that any geodesic α of $\Pi^{-1}(\beta)$ is a helix in $\mathbb{S}^3$ with curvature and torsion given, respectively, by

$$\kappa = \frac{\rho + 2g}{1 + g^2}, \qquad \tau = \frac{1 - \rho g - g^2}{1 + g^2}.$$

Then a geodesic of a circular Hopf tube is a helix in the three sphere. The converse of this fact also holds. Given a helix, say, α, in $\mathbb{S}^3$, with curvature $\kappa > 0$ and torsion τ, we consider in $\mathbb{S}^2$ the circle β with curvature

$$\rho = \frac{\kappa^2 + \tau^2 - 1}{\kappa}.$$

Now, in the Hopf tube $\mathbf{T}_\beta = \Pi^{-1}(\beta)$ choose the geodesic γ with slope

$$g = \frac{1 - \tau}{\kappa}.$$

It is clear that γ is a helix in $\mathbb{S}^3$. Moreover an easy exercise shows that both γ and α have the same curvature and the same torsion, so they are congruent in $\mathbb{S}^3$.

Circular Hopf tubes are actually Hopf tori. Furthermore, the constancy of the curvature of the section implies that circular Hopf tori have constant mean curvature in $\mathbb{S}^3$. In particular, if one constructs the Hopf torus whose section is a geodesic in $\mathbb{S}^2$ (curvature zero), one gets a minimal torus in $\mathbb{S}^3$, which is known as the Clifford torus.

This allows us to solve geometrically the problem of finding out the one parameter class of helices providing the dynamics of the constrained model. Just translate the constraint to (ρ, g) and then see the trajectories as geodesics of circular Hopf tori.

Using Hopf Map to Solve the Free Model Dynamics

Let β be a curve in $\mathbb{S}^2$ which we assume to be parametrized by its arc length. Let

$$\mathbf{F} := \{ T = \beta', \quad N = J(T), \quad \kappa \},$$

be the Frenet apparatus and

$$\nabla_T T = \kappa N, \qquad \nabla_T N = -\kappa T$$

the Frenet equations. On the other hand, a horizontal lift $\overline{\beta}$ is a curve in $\mathbb{S}^3$ which is automatically arc length parametrized because the Hopf map is a Riemannian submersion. The corresponding Frenet apparatus will be denoted by

$$\overline{\mathbf{F}} := \left[\{ \overline{T} = \overline{\beta}', \quad \xi, \quad \eta \}, \quad (\rho, \tau) \right],$$

and the corresponding Frenet equations

$$
\begin{aligned}
\nabla_{\overline{T}} \overline{T} &= \rho\, \xi, \\
\nabla_{\overline{T}} \xi &= -\kappa \overline{T} + \tau \eta, \\
\nabla_{\overline{T}} \eta &= -\tau \xi.
\end{aligned}
$$

Now, use the O'Neill equations to get the relationship between $\mathbf{F}$ and $\overline{\mathbf{F}}$

- **Unit normals:** $\xi = \overline{N}$, ξ is horizontal.
- **Unit binormal:** $\eta = V$, tangent to orbits.
- **Curvatures:** $\rho = \kappa \circ \Pi = \overline{\kappa}$.
- **Torsion:** $\tau^2 = 1$.

Summing up, the horizontal lifts, via the Hopf map, of curves in $\mathbb{S}^2$ provide trajectories for particles in the free model $(\mathbb{S}^3, \mathbf{C}, \mathscr{F}_0)$. Even more, they provides the whole dynamics, which can be showed using a congruence argument for curves in $\mathbb{S}^3$.

CLOSED TRAJECTORIES

In any theory of curves, there are two main problems which, hardly ever, are trivial. On one hand, the solving natural equations, that is, given two suitable functions of one variable (potentially curvature and torsion) find an arclength parametrized curve for which the two given functions work as the curvature and the torsion. Sometimes, this problem has been achieved using different procedures, either by quadratures, or solving a Ricatti equation, or by geometrical integration, or by numerical approaches.

On the other hand, the closed curve problem stated by Efimov and Fenchel. Curvature and torsion of any arclength parametrized closed curve are obviously periodic functions of the arclength. However, this is a necessary but not a sufficient condition. Now the problem is stated as follows: find explicitly necessary and sufficient conditions to determine when, given two periodic functions with the same period, the integral curve is closed.

The solving natural equations for the trajectories, of order one rigidity models, have been geometrically solved using the Hopf map. Precisely the Hopf map is a Riemannian submersion. The closed curve problem for trajectories can be also solved using the following strategy.

- Trajectories, no matter the model is constrained or free, are geodesic of Hopf tubes.
- Hopf tubes are flat surfaces and, in particular, when section is closed, we have geodesics of flat tori.
- Closed geodesics of flat tori are completely determined when we know the isometry type of those tori.
- The isometry type of the Hopf tori is again encoded in the Hopf map. To compute it, we use that the Hopf map is a circle bundle over a two sphere and then determine its holonomy number which carries the isometry of Hopf tori.

Using this program, we have the following quantization principle for closed trajectories (see [2] for details.

If β is a closed curve in $\mathbb{S}^2$, with length L and enclosed area A, then its Hopf torus, T_β, is isometric to $\mathbb{R}^2/\Gamma$, where Γ is the lattice in the Euclidean plane, $\mathbb{R}^2$, generated by $(0, 2\pi)$ and $(L, 2A)$.

The whole space of closed trajectories in the constrained model agrees with a rational one-parameter family of closed helices in $\mathbb{S}^3$. Geometrically, they are geodesics of circular Hopf tori which are obtained when the slope is quantized, a la Dirac, by a rational constraint.

The moduli space of closed trajectories in the free model is obtained when we lift, several times, closed curves in $\mathbb{S}^2$ enclosing an area that is a rational multiple of π. More precisely, a curve α in $\mathbb{S}^3$ is a closed trajectory of the total curvature model if and only if there exists a natural number n such that α is a horizontal lift, via Hopf, of the n-cover of a closed curve β in $\mathbb{S}^2$ enclosing an area $A = \frac{p}{n}\pi$, where p and n are prime numbers.

THE DYNAMICS IN THE ANTI DE SITTER THREE SPACE

To obtain the whole dynamics of both models in the anti se Sitter three space $\mathbf{AdS}_3$ we have to solve the following problem:

- **Constrained model.** Determine the one-parameter class of helices

$$\{(\kappa, \tau) \in \mathbb{R}^2 : \tau^2 - \varepsilon_2 m \kappa = 1\}.$$

- **Free model.** Determine the class of curves with torsion τ satisfying $\tau^2 = 1$.

The dynamics in the Lorentzian context is richer than in its Riemannian partner and this is because, now, we have a couple of Hopf mappings. Let $\mathbb{C}_1^2$ be the 2-dimensional complex lineal space $\mathbb{C}^2$ endowed with the Hermitian form $(z, w) = -z_1 \bar{w}_1 + z_2 \bar{w}_2$. Then $AdS_3 = \{z \in \mathbb{C}_1^2 / (z, z) = -1\}$. The hyperbolic plane $\mathbb{H}_0^2$ and the pseudo-hyperbolic plane (anti De Sitter plane) $AdS_2 = \mathbb{H}_1^2$ can be obtained as orbit spaces from two natural actions on AdS_3. In fact, $\mathbb{S}^1$ (the unit circle in $\mathbb{R}^2$) and $\mathbb{H}^1$ (the unit circle in $\mathbb{R}_1^2$), respectively, act on AdS_3 by $(a, (z_1, z_2)) = (az_1, az_2)$, where $a \in \mathbb{S}^1$ or $a \in \mathbb{H}^1$, respectively. Then, we obtain two natural Hopf fibrations $\pi_r : AdS_3 \to \mathbb{H}_r^2$, $r = 0, 1$, with fibers $\mathbb{S}^1$ and $\mathbb{H}^1$, respectively. Actually they become semi-Riemannian submersions when considering in $\mathbb{H}_r^2$, $r = 1, 2$, those metrics with constant curvature -4.

As in the Riemannian case, the study of helices in $\mathbf{AdS}_3$ is strongly related to the geodesics of the following flat surfaces

$$\pi_r^{-1}(\beta) \equiv \Phi(s, t) = \begin{cases} \cos(t)\bar{\beta}(s) + \sin(t)i\bar{\beta}(s), & \text{if} \quad r = 0, \\ \cosh(t)\bar{\beta}(s) + \sinh(t)i\bar{\beta}(s), & \text{if} \quad r = 1. \end{cases}$$

The former case, $r = 0$, are known as Hopf tubes, while the latter yields the so called B-scrolls in the sense of Graves and Nomizu (see [30]). In fact, helices in anti de Sitter three space are nothing but geodesics in either a Hopf tube or a B-scroll over constant curvatures curves. Therefore, they can be described, up to motions in $\mathbf{AdS}_3$, non only by curvature and torsion (κ, τ), which are both constant, but also by the so called cylindrical parameters. They are the constant curvature ρ, of the section of either the Hopf tube or the B-scroll, and the slope g, measured with respect to the fibers.

In the Lorentzian setting, however, another pair of parameters appear to describe the trajectories. They are the squared mass, obtained from the linear momentum P, and the spin, which are related with the geometry of trajectories by

$$M^2 = <P, P> = \frac{(\tau^2 + c)(\tau^2 - \kappa^2 + c)}{\kappa^2}, \qquad S^2 \frac{\tau^2 \kappa^2}{(\tau^2 + c)(\tau^2 - \kappa^2 + c)}.$$

All this nice dynamics can be summarized as follows

The Lagrangian $\mathscr{F}_m$, with $m \neq 0$, provides a consistent formulation to describe the dynamics of massive spinning particles in AdS_3. These evolve generating worldlines that are helices in AdS_3. The complete solution of the motion equations consists of a

one-parameter family of non-congruent helices. The moduli space of solutions may be described by three different (but equivalent) pairs of dependent real moduli:

1. The curvature κ and the torsion τ of the particle worldline, whose dependence defines a piece of parabola.
2. The mass M and the spin S of the particle, whose dependence gives the Regge trajectory.
3. The cylindrical coordinates (g, ρ) of the particle worldline regarded as a geodesic of either a Hopf tube or a B-scroll, with an obvious constraint.

To determine the dynamics of $\mathscr{F}_0$ we have to compute the moduli space of curves in $\mathbf{AdS}_3$ with torsion $\tau^2 = 1$. It can be completely solved mainly due to (i) The high rigidity of the standard gravitational field on AdS_3; and (ii) The nice geometry associated with the Hopf mappings allow us to obtain the whole moduli space of massless spinning particles for the Plyushchay model.

It should be first observed that any horizontal lift via π_r of a curve in $\mathbb{H}_r^2$, $r = 0, 1$, has torsion $\tau = \pm 1$, which automatically gives a worldline of a massless spinning particle evolving in $\mathbf{AdS}_3$. Conversely, by assuming that α is the worldline of a massless spinning particle in $\mathbf{AdS}_3$, then its torsion is $\tau = \pm 1$. Let κ^* be its curvature function and let γ be a curve in $\mathbb{H}_r^2$ whose curvature function is $\kappa = \pi_r \circ \kappa^*$. Choose finally a horizontal lift, say $\bar{\gamma}$, of γ. As α and $\bar{\gamma}$ have the same curvature κ^* and torsion $\tau = \pm 1$, then they must be congruent in $\mathbf{AdS}_3$. Therefore, we have determined the moduli space of solutions for the field equations associated with the Plyushchay model describing massless spining particles in $\mathbf{AdS}_3$.

It should be noted that, in contrast with the massive models, where two-parameter real moduli describe the space of solutions, now the only moduli moves along the space of smooth functions from $\mathbb{R}$ in $\mathbb{H}_r^2$. We can also obtain in both cases a quantization principle for closed trajectories in terms of a rational slope for massive model or the are enclosed by the section for non massive case (see [9] for a more general case).

WILLMORE HAS BEEN, AND HE IS STILL BEING, HIGHLY FASHIONABLE

It is obvious that, for surfaces $\mathbf{S} \in \mathbb{R}^3$ the two main geometrical invariants are the Gaussian curvature, $\mathbf{G}$, and the mean curvature, $\mathbf{H}$. According to the popular theorem *Egregium*, the Gaussian curvature is an intrinsic property of surfaces. It only depends on the metric, no matter how the surface is immersed in the space. Furthermore, for compact surfaces, the total Gaussian curvature only depends on the topology of the surface. This is the celebrated theorem of Gauss-Bonnet

$$\int_{\mathbf{S}} \mathbf{G} \, dA = 2\pi \chi(\mathbf{S}),$$

where $\chi(\mathbf{S})$ is the Euler characteristic of the surface and dA its area element.

On the contrary, the mean curvature $\mathbf{H}$ of a surface is extrinsic. It strongly depends on the way that surface is viewed in the space. The idea of integrating the squared mean curvature for compact surfaces is very old. However, a systematic study of this functional was proposed by T. J. Willmore in Oberwolfach (in the sixties last century), [53]. Since then, the action is popularly known as the Willmore functional. The associated variational problem, the Willmore variational problem, [54]

$$\mathscr{W}(\mathbf{S}) = \int_{\mathbf{S}} \mathbf{H}^2 \, dA.$$

has been widely studied and considered for many people in the later 45 years. Perhaps, a powerful reason to this popularity is the well known Willmore Conjecture. This may be justified from the following earlier result due to Willmore: **Round spheres are, globally, the least curved compact surfaces in Euclidean space.** More precisely

$$\mathbf{S} \subset \mathbb{R}^3, \quad \text{compact} \quad \Rightarrow \quad \mathscr{W}(\mathbf{S}) = \int_{\mathbf{S}} \mathbf{H}^2 \, dA \geq 4\pi.$$

Equality holding if and only if S is $\mathbb{S}^2(r)$, a round sphere no matter radius.

How Do We Prove That ?

Nowadays, we can give a beautiful prove of the Willmore theorem which is based on a powerful inequality due to Chern and Lashof, which says that

$$\mathbf{S} \subset \mathbb{R}^3, \quad \text{compact} \quad \Rightarrow \quad \int_{\mathbf{S}} |\mathbf{G}| \, dA \geq 2\pi(4 - \chi(\mathbf{S})).$$

Proof of Willmore theorem. Combine Chern-Lashof inequality with Gauss-Bonnet theorem to get

$$\int_{\mathbf{S}^+} \mathbf{G} \, dA + \int_{\mathbf{S}^-} \mathbf{G} \, dA = 2\pi\chi(\mathbf{S})$$

and

$$\int_{\mathbf{S}^+} \mathbf{G} \, dA - \int_{\mathbf{S}^-} \mathbf{G} \, dA \geq 2\pi(4 - \chi(\mathbf{S})),$$

where $\mathbf{S}^+ = \{p \in \mathbf{S} : \mathbf{G}(p) > 0\}$ and $\mathbf{S}^- = \mathbf{S} - \mathbf{S}^+$, to obtain

$$\int_{\mathbf{S}^+} \mathbf{G} \, dA \geq 4\pi.$$

On the other hand, we know that

$$\mathbf{H}^2 - \mathbf{G} \geq 0, \quad \text{equality holding in umbilical points}.$$

Then we get

$$\int_{\mathbf{S}} \mathbf{H}^2 \, dA \geq \int_{\mathbf{S}^+} \mathbf{H}^2 \, dA \geq \int_{\mathbf{S}^+} \mathbf{G} \, dA \geq 4\pi.$$

When Does Equality Hold ?

Certainly, for any round sphere one obtains the equality.

What About The Converse ?

If for a surface $\mathbf{S}$ the equality holds then

- The open $\mathbf{S}^+$ is made up of umbilical points and so it must be contained in a certain sphere $\mathbf{S}^+ \subset \mathbb{S}^2(r)$ so that

$$\mathbf{H} = \frac{1}{r} \quad \text{in} \quad \mathbf{S}^+.$$

- Its complementary $\mathbf{S} - \mathbf{S}^+$ is made up of minimal points, so that

$$\mathbf{H} = 0 \quad \text{in} \quad \mathbf{S}^-.$$

However, there are not compact minimal surfaces in $\mathbb{R}^3$, then $\mathbf{S}$ should be $\mathbb{S}^2(r)$.

Proof of Chern-Lashof inequality. As $\mathbf{S} \subset \mathbb{R}^3$ is compact, then it is orientable and there exists, globally defined, the Gauss map

$$N : \mathbf{S} \to \mathbb{S}^2(1).$$

Now $\int_{\mathbf{S}} |\mathbf{G}| \, dA$ measures the *average number of times* that $\mathbb{S}^2(1)$ is covered by the Gauss map.

Using Morse Theory

Given $\xi \in \mathbb{S}^2(1)$, it defines the corresponding height function

$$h_\xi : \mathbf{S} \to \mathbb{R} \qquad h_\xi(p) = \langle p, \xi \rangle.$$

The critical points of these functions are nicely characterized by

$$p \quad \text{critical point of} \quad h_\xi \quad \Leftrightarrow \quad \xi \perp T_p \mathbf{S}.$$

Functions all whose critical points are non-degenerate are called Morse functions. Then, as an application of the Sard theorem, we can see that

$$\mathbf{N} = \{\xi \in \mathbb{S}^2(1) : h_\xi \text{ is Morse}\} \quad \text{is an open } dense \text{ set in} \quad \mathbb{R}^3.$$

Morse inequalities. Given a Morse function $f : \mathbf{M} \to \mathbb{R}$ on a compact differentiable manifold, let $c_i(f)$ be the number of critical points with index i and let b_i be the i-th Betti number of $\mathbf{M}$. Then

$$b_i \leq c_i(f).$$

Observe that when p is a critical point of h_ξ then it is a critical point of $h_{-\xi}$ too. However, $N(p)$ is ξ or $-\xi$, so we get

$$\int_{\mathbf{S}} |\mathbf{G}| \, dA = \frac{1}{2} \int_{\xi \in \mathbf{N}} c(h_\xi) \, d\sigma^2,$$

where $c(h_\xi)$ is the number of critical points of h_ξ and $d\sigma^2$ is the area density on the unit two-sphere. We have used, of course, Sard's theorem. Now from Morse inequalities we find

$$\int_{\mathbf{S}} |\mathbf{G}|\, dA \geq \frac{1}{2}(b_0 + b_1 + b_2) \int_{\xi \in \mathbf{N}} d\sigma^2 = \frac{1}{2}(b_0 + b_1 + b_2) \int_{\xi \in \mathbb{S}^2(1)} d\sigma^2 =$$

$$= 2\pi(b_0 + b_1 + b_2) = 2\pi(2 + b_1) = 2\pi(4 - \chi(\mathbf{S})).$$

This finishes the proof of Chern-Lashof inequality as well as theorem of Willmore.

WILLMORE BEYOND EUCLIDEAN THREE SPACE

The Willmore functional can be defined for compact surfaces $\mathbf{S}$ immersed in a Riemannian space $(M^n; g = \langle,\rangle)$. However, in this case, Willmore density should be modified by adding the sectional curvature function $\mathbf{R}$, measured in the ambient space, restricted to the tangent plane of the surface, i.e.,

$$\mathscr{W}(\mathbf{S}) = \int_{\mathbf{S}} \left(H^2 + \mathbf{R} \right) dA.$$

Part of the beauty of this action comes from its invariance under conformal changes in the surrounding metric $g = \langle,\rangle$. Therefore, actually the Willmore program is stated in the conformal class $[g]$ of the original metric.

The Willmore Conjecture

The Euler-Lagrange equations, providing the critical points also known as Willmore surfaces, for Willmore functional in spaces with constant curvature, were computed by J. L. Weiner, [50]. In particular, in $\mathbb{R}^3$ or in the unit three sphere $\mathbb{S}^3$, the Willmore surfaces are the solutions of

$$\Delta \mathbf{H} = 2(\mathbf{H}^2 - \mathbf{G})\mathbf{H}, \qquad impossible\, to\, be\, solved.$$

This equation shows that minimal surfaces, $\mathbf{H} = 0$, are Willmore surfaces. This statement has no meaning in Euclidean space because the non existence of compact minimal surfaces. However, in the unit three sphere the class of compact minimal surfaces is very big. Using a stereographic projection, which is a conformal map, we can project compact minimal surfaces into $\mathbb{S}^3$ to obtain a wide class of Willmore surfaces in the Euclidean space.

The Clifford Torus as a Test Surface

The Clifford torus can be viewed as a Hopf one in $\mathbb{S}^3$, just that associated with a geodesic in the two sphere

$$\mathbf{T}_\beta = \Pi^{-1}(\beta), \qquad \beta = \text{geodesic in } \mathbb{S}^2(1/2).$$

This is enough, up to computations that we will do anywhere later, to conclude that the Clifford torus is minimal in $\mathbb{S}^3$. However, we will regard the Clifford torus explicitly constructed as follows.

- Start with a two-parameter class of isometric embeddings

$$\Phi_{ab}:\mathbb{R}^2 \to \mathbb{C}^2, \qquad \Phi_{ab}(s,t) = \left(a\,e^{\frac{is}{a}}, b\,e^{\frac{it}{b}}.\right)$$

- Choose $a^2 + b^2 = 1$ to obtain spherical embeddings in the unit three sphere

$$\Phi_{ab}(\mathbb{R}^2) \subset \mathbb{S}^3(1).$$

- It is obvious that each embedding can be induced to a flat torus

$$\Phi_{ab}:\mathbf{T}_{ab} \to \mathbb{S}^3(1) \subset \mathbb{C}^2,$$

$$\mathbf{T}_{ab} = \mathbb{R}^2/\Gamma_{ab}, \qquad \Gamma_{ab} = \text{Span}\{(2\pi a,0);(0,2\pi b)\}.$$

- Now, the Clifford torus is obtained, in the above construction, choosing a squared lattice $a = b = \sqrt{2}/2$ to write down

$$\Phi:\mathbf{T} \to \mathbb{S}^3(1) \subset \mathbb{C}^2, \qquad \Phi(s,t) = \frac{\sqrt{2}}{2}\left(e^{i\sqrt{2}s}, e^{i\sqrt{2}t}.\right)$$

For isometric immersions in Euclidean space, we have

$$\psi:\mathbf{M}^n \to \mathbb{R}^m, \qquad \Delta\psi = n\vec{\mathbf{H}},$$

Δ being the Euclidean Laplacian and $\mathbf{H}$ the mean curvature vector field. As for the Clifford torus we have

$$\vec{\mathbf{H}} = \frac{1}{2}\Delta\phi = 2\phi, \qquad \text{which is normal to} \quad \mathbb{S}^3(1).$$

Then it is minimal in $\mathbb{S}^3(1)$.

Several Conjectures on the Clifford Torus

The Clifford torus is a main **star** in many different contexts. Several conjectures have been formulated, and still are open, on this surface. I wish to mention the most important ones.

The Lawson Conjecture: The Clifford torus is the only minimal embedding of a genus one surface in the three sphere.

The Yau Conjecture: Every minimal embedding of a torus in the three sphere is constructed with eigenfunctions associated with the first Laplacian eigenvalue.

The Willmore Conjecture: A round sphere is a minimum of the Willmore functional on the whole class of compact surfaces, no matter the topology.

Now, the natural problem is to ask for minima of the Willmore functional when one fixes the topology of surfaces. The genus one topology provides the Willmore conjecture. The Willmore value in the Clifford torus is

$$\mathscr{W}(\mathbf{T}) = \int_{\mathbf{T}} \left(\mathbf{H}^2 + 1\right) dA = \left(2\pi \frac{\sqrt{2}}{2}\right)^2 = 2\pi^2.$$

Under a suitable stereographic projection, the image of the Clifford torus can be viewed as a revolution torus with radii in the ratio $\sqrt{2}$. However, it can be conformally deformed, for example changing the projection pole, and so viewed as, for example, a cyclide of Dupin. Let $\mathbf{S}$ be a genus one surface and $\mathbf{I}(\mathbf{S}\mathbb{R}^3)$ the space of immersions, $\psi : \mathbf{S} \to \mathbb{R}^3$. The Willmore functional can be viewed as

$$\mathscr{W} : \mathbf{I}(\mathbf{S}, \mathbb{R}^3) \to \mathbb{R}, \qquad \mathscr{W}(\psi) = \int_{\mathbf{S}} \mathbf{H}_\psi^2 \, dA_\psi.$$

Now, the Willmore conjecture is stated as follows

$$\mathscr{W}(\psi) \geq 2\pi^2; \qquad \text{equality holding} \quad \Leftrightarrow \quad \textbf{Clifford.}$$

These Three Conjectures Are Still Unsolved.

WILLMORE VIEWPOINT IN PHYSICS

Sometimes one finds tools and models in mathematics which apply in a wide variety of nonlinear phenomena in physics. Therefore, a single model can be of interest in many different contexts providing nice connections between two, or more, apparently unrelated physical phenomena. It is my plan here to describe several non linear physical phenomena where the Willmore functional occurs. In my opinion, this kind of universality may be strongly related to the fact that such models, and the equations governing them, very often have an underlying geometrical meaning.

Elastic Surfaces and Membranes

A membrane can be described as a two-dimensional thin structure that can endure the bending moment. It is usually thought of as a smooth surface since its thickness is much smaller than its lateral dimensions. I will mention, anyway, a few facts about this theory. It is a constant over the history, when one tries to provide a physical model at least variationally, the problem of searching for Lagrangian densities. We have sawn several examples when particles were exposed. The theory of elastic surfaces is not an exception. The search for Lagrangian to do the role of elastic energy has a long history. As far as I know, it was at the beginning of the XIX century when the first two serious proposals appeared. On one hand, Poisson wrote the free elastic energy for a solid membrane as

$$\mathscr{E}(\mathbf{S}) = \frac{k}{2} \int_{\mathbf{S}} (2\mathbf{H})^2 \, dA,$$

101

where k is the bedding modulus. It agrees, up to a constant, with the Willmore fuctional. Therefore, the equations governing the model coincides with those obtained in the Willmore program

$$\Delta \mathbf{H} = 2(\mathbf{H}^2 - \mathbf{G})\mathbf{H}.$$

On the other hand, in 1810, Sophie Germain (see [21]) proposed the following idea to construct elastic Lagrangians: **the elastic energy of a membrane is the Lagrangian associated with a density which is an even, symmetric function of the principal curvatures of the membrane**. In the last half century, physicists turned their interest to cell membranes and bi-layers (lipid, proteins and carbohydrates). Therefore, in the 1970's Canham, [22], and Helfrich [32], according to the Germain philosophy, proposed a quadratic, in the principal curvatures, elastic energy density, namely

$$\mathcal{E}(\mathbf{S}) = \int_{\mathbf{S}} \left(a + b \mathbf{H}^2 + c \mathbf{G} \right) dA$$

and elastic energy for bi-layers

$$\mathcal{E}(\mathbf{S}) = \int_{\mathbf{S}} \left(a + b(\mathbf{H} + b_o)^2 + c \mathbf{G} \right) dA.$$

The critical points of the elastic energy action are called equilibrium elastic surfaces. Stable equilibrium spherical shapes are known for a long time. Besides these spherical solutions, nowadays are known other analytic solutions: revolution tori with radii in the rate $\sqrt{2}$ and biconcave disks as red blood cells. There are also a lot of known numerical solutions of the corresponding field equations.

Notice that if we consider that fluctuations of elastic surfaces do not change the topology, then the total Gaussian curvature is constant from and Gauss-Bonnet and so the Canham-Helfrich elastic energy is

$$\mathcal{E}(\mathbf{S}) = \int_{\mathbf{S}} \left(a + b \mathbf{H}^2 \right) dA.$$

Moreover, we may assume that $b \neq 0$, otherwise we are talking about minimal surfaces which is not real. Therefore, the Canham-Helfrich elastic energy looks like a kind of Willmore functional.

Recently, lipid bi-layers with free exposed edges have been observed. This implies that we have to study the equilibrium equation with certain boundary conditions. Moreover, to study cell membranes, it is necessary to introduce in the elastic density certain terms to consider the so called membrane skeleton which is like a two dimensional rubber membrane.

Bosonic String Theories

String theory, or better string theories, constitutes a new way to understand the physical world, the reality. In this sense, the building block are one-dimensional extended objects called strings rather than the conventional zero-dimensional point particles. A string theory takes place in a gravitational space, $(M^n, g = \langle,\rangle)$, where certain curves dynamically evolve to generate surfaces, worldsheets. In contrast with the theories of membranes, where elastic surfaces are required to be embedded, now worldsheets are allowed to self-intersect and so surfaces are immersed. Now, the string are those curves which sweep out surfaces, worldsheets, that are solutions of the field equations associated with a string action. Certainly, this idea is extended to higher dimensions to obtain branes that sweep out worldvolumes.

How Do We Choose the String Action ?

Once we have the target space, $(M^n, g = \langle,\rangle)$, it is necessary on one hand to decide the space of elementary fields, curve configurations and then the more important step, the choice of the string action. Historically, the Nambu-Goto action was the first choice to construct a bosonic string theory. This action measures, up to a coupling constant, the area of the worldsheets **S** which are the surfaces generated in the string evolution, that is,

$$\mathcal{N}\mathcal{G}(\mathbf{S}) = \mu \int_{\mathbf{S}} dA.$$

This choice clearly obeys two reasons. First it is the simplest invariant action and then it imitates the conventional approach of the classical particle.

The Nambu-Goto bosonic string theory provides worldsheets which are minimal surfaces in the corresponding gravitational space. However, this action is not the fundamental one that physicists use when they develop quantized versions of string theory. The fundamental action in a bosonic string theory was constructed by S. Deser-B. Zumino, [19], and independently by L. Brink-P. DiVechia-P. S. Howe, [17]. However, this action is usually known as the Kleinert-Polyakov action because they made use of it in quantizing the string theory (they constructed the quantum chromo dynamics **QCD**) [33, 46, 47]. The main idea in the construction of the Kleinert-Polyakov action consists in the inclusion, as a fundamental ingredient, in the string action the extrinsic geometry of worldsheets. The way to involve that ingredient in the string density is through the mean curvature of worldsheets

$$\mathcal{P}\mathcal{K}(\mathbf{S}) = \mu \int_{\mathbf{S}} dA + \nu \int_{\mathbf{S}} \mathbf{H}^2 dA.$$

Later, in 1989, P. B. Wiegmann, [51], showed that every string action to be quantized must include in the string density the extrinsic geometry of the worldsheets.

Sigma Models

A classical field theory is a physical theory that describes the study of how one or more physical fields interact with matter. Of course physical fields means local data on the space-time (function, vector fields, forms, sections of bundles, metrics, connections...). These constitutes the dynamical variables of the theory and usually are called classical fields or elementary fields of the classical field theory. Now, when classical field theories incorporate quantum theories become into quantum field theories, **QFT**.

A non linear sigma model is a field theory in which the elementary fields, or dynamical variables, are maps

$$\varphi : \mathbf{E} \to \mathbf{M}$$

where

- The starting space, $\mathbf{E}$, is a manifold, called the source space, and its dimension is the dimension of the sigma model.
- The arriving space $\mathbf{M}$ is a Riemannian space or a space-time, called the target space, and its isometry group provides the symmetry of the sigma model.

The Lagrangian governing the dynamics of the sigma model measures the total energy of the mappings

$$\mathscr{E}(\varphi) = \int_{\mathbf{E}} |d\varphi|^2 \, dv_\varphi.$$

Now, the classical solutions of the sigma model, that is the solutions of the field equations, constitute the space of configurations where the solitons of the sigma model are localized.

The $\mathbf{O}(3)$ non linear sigma model is ubiquitous in Physics. It is used in a wide of ranging of fields from Condensed-Matter-Physics to High-Energy-Physics. In my opinion, this kind of universality is strongly related with the fact that the equations governing the sigma model have a deep, underlying, geometrical meaning. As the dimension is two and the symmetry $\mathbf{O}(3)$, the dynamical variables of the sigma model are maps from something of dimension two, a surface $\mathbf{E} = \mathbf{S}$, in a unit sphere, $\mathbf{M} = \mathbb{S}^2 \subset \mathbb{R}^3$. Certainly, it brings to mind the Gauss map of surfaces. So, this is the starting point to use the theory of surfaces, and more precisely its extrinsic face, in the study of this sigma model. According to this approach, the dynamical variables of the sigma model are viewed as Gauss maps associated with the immersions of a surface $\mathbf{S}$ in $\mathbb{R}^3$. In other words, let $\mathbf{S}$ be a surface and let $\mathbf{I}(\mathbf{S}, \mathbb{R}^3)$ be the space of immersions in $\mathbb{R}^3$. Now, for each $\phi \in \mathbf{I}(\mathbf{S}, \mathbb{R}^3)$, one has its Gauss map $N_\phi : \mathbf{S} \to \mathbb{S}^2$. Now, from this point of view, the Lagrangian governing the dynamics in this sigma model, that is, the total energy of a Gauss map is

$$\mathscr{E} : \mathbf{I}(\mathbf{S}, \mathbb{R}^3) \to \mathbb{R}, \quad \mathscr{E}(\phi) = \int_{\mathbf{S}} |dN_\phi|^2 \, dA_\phi.$$

Obviously, the most direct way to get configuration shapes of this sigma model is to compute the field equations and try to integrate them. Certainly, these equations have been computed and nothing more to say. However, an alternative way is to use procedures based in symmetries, geometry, numerical methods and so on, to obtain some solutions. For example

- If we consider solutions with conformal Gauss map (round spheres and minimal surfaces) then one obtains the famous solitons discovered by Belavin and Polyakov, [16].
- Assuming that surfaces have constant mean curvature, then one obtains the solitons of Purkait and Ray, [48], which correspond to the family of constant mean curvature helicoids discovered by Do Carmo and Dajczer, [20].

The Gauss formula for surfaces in Euclidean space gives

$$|dN_\phi|^2 = 4\mathbf{H}_\phi^2 - 2\,\mathbf{G}_\phi.$$

Therefore, if we consider the free sigma model, that is, we assume that $\mathbf{S}$ is closed (compact and boundary free), as it is obvious, admissible fluctuations do not change the topology, then this sigma model turns out to be equivalent, once more, to the Willmore program

$$\int_{\mathbf{S}} |dN_\phi|^2 dA_\phi = 4 \int_{\mathbf{S}} \mathbf{H}_\phi^2\, dA_\phi - 4\pi\chi(\mathbf{S}).$$

WILLMORE SURFACES

We have exhibited a series of physical theories where the Willmore program plays an important role. The underlaying geometry of these physical contexts is strongly related with the Willmore variational problem. Now, we could consider other models different from those studied before. In fact, for simplicity and for the first time, it is enough to study the free case, that is, surfaces providing the configurations which are closed (compact and without boundary) and living in Riemannian spaces. However, it is worth thinking on the following problems:

- Bring the Willmore program to surfaces in Lorentzian spaces and, more generally, in semi-Riemannian ones.
- Also, we can consider the constrained problem. In this case surfaces have non-empty boundary and the variational problem is stated through certain boundary conditions.
- Submanifolds with dimensions greater than two can be considered too. In this case, the program is known as the Willmore-Chen variational problem. This model has been used, for example, in searching for branes in superstrings theories and M-theory.

Perhaps, due to the Willmore conjecture, an exciting fever to study Willmore tori has flooding the specialized literature with lot of papers. It is well known that every torus is conformal to a flat one. On the other hand, the space of flat tori is also known, so the moduli space of conformal classes on a torus is well determined. That yielded to solve the Willmore conjecture on certain conformal classes. However, the partial answers are still very poor. The searching for Willmore tori could be compared with the gold fever. I would like to tell you a trailer of this movie.

The Principle of Symmetric Criticality: A Version Due To R. S. Palais.

An elegant way to carry out that idea is encoded in the so called principle of symmetric criticality. This principle has been **used** in a lot of applications, from the Calculus of Variations to Physics, without being explicitly noticed. Even in some where the principle does not work and so it can not be applied. An early typical example of the implicit use of this principle can be found in the H. Weyl formulation of the Schwarzschild solution to the Einstein field equations. The Einstein field theory version of general relativity deals with dynamical variables which are metrics (Lorentzian metrics) in a certain space, say of dimension four M^4. Now, the theory is governed by the so called Einstein-Hilbert action, which measures the total scalar curvature of metrics. The field equations of this action provides the popular Einstein field equations. In particular, one has the Einstein vacuum field equation, whose better solution was proposed by Schwarzschild in 1916. Shortly after Einstein formulated general relativity theory. The Schwarzschild solution models the gravitational field outside an isolated, static, spherically symmetric star.

A space-time M^4 is static if it admits a nowhere zero time-like Killing vector field V with integrable orthogonal distribution. Therefore, the leaves of $\langle X \rangle^\perp$ are space-like hypersurfaces, isometric to each other. M^4 is said to be spherically symmetric if there exists an isometric action of $\mathbf{SO}(3)$ such that orbits are either a space-like surface or a single point. Two-dimensional orbits have necessarily constant curvature. Both staticity and spherical symmetry are satisfied by

$$M = \mathbb{R} \times \mathbf{I} \times \mathbb{S}^2, \qquad \mathbf{I} \subset \mathbb{R},$$

endowed with any warped product metric

$$g = -F^2(s)\,dt^2 + ds^2 + G^2(s)\,d\sigma^2.$$

Let $\mathscr{M}$ be the whole space of Lorentz metrics on M^4 and let $\mathscr{M}_S$ be the subspace of Lorentz metrics on M^4 which are spherically symmetric.

The Weyl formulation works as follows

- Consider the total scalar curvature action restricted to the space $\mathscr{M}_S$.
- Compute its first variation.
- Compute F and G in order to vanish the first variation.
- Then get the Schwarzschild solution.

The Weyl formulation presents a serious problem. In fact, the Schwarzschild metric so obtained is a critical point of a part of the action. That is, the vanishing of the first variation of the action restricted to $\mathscr{M}_S$ is, certainly, a necessary condition to be a solution. However, the converse needs some extra work. What happens on the transversal space of $\mathscr{M}_S$?

The principle of symmetric criticality tries to look for sufficient conditions to make true the converse way, i.e., any critical symmetric point is a symmetric critical point.

The setting we need to establish the principle can be described as follows. First we have a space $\mathscr{N}$ on which a group G acts by diffeomorphisms. We also have a functional $\mathscr{W} : \mathscr{N} \to \mathbb{R}$ which is G-invariant

$$\mathscr{W}(a \cdot \varphi) = \mathscr{W}(\varphi), \quad \forall a \in G.$$

Then, we have the set of symmetric points

$$\mathscr{N}_G = \{ \varphi \in \mathscr{N} \ : \ a \cdot \varphi = \varphi, \forall a \in G \}.$$

In this framework, we can consider the following ingredients

- The set of critical points Σ of $\mathscr{W} : \mathscr{N} \to \mathbb{R}$.

- The set of critical points Σ_G of the restriction of $\mathscr{W}$ to the set $\mathscr{N}_G$ of symmetric points.

The latter requires that $\mathscr{N}_G$ be a differentiable manifold which is assured when, for example, it is a submanifold of $\mathscr{N}$. A sufficient condition to guarantee this fact is to assume that the group G is compact. Even more, under this assumption the following equality holds

$$\Sigma \cap \mathscr{N}_G = \Sigma_G,$$

which is nothing but the principle of symmetric criticality. It should be observed that this principle does not work in general. To see that we next give a very simple example. Choose

$$\mathscr{N} = \mathbb{R}^2, \quad G = \mathbb{R}, \quad \psi_t(x,y) = (x + y^k t, y), \quad k \in \mathbb{N}.$$

The set of symmetric points is

$$\mathscr{N}_G = \{(x,0) : x \in \mathbb{R}\}.$$

Then we have

$$\mathscr{W}(x,y) = f(y), \quad \text{for any } f, \quad (\mathscr{W}/\mathscr{N}_G)(x,0) = f(0),$$

$$\Sigma_G = \mathscr{N}_G, \quad \text{while} \quad \Sigma \cap \mathscr{N}_G = \emptyset, \quad \text{if} \quad f'(0) \neq 0.$$

The Willmore Revolution Tori

The whole space of revolution tori, in Euclidean three space, which are Willmore surfaces can be obtained through a recipe involving the following main ingredients

- **The Palais principle of symmetric criticality.**
- **A suitable conformal change of the Euclidean metric.**
- **Several geometric spices.**

We will give, in detail, how this recipe works mainly due to its beauty and simplicity. Without loss of generality, we choose the axis of revolution as the z-axis. Then, as a first step, we prepare the target Euclidean space $(\mathbf{M}, \bar{g}_o) = \mathbb{R}^3 - z$-axis with an appropriate conformal change. We use cylindrical coordinates to see that the Euclidean space $(\mathbf{M}, \bar{g}_o)$ is the warped product of an Euclidean half-plane $(\mathbf{P}, g_o)$ with

$$\mathbf{P} = \{(0,y,z) \in \mathbb{R}^3 : y > 0\},$$

and a unit circle. More precisely,

$$\mathbf{M} = \mathbf{P} \times_f \mathbb{S}^1, \quad \bar{g}_o = g_o + f^2 dt^2,$$

where the warped function is given by

$$f : \mathbf{P} \to \mathbb{R}, \quad f(0,y,z) = y.$$

Now, the following natural conformal change

$$\bar{g} = \frac{1}{f^2} \bar{g}_o = \frac{1}{f^2} g_o + dt^2,$$

yields a Riemannian product. Observe $\left(\mathbf{P}, g = \frac{1}{f^2} g_o \right)$ is a hyperbolic Poincaré half-plane with curvature -1. Therefore, the Euclidean metric in $\mathbf{M}$ is conformal to that associated with a Riemannian product of a hyperbolic plane with a unit circle. Now, to get the Willmore revolution tori we will use this new metric $\bar{g}$ in the class of $[\bar{g}_o]$.

We now identify each of the elements involved in the Palais setting. For a genus one surface, a torus $\mathbf{T}$, let $\mathcal{N} = \mathbf{I}(\mathbf{T}, \mathbf{M})$ be the space of immersions in $\mathbf{M}$. The group $G = \mathbb{S}^1$ acts on $\mathbf{M}$ by rotations around the z-axis and this action can be induced, in an obvious way, on $\mathcal{N}$. Let $\mathcal{W}_o : \mathcal{N} \to \mathbb{R}$ be the Willmore functional associated with the Euclidean metric, which, certainly, is G-invariant. The set of symmetric points $\mathcal{N}_G$ agrees with the space of revolution tori with axis z and so it can be identified with the space $\mathbf{C}$ of closed curves in the half-plane $\mathbf{P}$. In this framework, we apply the principle of symmetric criticality, so that a revolution torus is Willmore if and only if it is so among the revolution tori. Therefore, we need to compute the Willmore functional on the space of revolution tori. The conformal invariance of the Willmore functional allows us to compute it using the metric $\bar{g}$. Hence, for any closed curve $\gamma \in \mathbf{C}$, the Willmore functional on the corresponding revolution torus $\mathbf{T}_\gamma$ writes down

$$\mathcal{W}_o(\mathbf{T}_\gamma) = \mathcal{W}(\mathbf{T}_\gamma) = \int_{\mathbf{T}} \left(\mathbf{H}^2 + \mathbf{R} \right) dA.$$

As $(\mathbf{M}, \bar{g})$ is a Riemannian product, then its sectional curvature over mixed sections vanishes identically and so $\mathbf{R} = 0$. On the other hand, parallels in $(\mathbf{M}, \bar{g})$ are geodesics and consequently

$$\mathbf{H} = \frac{1}{2} \kappa,$$

κ being the curvature function of γ in the hyperbolic plane $(\mathbf{P}, g)$. Therefore,

$$\mathcal{W}_o(\mathbf{T}_\gamma) = \mathcal{W}(\mathbf{T}_\gamma) = \frac{\pi}{2} \int_\gamma \kappa^2(s) \, ds.$$

Conclusion. Willmore revolution tori are obtained just rotating free elasticae in the hyperbolic plane, that is, closed curves which are critical points of the squared total curvature in the hyperbolic plane.

This class of curves have been widely considered in the literature. In particular, to compute the corresponding Euler-Lagrange equation, one can see that the curvature function of an elastica in the hyperbolic plane must be a solution of the following differential equation, [35]

$$2\frac{d^2 \kappa}{ds^2} + \kappa(\kappa^2 - 2) = 0.$$

This equation provides two constant solutions. First $\kappa = 0$, that corresponds with a geodesic which is not closed. Secondly a geodesic circle ε, with $\kappa^2 = 2$, which generates an anchor ring with radii in the ratio $\sqrt{2}$, that is, a conformal Clifford torus. However, this equation can be integrated using elliptic function. Therefore, besides constant solutions, we have

- orbitlike solutions

$$\kappa(s) = 2a\,\mathbf{dn}(as, p), \quad a \in \mathbb{R}, \quad 0 < p < 1;$$

- asymptotically geodesic solution:

$$\kappa(s) = 2\,\mathbf{sech}(s);$$

and
- wavelike solutions

$$\kappa(s) = a\,\mathbf{cn}(rs, p), \quad a, r \in \mathbb{R}, \quad \sqrt{2}/2 < p < 1.$$

The space of closed free elasticae in a hyperbolic plane was obtained in [35]. Besides the above mentioned geodesic circle, one obtains an integer two-parameter family $\{\gamma_{mn} : 1 < 2m/n < \sqrt{2}\}$ of closed orbitlike free elasticae. A wavelike free elastica never closes and the asymptotically geodesic elastica is obviously not closed. This provides the complete class of Willmore revolution tori

$$\mathbf{F} = \{\mathbf{T}_\varepsilon, \mathbf{T}_{\gamma_{mn}} : 1 < 2m/n < \sqrt{2}\}.$$

WILLMORE TORI AND KALUZA-KLEIN MECHANISM

Let $p : M \to B$ be a circle bundle endowed with a gauge potential ω. For any metric h on the base B and any positive smooth function $f : B \to \mathbb{R}$, one can define the so-called generalized Kaluza-Klein metric $\bar{h}_f$ on M as follows

$$\bar{h}_f = p^*(h) + (f \circ p)^2\,\omega^*(ds^2).$$

In particular, is f is chosen to be constant then one has the notion of Kaluza-Klein metric, also known as bundle-like metric. Let us recall some of the main properties of this class of metrics

- Choosing the local warping function f to be identically one, the corresponding Kaluza-Klein metric can be seen depending on two variables: the gravity h and the electromagnetic potential ω. Now, the Kaluza-Klein metric evolves either by gravity or by electromagnetic potential. Therefore the Einstein and Maxwell equations arise from the same variational principle. As the above construction also works for structural Lie groups endowed with a bi-invariant metric, the same remark holds for Einstein equations as well as Yang-Mills ones.

109

- $p : (M, \bar{h}_f) \rightarrow (B, h)$ is a Riemannian submersion, which has geodesic fibers if and only if f is constant, that is, $\bar{h}_f$ is bundle-like.
- The natural $\mathbb{S}^1$-action is carried out through isometries of $(M, \bar{h}_f)$.
- By writing $\tilde{h}_f = \frac{1}{(f \circ p)^2} \bar{h}_f$, then

$$\tilde{h}_f = p^* \left(\frac{1}{f^2 h} \right) + \omega^*(ds^2),$$

which means that $\tilde{h}_f$ is a bundle-like metric in the same conformal class of the original one $[\bar{h}_f]$. In particular, the associated Willmore problems are equivalent.
- For any curve γ immersed in B, its total lift $N_\gamma = p^{-1}(\gamma)$ provides a surface immersed in M which is $\mathbb{S}^1$-invariant. This surface is embedded if and only is the starting curve is simple. Furthermore, all of $\mathbb{S}^1$-invariant surfaces in M are obtained in this way.
- It is clear that $N_\gamma = \{a \cdot \bar{\gamma}(s) : a \in \mathbb{S}^1\}$, where $\bar{\gamma}$ is a horizontal lifts of γ. In particular, N_γ can be naturally parametrized with coordinate curves being horizontal lifts and fibers.
- If γ is closed, then N_γ is a torus. However, horizontal lifts do not close, in general, because the holonomy of the gauge potential could be non trivial.
- The mean curvature function $\mathbf{H}$ of $N_\gamma = p^{-1}(\gamma)$ in $(M, \tilde{h}_f)$ and the curvature function κ of γ in $(B, \frac{1}{f^2} h)$ are related by

$$\mathbf{H}^2 = \frac{1}{4} (\kappa^2 \circ p).$$

In the general setting associated with a generalized Kaluza-Klein metric, a natural problem can be stated as follows: given a closed curve γ, when $N_\gamma = p^{-1}(\gamma)$ is a Willmore torus in $(M, [\bar{h}_f])$?

Several answers are known relative to this problem. Let me give some of them.

(1) Let $(\mathbb{S}^2, h)$ be the round two sphere with curvature four and let $\Pi : \mathbb{S}^3 \rightarrow \mathbb{S}^2$ be the Hopf map. Set ω the gauge potential associated with the horizontal distribution defined as the orthogonal one to the Hopf vector field. Given a positive real number $r > 0$, we can define the Kaluza-Klein metric on $\mathbb{S}^3$

$$\bar{h}_t = \Pi^*(h) + r^2 \, \omega^*(ds^2).$$

This one-parameter class of metrics on the three sphere constitutes the so called canonical variation of the round unit three sphere, which is reached for $r = 1$. Geometrically, these three spheres, called Berger spheres, can be viewed as geodesics three spheres in the complex projective plane $\mathbb{CP}^2$ endowed with the usual Fubini-Study metric. These metrics are highly rigid and have constant scalar curvature. The following result has been proved in [41] for $r = 1$ and in [6] for any r.

For a closed curve γ in $\mathbb{S}^2$, the torus $\Pi^{-1}(\gamma)$ is Willmore in $[\bar{h}_t]$ if and only if γ is a critical point of the action

$$\mathscr{C}_r(\gamma) = \int_\gamma (\kappa^2 + 4r^2)\,ds$$

acting on the space of closed curves on the two sphere. These curves are constrained elastic curves, where the constant $4r^2$ can be regarded as a Lagrange multiplier. The existence of this kind of curves on round two spheres has been studied in [35].

(2) Let $\mathbb{S}^5$ and $\mathbb{CP}^2$ be the unit round five-sphere and the complex projective plane, respectively. Let $\Pi : \mathbb{S}^5 \to \mathbb{CP}^2$ be the corresponding Hopf map. This becomes into a Riemannian submersion when we endow $\mathbb{CP}^2$ with the Fubini-Study metric h with constant holomorphic sectional curvature four. It certainly provides a circle bundle with a standard gauge potential ω. Furthermore, the round metric $\bar{h}$ in $\mathbb{S}^5$ is the Kaluza-Klein one associated with that of Fubini-Study when choosing the local warped function to be one

$$\bar{h} = \Pi^*(h) + \omega^*(ds^2).$$

The study of Willmore Hopf tori in the conformal class $[\bar{h}]$ can be reduced to the study of closed elasticae in $\mathbb{CP}^2$ (see [15]). The authors gave there the complete classification of elasticae with constant slant in $\mathbb{CP}^2$. It essentially consists of three families of elastic curves. Two of them corresponding with torsion free elasticae, which are living in certain totally geodesic surfaces (a complex line and a real projective plane, respectively). In these cases the slants reach the extremal values. The third family provides a two-parameter class of helices lying fully in $\mathbb{CP}^2$. An interesting quantization principle is obtained to describe, up to congruences, the moduli space of closed helices. Actually, it becomes a rational constraint on the slant. As a consequence, the complete class of Willmore Hopf tori, with conformal constant mean curvature, on the five sphere is obtained (see [15] for more details).

(3) To understand the next application, we wish to do a couple of considerations. First, the Willmore program and the building of generalized Kaluza-Klein of metrics can be both extended to a Lorentzian setting. Secondly, it should be noticed that generalized Kaluza-Klein metrics can be locally regarded as warped product ones. In particular, warped product metrics are generalized Kaluza-Klein metrics over trivial spaces. Our next application is viewed in this framework. It is known that the hyperbolic plane $\mathbb{H}^2$ can be seen in the Lorentz Minkowski space $\mathbb{L}^3 = \mathbb{R} \times \mathbb{R}^2$ as

$$\mathbb{H}^2 = \{(x_o, x) \in \mathbb{R} \times \mathbb{R}^2 : -x_o^2 + \langle x, x \rangle = -1, x_o > 0\}.$$

Also, the anti de Sitter three space in $\mathbb{R}_2^4 = \mathbb{R}^2 \times \mathbb{R}^2$ is

$$\mathbf{AdS}_3 = \{(\xi, \eta) \in \mathbb{R}_2^4 : -\langle \xi, \xi \rangle + \langle \eta, \eta \rangle = -1\}.$$

Both quadrics, with the corresponding induced metrics g_o and h_o, have constant curvature -1.

Define the diffeomorphism $\Phi : \mathbb{H}^2 \times \mathbb{S}^1 \to \mathbf{AdS}_3$ by

$$\Phi((x_o, x), u) = (x_o u, x).$$

Consider now the positive smooth function $f : \mathbb{H}^2 \to \mathbb{R}$ given by

$$f(x_o, x) = x_o.$$

Then, it is not difficult to see that $\mathbb{H}^2 \times \mathbb{S}^1$, endowed with the warped product metric $g = g_o - f^2 \, ds^2$, is isometric, via the diffeomorphism Φ, to $(\mathbf{AdS}_3, h_o)$. That construction is summarized in the following formula

$$\mathbf{AdS}_3 = \mathbb{H}^2 \times_f \mathbb{S}^1.$$

Then we have provided a trivial bundle $\Pi : \mathbf{AsS}_3 \to \mathbb{H}^2$ and h_o is a generalized Kaluza-Klein metric on the anti de Sitter space associated with the hyperbolic metric g_o and the warped function f. We do an obvious conformal change

$$h = \frac{1}{f^2} h_o = \frac{1}{f^2} g_o - ds^2.$$

Then observe that $(\mathbb{H}^2, (1/f^2)g_o)$ is a once punctured unit two sphere and use this fact to construct Willmore tori in the anti de Sitter three space from eleasticae in the two sphere (see [10]).

ACKNOWLEDGMENTS

The author wishes to express his thanks to Prof. A. Ferrández and Prof. A. Romero for their valued suggestions that improved the preparation of this paper. The work was partially supported by Spanish MEC Grant MTM2007-60731 and J. Andalucía Regional Grant P06-FQM-01951

REFERENCES

1. Y. Aharonov and D. Bonm, *Phys. Rev.* **115**, 485–491 (1959).
2. J. Arroyo, M. Barros and O. J. Garay, *Proc Edinburgh Math. Soc.* **43**, 587–603 (2000).
3. J. Arroyo, M. Barros and O. J. Garay, *J. Phys. A: Math. and Gen.* **35**, 6815–6824 (2002).
4. J. Arroyo, O. J. Garay and J. Mencia, *J. Geom. Phys.* **51**, 101 (2004).
5. J. Arroyo, M. Barros and O. J. Garay, *Gen. Rel. Grav.* **36**, 1441–1451 (2004).
6. M. Barros, *Math. Proc. Camb. Phil. Soc.* **121**, 321–324 (1997).
7. M. Barros, *Gen. Rel. Grav.* **34**, 837–852 (2002).
8. M. Barros, J. L. Cabrerizo, M. Fernández, and A. Romero, *J. Math. Phys.* **46**, 112905-1–15 (2005).
9. M. Barros, A. Ferrandez, M. A. Javaloyes and P. Lucas, *Class. Quantun Grav.* **35**, 489–513 (2005).
10. M. Barros, A. Ferrandez, P. Lucas and M. A. Meroño, *J. Geom. Phys.* **28**, 45–66 (1999).
11. M. Barros, A. Ferrandez and P. Lucas, *Quart. J. Math. Oxford* **2**, 385–388 (1999).
12. M. Barros, A. Ferrandez, P. Lucas and M. A. Meroño, *Trans. A. M. S.* **352**, 3015–3027 (2000).
13. M. Barros, A. Ferrandez and P. Lucas, *J. Geom. Phys.* **40**, 1–12 (2001).
14. M. Barros, A. Ferrandez, M. A. Javaloyes and P. Lucas, *Int. J. Modern Phys.* **19**, (11), 1738–1745 (2004).
15. M. Barros, O. J. Garay and D. A. Singer, *Tôhoku Math. J.* **51**, 177–192 (1999).
16. A. A. Bellavin and A. M. Polyakov, *JETP. Lett.* **22**, 245 (1975).
17. L. Brink, P. DiVechia and P. Howe, *Phys. Lett. B* **65**, 471 (1976).
18. A. Besse, *Einstein Manifolds*, Springer-Verlag, 1987.

19. S. Deser and B. Zumino, *Phys. Lett. B* **65**, 369 (1976).
20. M. DoCarmo and M. Dajczer, *Tôhoku Math. J.* **34**, 425–435 (1982).
21. L. L. Bucciarelli and N. Dworsky, *Sophie Germain: An Essay in the History of the Theory of Elasticity*, D. Reidel, 1980.
22. P. B. Canham, *Teor. Biol.* **26**, 61 (1970).
23. R. G. Chambers, *Phys. Rev. Lett.* **5**, 3–5 (1960).
24. P. A. M. Dirac, *Proc. Roy. Soc. A* **133**, 60–72 (1931).
25. P. A. M. Dirac, *Phys. Rev.* **74**, 817-830 (1948).
26. Z. Fang et al, *Science* **302**, 92 (2003).
27. J. Feoli, V. V. Nesterenko and G. Scarpetta, *J. Math. Phys.* **36**, 5552 (1995).
28. J. Feoli, V. V. Nesterenko and G. Scarpetta, *Class. Quantum Grav.* **13**, 1201 (1996).
29. J. Feoli, V. V. Nesterenko and G. Scarpetta, *Nuclear Phys. B* **705**, 577–592 (2005).
30. L. Graves, *Trans. Amer. Math. Soc.* **252**, 367–392 (1979).
31. H. Hasimoto, *J. Fluid Mech.* **51**, 477 (1972).
32. W. Helfrich, *Z. Natur.* **28c**, 693 (1973).
33. H. Kleinert, *Phys. Lett. B* **174**, 335–338 (1986).
34. H. Hopf, *Math. Ann.* **104**, 637–665 (1931).
35. J. Langer, and D. A. Singer, *J. Diff. Geom.* **20**, 1–22 (1984).
36. J. Langer, and D. A. Singer, *Ann. Global Anal. Geom.* **5**, 133 (1987).
37. J. Langer, and D. A. Singer, *SIAM Review* **38**, 605 (1996).
38. A. Nersessian, R. Manvelyan and G. J.W. Müller, *Nuclear Physics Proc. Suppl.* **88**, 381–384 (2000).
39. B. O'Neill, *Semi Riemannian Geometry*, Academic Press, 1983.
40. R. S. Palais, *Global Analysis Proc. Sympos. Pure Math.* **15**, 239–250 (1970).
41. U. Pinkall, *Invent. Math.* **81**, 379–386 (1985).
42. M. S. Plyushchay, *Mod. Phys. Lett. A* **4**, 837 (1989).
43. M. S. Plyushchay, *Phys. Lett. B* **243**, 383–388 (1990).
44. M. S. Plyushchay, *Phys. Lett. B* **262**, 71–78 (1991).
45. M. S. Plyushchay, *Nuclear Phys. B* **362**, 54–72 (1991).
46. A. M. Polyakov, *Nucl. Phys. B* **268**, 406–412 (1986).
47. A. M. Polyakov, *Nucl. Phys. B* **486**, 23–33 (1997).
48. S. Purkait and D. Ray, *Phys. Lett. A* **116**, 247 (1986).
49. H. K. Urbantke, *J. Geom. Phys.* **46**, 125–150 (2003).
50. J. L. Weiner, *Indiana Univ. Math. J.* **27**, 19–35 (1978).
51. P. B. Wiegmann, *Phys. Lett. B* **323**, 311 (1989).
52. F. Wilczek, *Phys. Rev. Lett.* **49**, 957–959 (1982).
53. T. J. Willmore, *Ann. Sti. Univ. Al I Cuza* **11**, 493–496 (1965).
54. T. J. Willmore, *J. London Math. Soc.* **3**, 307–310 (1971).
55. Y. Zhang, Y. Tan, H. L. Stormer and P. Kim, *Nature* **438**, 201 (2005).

The Geometry and Topology of Liquid Crystals

Christian Santangelo

Department of Physics, University of Massachusetts, Amherst, MA 01003

Abstract. Liquid crystals are phases of matter with properties intermediate to liquids and crystals. In these lectures, I will consider nematic and smectic liquid crystalline phases which are geometrically frustrated by either chiral molecular interactions, boundary conditions, or background spatial curvature. To resolve this frustration, these liquid crystals allow the introduction of topological defects in their ground state which may then organize into an ordered configuration. In particular, we will consider nematic blue phases, smectic twist-grain boundary phases, and focal domains in this light.

Keywords: liquid crystals, smectic, nematics, blue phases, dislocations, nonlinear elasticity, minimal surfaces
PACS: 61.30.Dk,61.30.Hn,61.30.Jf,61.30.Mp

1. INTRODUCTION

Liquid crystals are phases of matter with properties 'between' those of traditional liquids and crystals [1]. They are typically formed by molecules, dubbed mesogens, with a rod-like or disk-like structure. This structure endows the molecules with additional degrees of freedom and anisotropic interactions, allowing a rich variety of ordered and partially-ordered molecular configurations. Liquid crystals are important technologically: they form the basis for modern computer and television displays. They are also of purely theoretical interest because they are a playground to study the role of symmetry, geometry and topological defects in real materials. Liquid crystals occupy a rare role among materials studied by theoretical physicists: if one can dream up the phase, somewhere someone will eventually discover a material that exhibits it (disclaimer: *your actual results may vary*).

There is a zoo of already discovered liquid crystals, and it would be foolish to try to list every possible phase. Instead, we will concern ourselves with only a few, the longest known and best studied liquid-crystalline phases. The most famous of all possible liquid-crystalline phases is the nematic liquid crystal, in which rod-like or disk-like mesogens align. This alignment breaks the rotational symmetry of the isotropic liquid, which occurs at higher temperature or lower density, but not translational symmetry. In the nematic liquid crystal, however, this rotational symmetry breaking is partial: there is no way to distinguish one end of a molecule from another. Subsequently, rotating all the molecules by π returns them to their original state.

Cholesteric phases are locally nematics. However, the mesogens are chiral (they are inequivalent to their mirror image) and their interaction prefers to be at an angle relative to their neighbors. This results in a large scale rotation of the mesogens along one of the spatial directions or, possibly, to a nematic blue phase. Both cases will be discussed in these lectures.

CP1002, *Curvature and Variational Modeling in Physics and Biophysics*
edited by O. J. Garay, E. García-Río, and R. Vázquez-Lorenzo
© 2008 American Institute of Physics 978-0-7354-0521-9/08/$23.00

The mesogens of a smectic liquid crystal, on the other hand, not only align as in a nematic but develop a density modulation – they form uniformly-spaced fluid layers. Smectic liquid crystals come in a number of varieties: smectics-A, in which the molecules align with the layer normals, and smectics-C, in which the molecules align in a direction different than the layer normals, are the most common of these. In addition, there are columnar phases made of disk-like molecules and columnar phases and lamellar phases sharing the symmetries of a smectic-A but made from more structureless block copolymers.

In these lectures, we will be primarily be concerned with the use of geometrical tools to study systems that have been frustrated by the presence of molecular interactions incompatible with the large scale order of the liquid crystalline phase. These interactions are primarily due to chiral mesogens for which the locally preferred chiral order cannot be extended indefinitely. To resolve this frustration, the ground state adopts a configuration of topological defects. Because the liquid crystal "melts" at the core of the defect into a phase with larger symmetry, these defects play the role of addition boundary conditions. We are going to study a simple variational problem: minimize an energy to find the ground state configuration of a particular liquid crystal. Unlike the traditional formulation of a variational problem, however, we will be primarily interested in minimizing the configuration of defects that must occur. In other words, the boundary conditions themselves will be allowed to vary.

2. ELASTICITY OF LIQUID CRYSTALS

We begin by introducing some necessary background for our study of liquid crystalline phases. The key feature of liquid crystalline phases is that they break the full translational and rotational symmetry of the isotropic liquid. The necessary theoretical apparatus to describe this symmetry-breaking is the 'order parameter.' This is best introduced by example: the smectic-C phase is a layered liquid crystal in which each molecule is tilted with respect to the layer normal. The in-plane direction of the tilt is labelled by a vector, $\mathbf{c}$, of fixed magnitude. This vector selects a preferred direction, though the original rotation invariance is preserved in the sense that $\mathbf{c}$ can point in any particular direction with equal energy. Put another way, the various directions of $\mathbf{c}$ specify the degeneracy of the ground state.

From these humble beginnings one can develop a phenomenological, yet powerful, theory of the elasticity of a liquid crystalline phase. Once the broken symmetry has been identified, we may imagine assigning a vector to every point in the material, $\mathbf{c}(\mathbf{x})$. We are interested in understanding how the energy increases as $\mathbf{c}(\mathbf{x})$ changes spatially, although very slowly, compared to the relevant microscopic scales in the system.

2.1. Topological defects

Consider the vectors labelled by integers n, $\mathbf{c}_n = c_0[-\sin(n\theta), \cos(n\theta)]$ with respect to the two coordinates on a flat layer, (x, y), where the angle $\theta = \tan^{-1}(y/x)$ is the azimuthal polar coordinate. If we compute

$$T = \frac{1}{2\pi} \oint d\vec{l} \cdot \frac{\mathbf{c}_n}{|\mathbf{c}_n|} = n. \tag{1}$$

Since n is an integer, there is no continuous way to deform $\mathbf{c}_n$ to a constant vector field. This configuration describes a topological defect. It is a defect because $\mathbf{c}_n$ cannot exist at $x = y = 0$ (the defect core), though it exists everywhere else. The quantity T is the topological charge of the defect; two defects with different values of T are topologically distinct.

In fact, these defects occur more generally in all liquid crystalline systems regularly. One can thing of $\mathbf{c}$ as providing a map from closed loops surrounding the defect core to a point on the circle, S_1. Different values of T correspond to different maps from S_1 to S_1 that are not homotopically equivalent. This mathematical description turns out to be an extremely powerful [2], but by no means complete [3], way to describe the stability of defects in liquid crystalline materials. One key feature is that the order parameter itself determines which defects are topologically stable and which aren't.

However it is important to note that there can be defects that are unstable in the topological sense that, nevertheless, persist because of energetic reasons [4]. Moreover, there are distinct defects in smectic liquid crystals that are, nevertheless topologically indistinguishable [3].

As will be explained later in these lectures, defects are important because they act as boundary conditions for the minimization of a free energy. In essence, there is a core region, of size ξ, around $x = y = 0$ in which $\mathbf{c} = 0$. In this region, the phase has smectic-A order, the molecules align with the normal and do not select a preferred in-plane direction. There is an energy cost for melting the smectic-C in this region, since $\mathbf{c}$ has a preferred nonzero energy. Nevertheless, we can minimize the energy in the presence of a certain configuration of defects. Then the defects themselves become the primary and essential degrees of freedom - their existence and location must be further minimized. It is this second minimization problem which will be the primary focus of these lectures. Put another way, we are interested in energy minimization when the boundary conditions themselves are also subject to minimization.

To proceed further, we will discuss how free energy functionals are determined for liquid crystalline phases. We will then proceed by considering a nematic liquid crystal made from chiral molecules whose local interactions are frustrated by geometrical considerations. We will see that it will be necessary to introduce a network of topological defects to minimize the energy and resolve this frustration, at least partially.

2.2. Nematic liquid crystals

The nematic liquid crystal phase is the prototypical liquid crystal. Though the mesogens align, breaking rotational invariance, this breaking is not complete because there is no way to distinguish $\hat{n}$ from $-\hat{n}$. Thus the space of possible states of an order parameter is not the two-sphere S_2 but the real projective plane, RP_2, a sphere in which antipodal points have been identified. This encodes the fact that a rotation of all the molecules by π radians should return the system to its original state.

Do to this partially preserved rotational symmetry, the order parameter for a nematic is traditionally a symmetric, traceless, second rank tensor, $Q_{\alpha\beta}$. To characterize this further, imagine a state in which $Q_{\alpha\beta}$ was constant in space. In a diagonalizing basis, we have

$$Q_{\alpha\beta} = \begin{pmatrix} q_1 & 0 & 0 \\ 0 & q_2 & 0 \\ 0 & 0 & -q_1-q_2 \end{pmatrix}. \tag{2}$$

It follows that $Q_{\alpha\beta}$ selects a preferred orthonormal basis of directions, the one that diagonalizes it. The molecular direction, $\hat{n}$, for example, is the eigenvector of the third eigenvalue $-q_1 - q_2$. If $q_1 \neq q_2$, a second direction, orthogonal to $\hat{n}$ is selected. This describes a biaxial nematic liquid crystal. When $q_1 = q_2$, on the other hand, the remaining eigenvectors can be rotated around the $\hat{n}$ axis since $Q_{\alpha\beta}$ is diagonal in the subspace orthogonal to $\hat{n}$. This describes the uniaxial nematic, the typical situation experimentally. In the case of a uniaxial nematic, we can write

$$Q_{\alpha\beta} = S(\hat{n}_\alpha \hat{n}_\beta - \delta_{\alpha\beta}/3), \tag{3}$$

where $\delta_{\alpha\beta}$ is the Kronecker delta and S is a scalar quantifying the degree of nematic order.

The next step in determining the energy of a nematic liquid crystal under small deformations is to assume that $Q_{\alpha\beta}$ changes slowly with respect to microscopic, molecular length scales. If it changes slowly enough, we can expand the unknown elastic free energy $F(Q_{\alpha\beta})$ in powers of the *gradients* of $Q_{\alpha\beta}$. Our task is to write down every possible term up to some predetermined order in gradients and $Q_{\alpha\beta}$. The coefficients are to be determined by comparison to experiments or microscopic theories and can, in fact, differ depending on the molecule used.

For ease of analysis, one typically writes a free energy functional in terms of $\hat{n}$, however, taking care that $\hat{n} \to -\hat{n}$ is a symmetry respected by the energy. Writing all the terms consistent with the symmetries of the nematic phase, to second order in gradients of $\hat{n}$, one finds the Frank free energy,

$$F_n = \int dV \left\{ \frac{K_1}{2}(\nabla \cdot \hat{n})^2 + \frac{K_2}{2}[\hat{n} \cdot (\nabla \times \hat{n}) + q]^2 + \frac{K_3}{2}[\hat{n} \times (\nabla \times \hat{n})]^2 \right\}. \tag{4}$$

Here we have quartic terms in $\hat{n}$ because $\hat{n}$ is not small – in fact, it has unit magnitude. The first term is the energy of a splay deformation, (figure 1a), the second that of a twist deformation (figure 1b) and the third a bend deformation (figure 1c). The parameter q is the spontaneous twist – relevant for chiral mesogens in which the molecular interactions prefer the molecules to make an angle with respect to one another.

In addition to these, there is another term that is sometimes included in the free energy, and should always be included if we're being honest:

$$\begin{aligned} F_{SS} &= K_{24} \int dV\, \nabla \cdot [\hat{n}(\nabla \cdot \hat{n}) - (\hat{n} \cdot \nabla)\hat{n}] \\ &= K_{24} \int dV\, \left[(\partial_i \hat{n}^i)^2 - \partial_i \hat{n}^j \partial_j \hat{n}^i \right], \end{aligned} \tag{5}$$

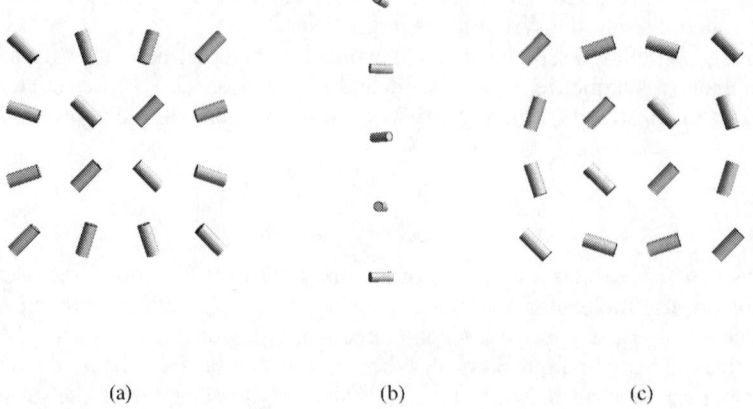

(a) (b) (c)

FIGURE 1. Graphical illustrations of the three terms of the Frank free energy: (a) splay, (b) twist, and (c) bend.

where $\partial_i = \partial/\partial x^i$ is the derivative in the ith direction. This term, the saddle-splay, is also quadratic in gradients – however, because it is a total derivative, it can be reduced to a boundary term and won't play a role in determining the Euler-Lagrange equations for minimizing equation (4). Nevertheless it is an important piece of the minimization problem if we allow the configuration of topological defects to change as well. In that case, the defect cores act as boundaries on which the saddle-splay term receives a contribution [5]. By changing the defect configuration, we also change this contribution.

2.3. Cholesteric liquid crystals and double-twist tubes

A molecule is chiral if its mirror image, $\mathbf{x} \to -\mathbf{x}$, cannot be rotated or translated back to the original structure. These molecules can introduce interaction that also does not preserve the symmetry under reflection. The physical interpretation of this is that chiral molecules prefer to lie at a particular angle with respect to their neighboring molecules. Within the framework of the Frank free energy, such interactions can be described by adopting a spontaneous twist, $q \neq 0$, after which the Frank free energy loses its invariance under reflections. In that case, we find a minimum of the Frank free energy by

$$\hat{n}(z) = [\cos(qz), \sin(qz), 0], \qquad (6)$$

for which $\nabla \cdot \hat{n} = 0$ and $\nabla \times \hat{n} = -q\hat{n}$. This ground state further breaks the symmetry of the nematic: whereas nematics are invariant under translations $z \to z + a$, now they are only invariant when $qa = 2\pi m$ for integers m. From the point of view of symmetry, the ground state of the cholesteric develops periodic order along the z direction, endowing it with the same symmetries as the smectic-A liquid crystal. Replacing the layers in the case of the cholesteric are the surfaces of constant $\hat{n}$.

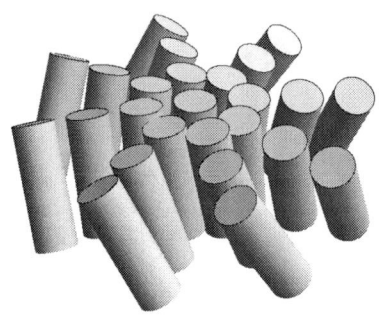

FIGURE 2. Mesogens in a cross-section of a double twist tube. The mesogens surrounding the central core have a fixed angle with respect to the center in all directions. This condition of double-twist does not persist away from the center.

If chiral molecules prefer to maintain a relative angle between their neighbors, the cholesteric does not obviously appear to be the lowest energy state possible, since molecules within a plane of constant $\hat{n}$ are not rotated with respect to each other. An even lower energy state could be reached if the molecules twisted in all directions from each other. This geometrical statement can be reduced to a single equation that must be satisfied by $\hat{n}$ [2, 6]

$$\partial_i \hat{n}_j = -q e^{ijk} \hat{n}_k \tag{7}$$

everywhere in space. If an $\hat{n}$ could be found such that this "double-twist" condition holds true, we would then have $\nabla \cdot \hat{n} = 0$, $\hat{n} \cdot (\nabla \times \hat{n}) = -q$ and $\hat{n} \times (\nabla \times \hat{n}) = 0$. In addition, the saddle-splay gives

$$\partial_i \hat{n}^i \partial_j \hat{n}^j - \partial_i \hat{n}^j \partial_j \hat{n}^i = -q^2, \tag{8}$$

lowering the energy below that of the cholesteric ground state.

The role of the saddle-splay term is interesting. On the one hand, it contributes $-q^2$ to every point. On the other hand, it is a total derivative and its contribution only contributes a term on the boundary of the sample, a fact seemingly in conflict with the double-twist state. In fact, matters are far more complex than they first appeared. Though equation (7) can be satisfied locally, there is no guarantee a unit vector $\hat{n}$ can be found to satisfy it everywhere. To see this, consider a vector field $\mathbf{v}_k$ satisfying the double-twist condition of equation (7). If we Fourier transform the vector field, we find

$$\left(i q_i \delta_{jk} + q e^{ijk} \right) v_k(\mathbf{q}) = 0, \tag{9}$$

which implies immediately that $v_k = 0$. Hence, it is certainly not possible to find a unit vector $\hat{n}$ satisfying the double-twist condition everywhere. This inability to extend a local minimal configuration globally is known as geometric frustration and arises in a number of different contexts.

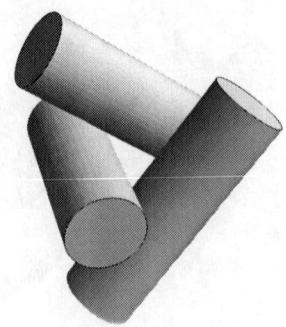

FIGURE 3. Three double-twist tubes in a perpendicular arrangement. The junction between them cannot be joined smoothly, resulting in a $+1/2$ disclination. This set of double-twist tubes can be joined into a periodic cubic array of tubes. Alternatively, this describes a network of $+1/2$ disclinations. Other arrangements are also possible and it is difficult to predict which structure one expects theoretically.

To gain further insight, we can attempt an explicit construction of a double-twist state through the following ansatz for $\hat{n}$ in cylindrical coordinates:

$$\hat{n}^r = 0 \quad \hat{n}^\theta = -\sin \psi(r) \quad \hat{n}^z = \cos \psi(r) \tag{10}$$

subject to the boundary conditions $\psi(0) = 0$. If $\psi(r) = 0$ everywhere, this describes a unit vector $\hat{n} = \hat{z}$. If $\partial_r \psi(r) \neq 0$, however, the nematic satisfies the condition of double twist along the line $r = 0$. This double-twist cannot be satisfied away from this line, however.

Furthermore, consider the free energy using the form of equation (10):

$$F = \int dV \left[\frac{K_2}{2} \left(q_0 - \partial_r \psi - \frac{1}{r} \sin \psi \cos \psi \right)^2 + \frac{K_3}{2} \frac{\sin^4 \psi}{r^2} - \frac{K_{24}}{r} \partial_r \left(\sin^2 \psi \right) \right]. \tag{11}$$

The last term in this expression is the saddle-splay. In a cylinder of radius R, it is evaluated at $r = R$ to give $-2\pi K_{24} \sin^2 \psi(R)$. Therefore, if K_{24} is large enough, this term prefers $\psi(R) \neq 0$, resulting in a double-twist tube.

2.4. Cholesteric blue phases

With such a structure in hand, we can construct an appropriate ground state satisfying the double-twist condition, at least approximately. This could be done, for example, as in figure 3, in which triples of double-twist tubes meet at right angles [2, 5]. Pairs of double-twist tubes can be joined continuously by choosing $\psi(R)$ appropriately (for example, double twist tubes can meet at right angles if $\psi(R) = \pi/4$). However, there is no way to seamlessly connect the director at the junctions of three tubes. Doing so necessarily forces a $+1/2$ disclination in the nematic (see figure 4). Moreover, an ordered space-filling array of double-twist tubes necessarily results in a cubic network of $+1/2$ disclinations. There are, of course, other possibilities for arranging this network of disclinations [2]. We could have also started this analysis by assuming the existence of a network of disclinations and evaluating the saddle-splay energy at the dislocation core [5].

Whether such a phase can exist in a real material is a delicate balance between the energy gain from satisfying the double-twist condition locally and the energy cost of inserting a network of disclinations [5]. In fact, such phases do exist and have been described since at least 1906 [7]. They are called "blue" phases precisely because of their tendency to scatter blue light, which is related to the periodicity of the structure. In fact, there are at least three blue phases: BPI and BPII have cubic symmetry, and BP3 is believed to be an isotropic, melted arrangement of defects [8, 9, 10].

In an interesting twist, Sethna *et al.* has shown that double-twist is not geometrically frustrated on the three-dimensional sphere of radius $1/q$ [6, 11],

$$x_0^2 + x_1^2 + x_2^2 + x_3^2 = 1/q \qquad (12)$$

for coordinates (x_0, x_1, x_2, x_3) in four dimensional Euclidean space. We can define an orthonormal set of basis vectors

$$
\begin{aligned}
e_1 &= q(-x_1, x_0, x_3, -x_2) \\
e_2 &= q(-x_2, -x_3, x_0, x_1) \\
e_3 &= q(-x_3, x_2, -x_1, x_0).
\end{aligned}
$$

This basis has the property that $e_i \partial_j e_k = q\varepsilon_{ijk}$ where ε_{ijk} is the completely antisymmetric tensor. The covariant derivative of $\hat{n}$ is written as

$$\nabla_i \hat{n}_j = \partial_i \hat{n}_j + \hat{n}^k e_j \partial_i e_k = \partial_i \hat{n}_j - q\varepsilon_{ijk} \hat{n}^k. \qquad (13)$$

It is clear from that that constant $\hat{n}$ satisfies the double-twist condition and that this texture is completely *unfrustrated*. Whether there is a consistent procedure to "uncurve" this space to achieve the types of structures observed in experiments remains unclear [6, 12].

2.5. Smectic liquid crystals

We start by recalling the cholesteric liquid crystal, characterized by $\hat{n} = [\cos(qz), \sin(qz), 0]$, which has the same symmetry as the smectic-A liquid crystal. To describe small deformations from the ground state, we introduce a scalar field, u. We then write

$$\hat{n} = [\cos(qz + qu), \sin(qz + qu), 0]. \qquad (14)$$

The smectic free energy can be written in terms of $u(\mathbf{x})$. To do so, we follow the same prescription as with the nematic, we expand the free energy functional in powers of $\partial_i u$.

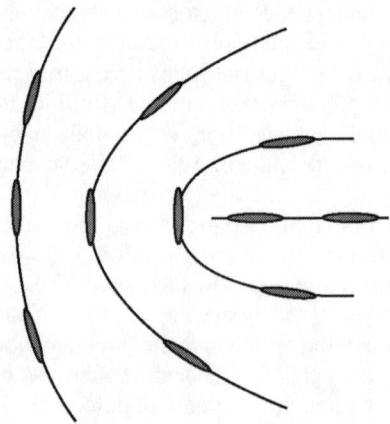

FIGURE 4. A $+1/2$ disclination in a nematic viewed in cross-section. The molecules are shown in red.

To lowest order, we find

$$F = \frac{B}{2} \int dV \left[(\partial_z u)^2 + \lambda^2 \left(\nabla_\perp^2 u \right)^2 \right],$$ (15)

where $\nabla_\perp = \hat{x}\partial_x + \hat{y}\partial_y$, λ has units of length, and B is known as the bulk modulus.

The first term of equation (15) describes changes in layer spacing; the second term, as we will see, describes the curvature of the layers. There are no terms proportional to $\nabla_\perp u$ because they are inconsistent with the rotation invariance (to lowest order). A uniform rotation of the layers by an angle θ (around the y axis) can be written

$$u(x,y,z) = -[\cos\theta - 1]z + \sin\theta x \approx \theta x + \mathcal{O}(\theta^2).$$ (16)

For such an infinitessimal rotation $\partial_x u \neq 0$ but $\partial_x^2 u = 0$. Therefore, rotation invariance ensures that terms linear in derivatives of u vanish but not those quadratic in derivatives.

For smectics described by density modulations, on the other hand, $\rho = \rho_0 + \rho_1 \cos(qz - qu) = \rho_0 + \mathrm{Re}\rho_1 e^{iqz-iqu}$. We write the "phase" $\phi(\mathbf{x}) = z - u$ and define an order parameter $\psi = \rho_1 e^{iq\phi(\mathbf{x})}$ (the phase ϕ might also give the direction of the nematic director in a cholesteric phase). Of course, properly speaking, the phase should be $qz - qu$. However, we will find it useful later to use this convention and it may as well be introduced at this point. The free energy functional is written in terms of ψ and the molecular direction, $\hat{n}$. If we expand the free energy functional in powers of ψ and $\hat{n}$, all terms being consistent with the symmetries of a smectic liquid crystal, we find [1]

$$F = \int d^3x \left\{ |(\nabla - iq\hat{n})\psi|^2 + r|\psi|^2 + u|\psi|^4 \right\} + F_N,$$ (17)

where F_N is the Frank free energy for nematics. It is useful to clarify the coupling between ψ and $\hat{n}$ that arises in equation (17). First, we note that

$$\nabla\psi = iq\nabla\phi\,\psi.$$ (18)

122

The vector $\nabla \phi$ points along the layer normal. Simultaneous rotations of the layers, in terms of ψ, and the director $\hat{n}$ must leave the energy invariant. Individual rotations of either, however, will not. The form of the first term of equation (17) is necessary in order to maintain this full rotation invariance.

How does a free energy functional such as the one in equation (17) work? Notice that, when $r > 0$, the minimum of equation (17) is $\psi = 0$ and the free energy gives the nematic free energy. When $r < 0$, however, the minimum occurs at $\psi \neq 0$ and the free energy necessarily describes a smectic phase since the particle density will now be modulated into layers. The parameter r is usually taken to depend on a "knob", such as temperature, controlled by an experimenter. There is a great deal of fascinating physics (and math) that emerges by studying the properties of theories near $r = 0$. However, our tack shall be different, since we're interested in how the energy changes when deforming the smectic phase.

Let's suppose that $r < 0$ so that $|\psi| \neq 0$ at the minimum of the free energy. Defining $\delta \hat{n} = \hat{n} - \hat{z}$ and assuming $|\psi|^2$ is constant, we expand equation (17) to quadratic order in u and $\delta \hat{n}$, finding

$$F = \frac{1}{2} \int d^3x \left\{ B (\partial_z u)^2 + B' (\nabla_\perp u - \delta \hat{n})^2 + K_1 (\nabla \cdot \delta \hat{n})^2 \right. \tag{19}$$
$$\left. + K_2 [\hat{z} \cdot (\nabla \times \delta \hat{n}) + q]^2 \right\}.$$

The first term is the layer compression term from earlier. The last two terms come from the nematic splay and twist. The second term enforces the smectic-A symmetry, in which the layer normal $\hat{N} = \nabla \phi / |\nabla \phi| \approx \hat{z} - \nabla_\perp u$ aligns with the director $\hat{n}$.

The free energy of equation (17) is analogous to the Landau-de Gennes free energy functional for the superconductor-normal metal transition. The similarity arises because a superconductor also has a complex, scalar order parameter, which is coupled to the vector potential, $\mathbf{A}$ in the following manner [13]:

$$F = \int dV \left\{ \frac{1}{2m} \left| \left(i\hbar \nabla - \frac{e}{c} \mathbf{A} \right) \psi \right|^2 + r|\psi|^2 + u|\psi|^4 \right. \tag{20}$$
$$\left. + \frac{1}{8\pi} (\nabla \times \mathbf{A})^2 - \frac{1}{4\pi} \mathbf{H} \cdot (\nabla \times \mathbf{A}) \right\}.$$

The last two terms arise from the magnetic field energy in the superconductor, $\mathbf{B} = \nabla \times \mathbf{A}$ and from the coupling to an external applied field, $\mathbf{H}$. Much like the case of smectics, the coupling between ψ and $\mathbf{A}$ is dictated by a symmetry – gauge invariance: $\mathbf{A} \rightarrow \mathbf{A} + \nabla f$ and $\psi \rightarrow e^{if} \psi$.

Much has been made about this similarities between these two free energy functionals. There are some differences, however. On the one hand, the gauge symmetry of superconductors is a *local* symmetry in an internal, non-physical space. The rotation symmetry of a smectic is global and related to a physical rotation of the layers themselves. These differences, it will turn out, have observable and qualitative consequences. Nevertheless, there are definite similarities and it is instructive to explore them.

When $r > 0$, $\psi = 0$ and the last term terms have a minimum when $\nabla \times \mathbf{A} = \mathbf{H}$. When $r < 0$ and $\mathbf{H} = 0$, however, $\psi = \sqrt{2u/|r|}e^{i\theta(\mathbf{x})}$ for an arbitrary phase $\theta(\mathbf{x})$. If we use this form in the free energy, we find

$$F = \int dV \left[\frac{\hbar^2}{2m} (\nabla\theta - \mathbf{A})^2 + (\nabla \times \mathbf{A} - \mathbf{H})^2 \right] \tag{21}$$

up to a constant term depending on $\mathbf{H}$. From this form it is clear that the two terms are frustrated. If the first term is minimized exactly, $\mathbf{A} = \nabla\theta$ and $\nabla \times \mathbf{A} = 0$. Thus, the last term is nonzero. On the other hand, minimizing the last term suggests that the first term will be nonzero. This frustration is the source of the celebrated Meissner effect, in which magnetic flux is expelled from a superconductor [13].

A similar frustration is also present in the case of smectics-A. If the last term of equation (19) is minimized, $\nabla \times \delta\hat{n} \neq 0$ implying that $\nabla_\perp u \neq \delta\hat{n}$. On the other hand, setting $\nabla_\perp u = \delta\hat{n}$ implies that the $\nabla \times \delta\hat{n} = 0$. These two frustrated terms result in a geometric analogy to the Meissner effect – smectics expel twist! That is, the existence of layers is incompatible with the nematic twist of the layer normals. We could have also arrived at this result by considering the conditions under which a vector field $\hat{n}$ can be integrated to a set of layers – that is, when can $\hat{n} \propto \nabla\phi$ for a scalar field ϕ? As it happens, the vanishing of twist is sufficient for the existence of layers.

We can take the analogy further. There are type I superconductors which expel magnetic flux up to a critical external field, H_c, after which they become normal metals [13]. There are type II superconductors in which, for large enough fields, some magnetic flux can penetrate in the form of a lattice of vortices. That is, the flux penetrates by melting a local region (a tube) of the superconductor. Outside this tube, $|\psi| > 0$ whereas inside, $\psi = 0$. Since the superconducting region is now no longer simply connected, we can have a vortex in which

$$\oint d\mathbf{l} \cdot \nabla\theta = 2\pi n \tag{22}$$

for integers n since phases need only agree by an integer multiple of 2π. Stokes' theorem tells us that $\nabla \times \nabla\theta \neq 0$, allowing us to find a minimum that optimizes between the two frustrated terms in equation (21).

Similarly, a smectic-A constructed from chiral molecules can exhibit either type I or type II behavior [14]. For a type II smectic-A, twist may penetrate in local regions in which the smectic phase melts. Outside these melted regions, we have

$$\oint d\mathbf{l} \cdot \nabla u = na. \tag{23}$$

This condition describes a type of defect known as a dislocation (see figure 6) which we will explore in the next section.

2.6. Nonlinear theory of smectics

One difficulty with the expansion of the smectic energy in powers of u is that, to quadratic order, the free energy is only infinitesimally rotation invariant. One might worry that an approximation that breaks a a symmetry of the free energy might be prob-

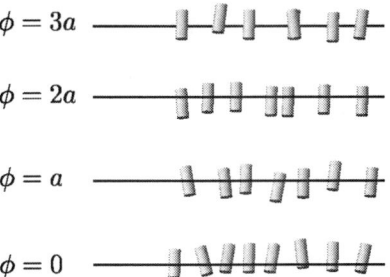

$\phi = 3a$

$\phi = 2a$

$\phi = a$

$\phi = 0$

FIGURE 5. A cross-section of a smectic liquid crystal. The layers can be described as the level sets of a function ϕ, indicated with dark lines. In a smectic-A, the molecules align with the layer normal, $\nabla\phi/|\nabla\phi|$.

lematic, and indeed we will find significant problems. To develop a fully rotationally-invariant theory, we can consider $\psi = |\psi|e^{i\phi(\mathbf{x})/a}$, where a is the equilibrium layer spacing. Layers are described by the level sets of $\phi(\mathbf{x})$ – that is, by the equation $\phi(\mathbf{x}) = na$ for integers n. We either substitute this into equation (17) or write down terms to lowest order in $\nabla\phi$ consistent with rotation invariance. A candidate energy is thus

$$F = \frac{B}{2}\int dV \left\{ \left[\frac{1-(\nabla\phi)^2}{2} \right]^2 + \lambda^2 (\nabla\cdot\hat{n})^2 + \lambda_3^2 [\hat{n}\times(\nabla\times\hat{n})]^2 \right.$$

$$\left. + \bar{\lambda}\int dV \, \nabla\cdot[\hat{n}(\nabla\cdot\hat{n}) - (\hat{n}\cdot\nabla)\hat{n}] . \right\} \tag{24}$$

The last three terms are the splay, bend, and saddle-splay terms of the nematic free energy. We define $\bar{\lambda} = K_{24}/B$, $\lambda^2 = \sqrt{K_1/B}$, and $\lambda_3 = \sqrt{K_3/B}$. The first term is the compression energy term, which prefers to have $|\nabla\phi| = 1$ (now it should be clearer why ϕ is defined to have units of length).

The first term accounts for the energy of changing the spacing of the layers from their equilibrium spacing. This term can be computed directly from equation (17). However, none of these terms are necessarily uniquely specified. This form is constrained by symmetries but little else. One might also consider a compression term of the form $F = \int dV \, e^2$, where e is the strain,

$$e = 1 - |\nabla\phi|. \tag{25}$$

Notice that, as long as the smectic phase is well defined, $\nabla\phi \propto \hat{n} \neq 0$ and there is no problem with the nonanalyticity of $|\nabla\phi|$. If $e' = [1-(\nabla\phi)^2]/2$, then [15, 16]

$$e' = \frac{1}{2}[1-(1-e)2] = e - \frac{e^2}{2}. \tag{26}$$

Thus, e and e' differ by higher order terms in the strain which can be neglected if the strain is small.

Expanding e and e' in powers of u, we find $e \approx e' \approx \partial_z u - (\nabla_\perp u)^2$. This nonlinearity in u is substantially different than higher order terms in e. In particular, since smectics are anisotropic, there is no guarantee that $\partial_z u$ and $\nabla_\perp u$ are the same order. As we will see, the presence of this higher order term in u can substantially affect many of the properties of smectics.

The bend term is typically not included since it is higher order than the compression strain. To see this, we compute [17]

$$\nabla |\nabla \phi| = \hat{n}\hat{n} \cdot \nabla |\nabla \phi| + |\nabla \phi| \hat{n} \cdot \nabla \hat{n}. \tag{27}$$

The last term is proportional to the bend, $\hat{n} \times (\nabla \times \hat{n})$, while the first term is $-\nabla e$. It is typical, therefore, to neglect the bend when the deformation from ideal, flat layers is small.

Similarly, we could replace $\nabla \cdot \hat{n}$ with $\nabla^2 \phi$, also a rotation invariant (and a term that arises from equation (17). Again, these two differ by higher order terms. There is a compelling reason to prefer $\nabla \cdot \hat{n}$, however: it is proportional to the mean curvature of the layers, allowing us to exploit geometry to find smectic minima. In fact, [18, 19]

$$H = -\frac{1}{2}\nabla \cdot \hat{n} \tag{28}$$

$$K = \frac{1}{2}\nabla \cdot [\hat{n}(\nabla \cdot \hat{n}) - (\hat{n} \cdot \nabla)\hat{n}], \tag{29}$$

where $H = (c_1 + c_2)/2$ is the mean curvature, $K = c_1 c_2$ is the Gaussian curvature, and c_1 and c_2 are the principle curvatures found by diagonalizing $h^i_j = g^{ik} h_{kj}$, where h_{jk} are the components of the second fundamental form of the layers and g_{ik} are the components of the first fundamental form. Thus, for equally spaced layers with spacing a the nematic splay term becomes [19, 16]

$$\int dV \ (\nabla \cdot \hat{n})^2 = a \sum_n \int dA_n H_n^2, \tag{30}$$

where the index n indicates that dA_n is the area element of layer n and H_n is the mean curvature of layer n. The saddle-splay term, on the other hand, becomes [19, 16]

$$\int dV \ \nabla \cdot [\hat{n}(\nabla \cdot \hat{n}) - (\hat{n} \cdot \nabla)\hat{n}] = a \sum_n dA_n K_n, \tag{31}$$

which, by the Gauss-Bonnet theorem, can be reduced to a topological invariant plus an integral on the boundary of layer n.

Though there is a ground state minimizing the free energy, flat, equally-spaced layers, the saddle-splay term or boundary conditions can frustrate this simple structure. For example, in the presence of defects, the layers necessarily bend. In that case, it will prove impossible to minimize both the bending and compression parts of the energy.

126

3. SCREW DISLOCATIONS IN SMECTICS

3.1. Linear theory

To begin our study of dislocations, we specialize to the quadratic, smectic-A free energy:

$$F = \frac{B}{2} \int dV \left[(\partial_z u)^2 + \lambda^2 \left(\nabla_\perp^2 u \right)^2 \right].$$

(32)

Of course, I have argued that this will give the wrong results for dislocations but it will be instructive to *see* this occur. For a closed loop surrounding a dislocation (which is the boundary of some area M through which the dislocation penetrates), we have

$$\oint_{\partial M} d\mathbf{l} \cdot \nabla u = na$$

(33)

which we can take to be the definition of a dislocation. We apply Stokes' theorem to this equation find

$$\int_M dA \, \nabla \times \text{“}\nabla u\text{”} = \int_M dA \, an\delta^2(\mathbf{x}_\perp - \mathbf{x}_{\perp,0}).$$

(34)

where $\mathbf{x}_\perp$ gives the coordinates perpendicular to the axis of the dislocation, and $\mathbf{x}_{\perp,0}$ is the location of the dislocation in these coordinates.

Equation (34) makes use of an object, "∇u", which is clearly not the gradient of a scalar since, if it were, it would have vanishing curl. One can think of it, perhaps, as a one-form, a generalization which may have some use when considering smectics on curved surfaces [20]. For the case we are considering here, we make sense of "∇u" by decomposing u into a smooth scalar, u_{sm}, and a singular vector, $\mathbf{u}_{si}$. Thus we write "∇u" $= \nabla u_{sm} + \mathbf{u}_{si}$, where $\nabla \times \nabla u_{sm} = 0$. It is the vector $\mathbf{u}_{si}$ which encodes the topology of the dislocation, whereas u_{sm} tells us something about how the layers away from the dislocation deform. Both pieces are important, of course, and separating them this way simplifies our calculations dramatically (at least in the quadratic theory!).

We define the defect density by $\nabla \times \mathbf{u}_{si} = \mathbf{m} \neq 0$. Since the free energy is quadratic, we proceed to perform our calculations in Fourier space, finding

$$\mathbf{u}_{si} = \frac{-iq \times \mathbf{m}(q)}{q^2}$$

(35)

$$u_{sm} = \frac{q_z \left(1 - \lambda^2 q_\perp^2 \right) (q_x m_y - q_y m_x)}{q^2 \left(q_z^2 + \lambda^2 q_\perp^4 \right)},$$

(36)

when we minimize F with respect to u_{sm}. These expressions can be substituted back into the free energy to yield an expression in terms of only the defect density $\mathbf{m}$,

$$F = \frac{B\lambda^2}{2} \int \frac{d^3q}{(2\pi)^3} \frac{|\hat{z} \cdot (\mathbf{q} \times \mathbf{m}(\mathbf{q}))|^2}{q_z^2 + \lambda^2 q_\perp^4}.$$

(37)

We distinguish between two types of dislocations. A screw dislocation is defined by $\mathbf{m}(\mathbf{q}) = 2\pi na\hat{z}\delta(q_z)$ (see figure 6a) and an edge dislocation is defined by $\mathbf{m}(\mathbf{q}) =$

127

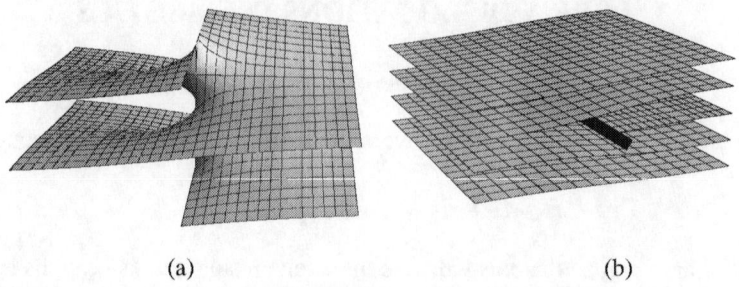

(a) (b)

FIGURE 6.   (a) Screw dislocation and (b) edge dislocation.

$2\pi na\hat{y}\delta(q_y)$ (see figure 6b). Though they are topologically indistinguishable, differing only in the direction of their defect axis relative to the direction of **m**, we will find that they have vastly different properties in smectics in both the quadratic and especially fully nonlinear theories.

In particular, we use equation (37) to compute that the free energy of a single screw dislocation is $F_{screw} = 0$. This result is true even for multiple dislocations, implying that screw dislocations do not interact to quadratic order in u. For edge dislocations, the energy per length of dislocation is

$$\tau_{edge} = \frac{Bn^2 a^2 \sqrt{\pi}}{\sqrt{2}} \sqrt{\frac{\lambda}{\xi}} \tag{38}$$

where we have introduced a microscopic length scale ξ to cutoff a divergent integral at the core. In fact, this length is physical, being related to the size of the region where the smectic is completely melted. The different geometrical character of the two dislocations can be seen readily in figure 6, from which this difference in energetics arises.

It is typical to introduce a Burgers scalar, $b = na$, in smectics, which is analogous to a similar quantity used to study dislocations in crystals. Notice, however, that in smectics the burgers scalar is just that – a scalar – whereas in crystals it is a vector quantity. In other contexts, this subtle difference is typically glossed over, and $\mathbf{b} = b\hat{z}$ by definition. It is also important to note that general dislocations can have axes at any angle to the direction of **m**. These dislocations will typically have both a screw and edge dislocation character. Unfortunately, little is known about such mixed screw-edge dislocations even in the quadratic theory.

A screw dislocation in real space, with defect axis centered at $x = y = 0$, is given by

$$u_{screw} = \frac{b}{2\pi} \tan^{-1}\left(\frac{y}{x}\right). \tag{39}$$

It is straightforward to compute that, for equation (39), $H = 0$. In fact, when $b = 2$, this gives a level set representation of the helicoid, a classical minimal surface [21]. The layer normal is

$$\hat{N} = \frac{\hat{z} - \frac{b}{2\pi r}\hat{\theta}}{\sqrt{1 + \left(\frac{b}{2\pi r}\right)^2}} \tag{40}$$

which implies $\hat{N} \cdot (\nabla \times \hat{N}) \propto \delta^2(\mathbf{x}_\perp)$. Thus, a screw dislocation allows twist to penetrate along its defect axis. Physically speaking, the smectic does not exist at the core of the dislocation. In a proper theory of dislocations, we would find that the smectic has melted back to a nematic (or an isotropic fluid) at the dislocation core, which has some microscopic size ξ. Within a nematic core, it is perfectly reasonable to have nematic twist, which will couple to the layers outside of the core.

Interestingly, equation (39) is also an extremum of the nonlinear smectic free energy. The nonlinear Euler-Lagrange equation is

$$\nabla \cdot \left[|\nabla\phi|\hat{N}e' + \lambda^2 \frac{P\nabla H}{|\nabla\phi|} \right] = 0, \tag{41}$$

where P is an operator that projects a vector onto the subspace orthogonal to the layer normal, $\hat{N}$. Substituting this solution back into the free energy yields

$$F_{screw} = \frac{B}{4}\frac{b^2}{\xi^2} \neq 0. \tag{42}$$

While the energy in the quadratic theory is zero, the energy of a screw dislocation diverges as a power in the nonlinear energy and must be cut off by a microscopic core size, ξ.

3.2. Twist-grain boundary phases

3.2.1. Chiral Phases of Smectics

How far can the analogy between smectics and superconductors be pushed? Screw dislocations are natural candidates for an analogue to vortices in superconductors, in which magnetic flux penetrates. The ground state of a type II superconductor in a strong enough magnetic field is a triangular lattice of parallel, magnetic flux tubes. Is there an analogous state in a smectic-A?

At first glance, an argument by Sethna suggests there is not (see reference [14]). Imagine a lattice of parallel screw dislocations in a smectic, all with burgers scalar b. Integrating ∇u along a loop containing N dislocations gives

$$\oint d\mathbf{l} \cdot \nabla u = Nb. \tag{43}$$

Since $\nabla u \propto \hat{\theta}$ for such a configuration, as $N \to \infty$, $|\nabla u| \to \infty$. This implies significant compression strains and an energetically forbidden configuration of layers.

In fact, there is a loophole in this argument, apparent if one considers a phase field ϕ that is a sum of parallel screw dislocations along only the x direction [22, 23]:

$$\phi = \gamma z - \frac{b}{2\pi} \sum_{n=-\infty}^{\infty} \tan^{-1}\left(\frac{y}{x - n\ell_d}\right). \tag{44}$$

We will set the constant γ shortly using the appropriate boundary conditions. This can be summed exactly by first noting that $\mathrm{Im}\ln(x + iy) = \tan^{-1}(y/x)$ [17]. Then we have

$$
\begin{aligned}
\phi &= \gamma z - \frac{b}{2\pi}\mathrm{Im}\ln(x + n\ell_d + iy) \\
&= \gamma z - \frac{b}{2\pi}\mathrm{Im}\ln\left[1 - \frac{\pi(x + iy)/\ell_d}{n\pi}\right] - \frac{b}{2\pi}\ln(x + iy) + \text{constant}. \tag{45}
\end{aligned}
$$

Utilizing the identity $\sin w = w\prod_{n=1}^{\infty}[1 - w^2/(n^2\pi^2)]$, we find

$$\phi = \gamma z - \frac{b}{2\pi}\mathrm{Im}\ln\sin\left[\frac{\pi(x + iy)}{\ell_d}\right]. \tag{46}$$

In the limit that $y \to \infty$, $\nabla\phi \to -b/(2\ell_d)\hat{x}$. When $y \to -\infty$, $\nabla\phi \to b/(2\ell_d)\hat{x}$. Enforcing the condition that the compression strain, $e = [1 - (\nabla\phi)^2]/$ vanishes at $y = \pm\infty$,

$$\gamma^2 = 1 - \left(\frac{b}{2\ell_d}\right)^2. \tag{47}$$

Notice that as $y \to \pm\infty$,

$$\nabla\phi \propto \hat{N} \to \sqrt{1 - \left(\frac{b}{2\ell_d}\right)^2}\,\hat{z} \mp \frac{b}{2\ell_d}\hat{x}. \tag{48}$$

Hence the layers at $y = \pm\infty$ are flat with normals rotated by an angle α given by $\sin(\alpha/2) = b/(2\ell_d)$. The subsequent structure of a single twist-grain boundary is shown in figure 7, and was originally proposed by Renn and Lubensky [14]. This structure is known as a twist-grain boundary. This immediately suggests a potential ground state for a smectic with penetrating twist: rather than having a lattice of parallel screw dislocations, the screw dislocations are arranged in twist-grain boundaries. Since the layers rotate through a twist-grain boundary, the screw dislocations in adjacent twist-grain boundaries rotate along with the layers.

Setting $\phi = 0$, we find an expression for the height function (or graph) of the surface [22]:

$$z(x,y) = \frac{b}{2\pi}\sec(\alpha/2)\tan^{-1}\left\{\frac{\tanh[(2\pi y/b)\sin(\alpha/2)]}{\tan[(2\pi x/b)\sin(\alpha/2)]}\right\}. \tag{49}$$

This can compared to the height function of another classical minimal surface, Scherk's first surface [22]:

$$z_S(x,y) = \sec(\alpha/2)\tan^{-1}\left\{\frac{\tanh[y\sin(\alpha/2)\cos(\alpha/2)]}{\tan[x\sin(\alpha/2)]}\right\}. \tag{50}$$

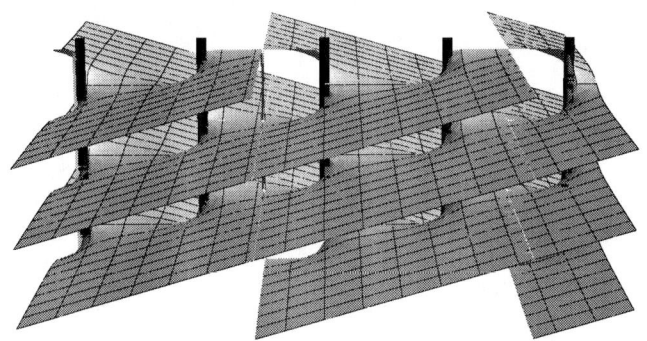

FIGURE 7. A single twist-grain boundary, with screw defects represented by dark lines. The twisting of the layers induces a subsequent twist of defects in the full twist-grain boundary phase.

These two expression agree if we rescale $y \to by\cos(\alpha/2)/(2\pi)$, $x \to bx/(2\pi)$, and $z \to b/(2\pi)z$. Amazingly, equation (49) describes a surface which is merely a rescaling away from a minimal surface. From this we draw the conclusion that Scherk's first surface is decomposable into a "sum" of screw dislocations [24], and that it can serve as a model for twist-grain boundaries as well. Though neither sums of screw dislocations nor Scherk's first surface are exact minima of the smectic free energy, the true structure lies somewhere between [25].

Finding an analytical expression for the entire twist-grain boundary phase is a much greater challenge since the dislocations are not parallel. If $\alpha = \pi/2$, however, we can exploit a surprising symmetry of equation (49). Rewriting the height function as a parametric equation [23],

$$\tan\left[\frac{2\pi z}{b}\cos(\alpha/2)\right]\tan\left[\frac{2\pi x}{b}\sin(\alpha/2)\right] = \tanh\left[\frac{2\pi y}{b}\sin(\alpha/2)\right]. \qquad (51)$$

When $\alpha = \pi/2$, taking $(x,z) \to (z,-x)$ (a $\pi/2$ rotation) and $b \to -b$, the surface is invariant.

This symmetry can be exploited for the particular case of $\pi/2$ twist-grain boundary phases. In that case, twist-grain boundaries can be rotated to be parallel, at the expense of introducing screw dislocations with negative burgers vectors. The parallel screw dislocations lie on a two dimensional lattice, $(n\ell_d + m\ell_d/2, m\ell_b)$ with $\text{sgn}(b) = (-1)^m$.

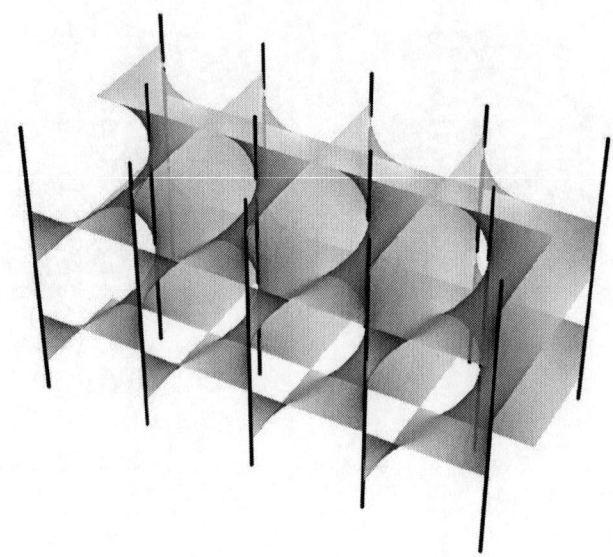

FIGURE 8. Schnerk's first surface, built from dislocations with $b = \pm 2$ and elliptic modulus $k^2 \approx -0.03033$. Reproduced from reference [26].

Then [23]

$$\phi = \gamma z - \frac{b}{2\pi} \text{Im} \sum_{m=-\infty}^{\infty} (-1)^m \ln \sin \left\{ \frac{\pi[(x+m/2)+i(y+m\ell_b)]}{\ell_d} \right\}. \tag{52}$$

Other lattices are, of course, possible, but this choice is particularly meaningful.

If we could sum these twist-grain boundaries, $\phi = \gamma z - \text{Im} \ln f(x+iy)$ where $f(x+iy)$ is a doubly-periodic function on the complex plane. This suggests that $f(x+iy)$ is an elliptic function, the generalization of trigonometric functions that are doubly-periodic on the complex plane. In fact, this sum can be performed exactly. Defining $\tau = 1 + 2i\ell_b/\ell_d$, we have [23]

$$\phi = \gamma z - \frac{b}{2\pi} \text{Im} \ln \text{sn} \left[\frac{2\mathbf{K}(k)x}{\ell_d} + i\frac{\text{Re}\mathbf{K}'(k)y}{\ell_b} \right], \tag{53}$$

where

$$\tau = i\frac{\mathbf{K}'(k)}{\mathbf{K}(k)} \tag{54}$$

and

$$\mathbf{K}(k) = \int_0^1 dx \frac{1}{\sqrt{(1-x^2)(1-k^2x^2)}} \tag{55}$$

is an elliptic integral and $\mathbf{K}'(k) = \mathbf{K}(\sqrt{1-k^2})$. The function sn is the elliptic generalization of sin in the sense that $\text{sn}(z,0) = \sin(z)$.

FIGURE 9. A unit cell of Schnerk's first surface is topologically equivalent to that the of Schwartz D surface. The mean curvature of the unit cell is projected below with intensity proportional to mean curvature. Reproduced from reference [23].

This surface, Schnerk's first surface, is triply-periodic with the unit cell is shown in figure 9. Though it is not minimal, and no simple transformation to make it minimal has been found, this unit cell is topologically identical to another classical minimal surface, the Schwartz D surface. Put another way, u differs from the Schwartz D surface only by a smooth function, not a singular function.

Schnerk's first surface is alternatively defined by a parametric equation,

$$\tan\left(\frac{2\pi\gamma z}{b}\right)\frac{\text{sc}(2\mathbf{K}x/\ell_d)}{\text{dn}(2\mathbf{K}x/\ell_d)} = -i\frac{\text{sc}(2\text{Re}\mathbf{K}'y/\ell_b)}{\text{dn}(2\text{Re}\mathbf{K}'y/\ell_b)}. \tag{56}$$

The function sc/dn is an elliptic generalization of the trigonomentric tan. This equation does not share the symmetries of Scherk's first surface, however, as can be verified by taking $(x,z) \rightarrow (z,-x)$. Nevertheless, it can be decomposed into any of *three* different sets of screw dislocations, identified as the poles and zeros of the elliptic and trigonometric tangents [26]. The zeros correspond to dislocations of one sign, the poles correspond to the other. In the yz-plane, the zeros are found at $(y,z) = (2n\ell_b, mb/(2\gamma))$ and poles are $(y,z) = [(2n+1)\ell_b, (2m+1)b/(4\gamma)]$, where n and m are integers. In the xz-plane, $(x,z) = [m\ell_d, nb/(2\gamma)]$ gives the location of the zeros and $(x,z) = [(2m+1)\ell_d/2, (2n+1)b/(4\gamma)]$ the location of the poles.

3.2.2. Is it stable?

Since there is a closed analytical expression, the energy of Schnerk's first surface can be determined as a function of the distance between grain boundaries, ℓ_b. The elliptic functions have well-known expansions in powers of $q = e^{iK'(k)/K(k)} = -e^{-2\pi\ell_b/\ell_d}$ [27]. Even for $\ell_b = \ell_d$, $q \approx -0.002$ so the grain boundaries do not have to be very far in order for q to be very small. From these expansions, it is tedious but straightforward to determine the small q behavior of the compression energy is given by [26]

$$\frac{\Delta F_c}{A} \sim \frac{BL_z\ell_d}{2\pi\ell_b} \left[C + \left(\frac{b}{2\pi\xi} \right)^2 + 2\ln\left(\frac{2\sqrt{2}\xi}{b} \right) \right] e^{-2\pi\ell_b/\ell_d}, \tag{57}$$

where $C > 0$ is of order unity, xi is the cutoff at the core size, and L_z is the dimension of the system along the z direction. The most important parts of this equation are that the interaction is attractive and the $\ell_b^{-1} e^{-2\pi\ell_b/\ell_d}$ dependence. A similar calculation using the bending energy is even more tedious but generally finds attractive behavior at close distances and repulsive behavior at longer distances ($\ell_b/\ell_d \geq 1$) [26]. Thus, there should be a stable $\ell_b \propto \ell_d$ for which Schnerk's first surface is at least a local minimum of the free energy.

3.2.3. Other twist angles

There is no known analytical expression for twist-grain boundary phases of twist angles other than $\pi/2$. However, the existence of defects along the twist-grain boundary pitch axis (in this case, the y direction) can be exploited to construct such structures topologically. The key is to twist the $\pi/2$ structure slightly to generate twist angles $\alpha < \pi/2$. For small twists, the defects along the pitch axis twist along with the surface, increasing their length and, presumably, their energy.

As the surface is twisted to larger and larger angles, it may happen that at particular twist angles a simpler, less twisted configuration of defects exists which also, presumably, decreases the overall energy of the structure. In that case, we will discover a topologically special set of angles which, by presumption but not calculation, have a lower energy than arbitrary angles. Note that, as you twist the structure, ℓ_d will adjust to match the twist angle of the grain boundaries.

Using the expressions for poles and zeros in the xz-plane and rotating these planes through adjacent grain boundaries, we note that certain special angles occur when the defects running through one grain boundary coincide with those in the adjacent grain boundary. When that occurs, a twist defect can be replaced with a straight one. Thus the special angles occur when a subset of the defects through grain boundaries align. The set of angles for which this occur are [26]

$$\tan(\alpha_n/2) = \frac{1}{\sqrt{2n+1}}. \tag{58}$$

This gives $\alpha = \pi/2$ when $n = 0$, or a ninety degree twist-grain boundary phase in which all defects can be rendered straight. This structure is achiral and identical to Schnerk's first surface.

When $n = 1$, we find a sixty degree twist-grain boundary phase in which all defects also align. However, the positive defects may align with negative defects. This mismatch can be corrected by the introduction of edge dislocations along lines of mismatched screw dislocations, as governed by the topological condition $\oint d\mathbf{l} \cdot \nabla\phi = b$. The edge dislocations themselves rotate by sixty degrees from grain boundary to grain boundary, endowing the structure with its chirality.

In fact, most experimental systems with large-angle twist-grain boundaries exhibit both ninety and sixty degree structures [28].

4. EDGE DISLOCATIONS IN SMECTICS

4.1. Linear theory

Let us consider again the minimization of the quadratic free energy

$$F = \frac{B}{2} \int dV \left[(\partial_z u)^2 + \lambda^2 \left(\nabla_\perp^2 u \right)^2 \right].$$

In this section, we are interested in edge dislocations, for which $\mathbf{m}(\mathbf{q}) = 2\pi b\hat{y}\delta(q_y)$. We have already exhibited the minimum of the smooth part, which gives the layer configurations of the defect far from the defect core by directly solving the Euler-Lagrange equation in Fourier space. Now, I will introduce a new method for finding the layer configuration of an edge dislocation. This method has the virtue that it can be extended to find layer configurations around edge dislocations in a completely rotationally-invariant theory.

First we rewrite the free energy, which is a sum of two squares, by completing the square. This yields

$$F = \frac{B}{2} \int dV \left[\left(\partial_z \mp \lambda \nabla_\perp^2 u \right)^2 \pm 2\lambda \partial_z u \nabla_\perp^2 u \right]. \tag{59}$$

Notice that we can complete the square with one of two possible signs. After some gentle massaging, the cross-term can be expressed as a total derivative, $2\lambda\nabla_\perp \cdot (\partial_z u \nabla_\perp u) - \lambda\partial_z(\nabla_\perp u)^2$, which can then be evaluated on the boundary of the smectic [29]. This boundary exists both far from the edge dislocations, at the edge of the smectic sample, at near the core of the dislocation. Moreover, this boundary term does not contribute to the Euler-Lagrange equation. Finally, we notice that there are a class of exact, global minima for a given set of boundary conditions that can be found by setting the perfect square to zero. In other words, by solving the equation

$$\partial_z u = \pm\lambda\nabla_\perp^2 u. \tag{60}$$

This equation is lower order in all derivatives than the full Euler-Lagrange equation. Therefore, any solution to either equation (60) is a global minimum, yet not all minima of the full equations will be solutions to the lower order equation. In particular, the lower order equations does not allow us to specify a full set of boundary conditions; therefore we must check explicitly that solutions to equation (60) satisfy all the boundary conditions that we have.

Solutions of this form are known as Bogomol'nyi-Prasad-Sommerfield (BPS) minima. General solutions, of course, do not satisfy equation (60). However, since the perfect square is always positive, we can compute a rigorous lower bound to the energy of the actual solution by using only the boundary term in the energy. It is precisely the minima that saturate this bound that are given by the solution of equation (60).

To see how this works explicitly, consider the case of an edge dislocation at $x = z = 0$. Then in the region $z > 0$, we solve $\partial_z u = \lambda \nabla^2_\perp u$, for which $u \to 0$ when $z \to \infty$. In the region $z < 0$, we solve the equation of opposite sign to ensure that $u \to 0$ as $z \to -\infty$. By "gluing" together the solutions above and below $z = 0$, we arrive at a complete solution for an edge dislocation. The boundary conditions for the $z > 0$ region will be $u = (b/2)\Theta(x)$ and in the $z < 0$ region, $u = -(b/2)\Theta(x)$. Here, $\Theta(x)$ is the Heaviside step function, defined by $\Theta(x) = 1$ when $x > 0$ and $\Theta(x) = 0$ when $x < 0$. The mismatch of sign will ensure that $\oint d\mathbf{l} \cdot \nabla u = 0$ for an integral taken around the dislocation core.

The solution is

$$u = \text{sgn}(z) \frac{b}{4} \left[\text{erf}\left(\frac{x}{\sqrt{4\lambda |z|}} \right) + 1 \right], \tag{61}$$

where $\text{erf}(u) = (2/\sqrt{\pi}) \int_0^u dt\, e^{-t^2}$ is the error function and $\text{sgn}(z)$ is $+1$ if $z > 0$ and -1 if $z < 0$. It is easier to understand the expressions for $\partial_z u$ and $\partial_x u$:

$$\partial_z u = -\frac{bx}{8\sqrt{\pi\lambda}|z|^{3/2}} \exp\left(-\frac{x^2}{4\lambda |z|} \right) \tag{62}$$

$$\nabla_\perp u = \frac{b\hat{x}}{4\sqrt{\pi\lambda}|z|^{1/2}} \exp\left(-\frac{x^2}{4\lambda |z|} \right).$$

These expressions show us that $(\nabla_\perp u)^2/\partial_z u \sim |z|^{1/2}/x$. Recall that the compression strain $e \approx \partial_z u - (1/2)(\nabla_\perp u)^2$ to second order in u. We see that, at least for edge dislocations, the quadratic term in u is *not* negligible compared to the first order term. The linear solution will fail, even when the derivatives of u are small far away from the dislocation core.

Some configurations that are not BPS minima would be those that involve two edge dislocations separated by some number of layers. In this case, the layers between would have to satisfy two nontrival boundary conditions, one at each dislocation. However, in this case, we can compute a lower bound on the interaction energy between the two just by evaluating an integral at the defect cores.

4.2. Nonlinear BPS minima

Before turning our attention to a completely rotation invariant theory, it is instructive to consider the somewhat rotationally-invariant nonlinear free energy [29]

$$F = \frac{B}{2} \int dV \left\{ \left[\partial_z u - \frac{1}{2} (\nabla_\perp u)^2 \right]^2 + \lambda^2 \left(\nabla_\perp^2 u \right)^2 \right\}. \tag{63}$$

This energy is invariant up to order θ^2 for uniform rotations by an angle θ. Furthermore, it has been shown that, in the presence of thermal fluctuations, this is the correct long wavelength theory for smectic liquid crystals [30, 31]. This energy also gives the minimal expression for the smectic free energy that is valid at long distances and small deformations.

Completing the square gives

$$F = \frac{B}{2} \int dV \left\{ \left[\partial_z u - \frac{1}{2} (\nabla_\perp u)^2 - \lambda \nabla_\perp^2 u \right]^2 + \frac{4\lambda}{3} \bar{K} u + \lambda \nabla \cdot \mathbf{A} \right\}, \tag{64}$$

where

$$\bar{K} = \frac{1}{2} \nabla_\perp \cdot [\nabla_\perp u \nabla_\perp^u - (\nabla_\perp u \cdot \nabla_\perp) \nabla_\perp u] \tag{65}$$

is the Gaussian curvature of the layers to quadratic order in u and

$$\mathbf{A} = - (\nabla_\perp u)^2 \hat{z} + 2 \partial_z u \nabla_\perp u - \frac{1}{3} (\nabla_\perp u)^2 \nabla_\perp u + \frac{2}{3} u (\nabla_\perp \psi + \hat{z} \times \nabla_\perp \phi). \tag{66}$$

The fields ψ and ϕ are arbitrary; their meaning is currently unclear.

When the layers bend in only the x direction, as they do for edge dislocations, $\bar{K} = 0$ and this free energy has the form of a perfect square plus a total derivative. This we find BPS minima by solving the lower order equation (in the region $z > 0$)

$$\partial_z u - \frac{1}{2} (\nabla_\perp u)^2 - \lambda \nabla_\perp^2 u = 0. \tag{67}$$

This can be solved exactly through the Hopf-Cole transformation, $u = 2\lambda \ln S$, yielding the *linear* equation

$$\partial_z S = \lambda \nabla_\perp^2 S. \tag{68}$$

The solutions have the form $u = 2\lambda \ln[1 + N u_{lin}(x/\sqrt{|z|})]$, where u_{lin} is a solution to the linear Euler-Lagrange equation from the quadratic theory and the normalization N is set by the boundary conditions $u = b/2$ when $x \to +\infty$ and $u = 0$ when $x \to -\infty$.

In the linear theory, there is a superposition principle between edge dislocations: a two edge dislocation solution is just the sum of two single edge dislocation solutions. BPS minima appear to have a nonlinear superposition principle. If the two dislocations lie in the $z = 0$ plane, a solution can be found by summing the solutions inside the logarithm! A four dislocation solution is shown in figure 10, comparing the nonlinear (solid) to linear (dashed) solutions.

137

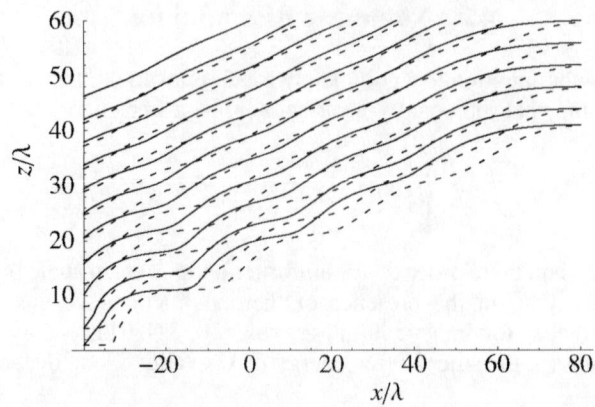

FIGURE 10. The layers of the nonlinear theory (dark lines) compared to the layers of the linear theory (dashed lines) for a configuration of four dislocations with burgers scalar b. The linear theory and nonlinear theory disagree dramatically. Reproduced from reference [29].

The energy for an edge dislocation comes from the term $-(\nabla_\perp u)^2 \hat{z}$ in $\mathbf{A}$ in particular. At $z = \pm\infty$, this vanishes. At the defect core, we use the boundary conditions at $z = 0$ both above and below to find an expression for the line tension

$$\tau_{edge} \frac{Bb^2\sqrt{\pi}}{\sqrt{2}}\sqrt{\frac{\lambda}{\xi}}. \tag{69}$$

This is the same form as in the linear solution up to a suitable rescaling of ξ, whose value is, after all, unknown. Thus, the nonlinear and linear energies agree! Moreover, the interactions between parallel dislocations is unmodified. This is in stark contrast to the case of screw dislocations, in which the nonlinear layer configuration agrees with the linear theory but the energies and interactions are dramatically modified.

The full Euler-Lagrange equation can be recast by defining [15]

$$\Gamma = \partial_z u - \frac{1}{2}(\nabla_\perp u)^2 - \lambda\nabla_\perp^2 u. \tag{70}$$

Then the extrema of the free energy are given by

$$\partial_z\Gamma + \nabla_\perp\left[(\nabla_\perp u)\Gamma\right] - \lambda\nabla_\perp\Gamma = -\frac{2}{3}\lambda\bar{K}, \tag{71}$$

and the Gaussian curvature acts as a source for Γ. The nonlinearity appears to generate an advection term for Γ. At this time, little is known about this modified form of the smectic Euler-Lagrange equations.

4.3. Geometry of BPS minima

By passing to a completely nonlinear, and geometrical, theory, we gain further insight into the BPS decomposition. Consider the rotationally invariant free energy, [15]

$$F = \frac{B}{2} \int dV \left[(1 - |\nabla \phi|)^2 + \lambda^2 (\nabla \cdot \hat{n})^2 \right]. \tag{72}$$

We first make the simple observation that $|\nabla \phi| = \hat{n} \cdot \nabla \phi$. Completing the square gives

$$F = \frac{B}{2} \int dV \left\{ [1 - \hat{n} \cdot \nabla \phi - \lambda \nabla \cdot \hat{n}]^2 + \lambda \nabla \cdot \hat{n} - 2\lambda \nabla \phi \cdot [\hat{n}(\nabla \cdot \hat{n}) - (\hat{n} \cdot \nabla)\hat{n}] \right\}. \tag{73}$$

The last term in this free energy can appear because, since derivatives of a unit vector are orthogonal to the vector and $\nabla \phi \propto \hat{n}$, $\nabla \phi \cdot (\hat{n} \cdot \nabla)\hat{n} = 0$. Defining the quantity

$$\mathbf{A}' = \hat{n}(\nabla \cdot \hat{n}) - (\hat{n} \cdot \nabla)\hat{n}, \tag{74}$$

we notice that the Gaussian curvature is $K = (\nabla \cdot \mathbf{A}')/2$. Hence,

$$F = \frac{B}{2} \int dV \left\{ [1 - \hat{n} \cdot \nabla \phi - \lambda \nabla \cdot \hat{n}]^2 + \lambda \nabla \cdot \hat{n} + 4\lambda \phi K - 2\lambda \nabla \cdot (\phi \mathbf{A}') \right\}. \tag{75}$$

This free energy has the BPS form when $K = 0$ – a perfect square plus a total derivative – as it does for edge dislocations where the curvature along the defect axis will be zero. We find BPS minima by solving the equation

$$1 - |\nabla \phi| = \pm \lambda \nabla \cdot \hat{n}. \tag{76}$$

Defining a function $\mathbf{r}_n(x, y)$ embedding the nth layer as a function of surface coordinates (x, y), this can be recast as an evolution equation for the layers [15]:

$$\partial_n \mathbf{r}_n(x, y) = \frac{\hat{n}}{1 - 2\lambda H} \tag{77}$$

where H is the mean curvature. This has the form of a generalized curvature flow, from which we generate all the layers from a single layer near the defect core.

The crucial step in deriving equation (73) was to rewrite

$$\int dV \, \hat{n} \cdot \nabla \phi \, (\nabla \cdot \hat{n}) = \frac{1}{2} \int dV \, \nabla \phi \cdot [\hat{n}(\nabla \cdot \hat{n}) - (\hat{n} \cdot \nabla)\hat{n}]. \tag{78}$$

This result is a generalization of a famous and old integral theorem of Minkowski's for closed surfaces [32],

$$\int dA \, H = \int dA \, (\mathbf{r} \cdot \hat{n}) K. \tag{79}$$

The function $\mathbf{r} \cdot \hat{n}$, where $\hat{n}$ is the outward surface normal, is called the support function, and acts as a generalized radius.

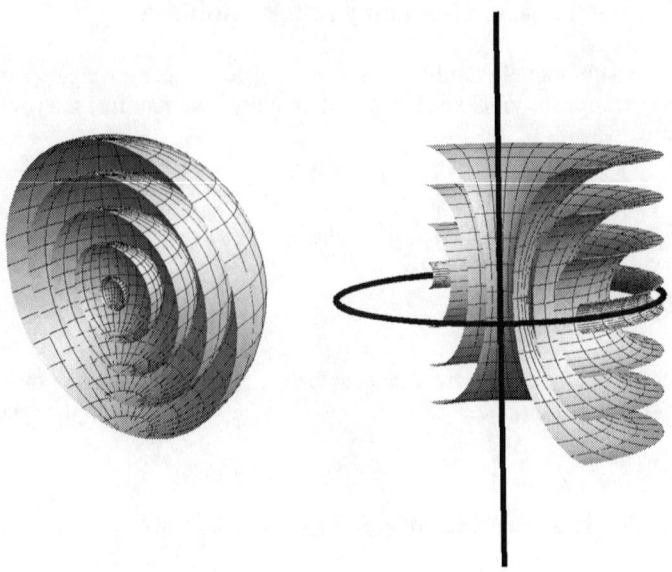

FIGURE 11. Focal conic domains are sets of equally spaced layers. Left: A focal conic domain with positive Gaussian curvature has a singular point at the center of the spheres. Right: A focal conic domain with negative Gaussian curvature has two singular curves, a line and circle. In the general case, these singular curves become a hyperbola and ellipse.

5. FOCAL DOMAINS

5.1. Three-dimensional case

In a typical smectic-A, the length scale λ measuring the ratio between the bending modulus and compression modulus is microscopic – hence over long distances smectics should be well-approximated by layers of equal spacing [1]. In this section, we will study these equally-spaced structures, known as focal domains.

We begin with our previous observation that bend is coupled to compression strain through the equation [17]

$$\nabla|\nabla\phi| = \hat{n} \cdot \nabla|\nabla\phi| + |\nabla\phi|(\hat{n} \cdot \nabla)\hat{n}. \tag{80}$$

Notice that

$$\hat{n} \times (\nabla \times \hat{n}) = -2(\hat{n} \cdot \nabla)\hat{n}, \tag{81}$$

so the last term is related to the nematic bend term of the Frank free energy. If the layers are equally spaced, $|\nabla\phi| = 1$, then

$$(\hat{n} \cdot \nabla)\hat{n} = 0, \tag{82}$$

from which we make a number of observations.

One thing we notice immediately is that the solutions of equation (82) are given by straight lines – that is, the layer normals of equally-spaced layers lie along straight lines. This implies that equally-spaced layers that simultaneously have splay (the layers bend) tend to focus the normals into singularities. Two obvious examples come to mind: (1) equally-spaced cylinders, which have a line singularity at their center (technically a $+1$ disclination), and (2) equally-spaced spheres (known as "focal conics with positive gaussian curvature" or FCD II [33]) with a point singularity at their centers. These do not, by any means, exhaust all possibilities and, in fact, regions of equally spaced layers can be connected to other regions of equal spacing without introducing a great deal of compression energy. The singularities of these focal domains tend to be conic sections (points, circles, lines, ellipses and hyperboli), hence they are often referred to as focal conic domains.

The most famous of all focal conic domains are Dupin cyclides (FCD I) [34]. A special case of such a structure is a torus, with a singular circle (a $+1$ disclination) and a singular line going through the center of the circle and perpendicular to the plane of the circle (see figure 11). By removing the outer half of the torus, the set of layers can be joined to a set of flat layers – hence these focal tori tend to be only half tori. In the general case, Dupin cyclides are formed from singular ellipses and hyperboli through one of the elliptic foci. These appear to be the inside of lopsided tori which are then attached to flat layers. A pair of such generalized tori can be "glued" together, from which one derives the rule that lines traveling through both foci of the ellipses meet at a single point [35].

There are a variety of nontrivial surfaces that have no bending: the minimal surfaces, for which $H = 0$. Can equally-spaced textures with low bending energy be made if we start with a minimal surface and fill in adjacent layers with equal spacing? Can this be arranged so that all the layers are minimal?

In fact, if we start with a surface with mean and Gaussian curvatures, H and K, and build a second surface with spacing a along their normals, we find the new curvatures [18]

$$H' = \frac{H + aK}{1 + 2aH + a^2 K} \tag{83}$$

$$K' = \frac{K}{1 + 2aH + a^2 K}. \tag{84}$$

If the first surface is minimal, $2H = (1/r_1 + 1/r_2) = 0$, where r_1 and r_2 are the principle radii of curvature. But this implies that r_1 and r_2 have opposite signs so $K \leq 0$. If $K = 0$ the layers are flat, but if $K < 0$, H' cannot also be zero. Written in differential form, the mean and Gaussian curvatures change along the normals according to [19]

$$\hat{n} \cdot \nabla H = K - 2H^2 \tag{85}$$

$$\hat{n} \cdot \nabla K = -2KH. \tag{86}$$

Gaussian curvature frustrates equally-spaced layers. Integrating these equations directly, one finds places where the normals focus into singularities and where the curvatures must therefore diverge.

FIGURE 12. A smectic focal domain modeled on the minimal Schwartz P surface. The layers are all equally spaced. Notice the curvature singularities that develop at the cusps and at the center. These curvature singularities are generic features of focal domains. Reproduced from reference [36].

Despite this frustration, minimal surface focal domains have been conjectured as a possible ground state for certain phases of smectics formed from bent-core (or banana molecules). The conjecture is that the packing of the molecular tails is less frustrated on layers with negative Gaussian curvature. If the modulus for Gaussian curvature $\bar{\lambda}$, related to the saddle-splay modulus of the Frank free energy, becomes negative, configurations with large negative K are preferred. Since the surfaces must fill space, they achieve their negative Gaussian curvature by assuming configurations similar to triply-periodic minimal surfaces, which must then be filled in with layers. DiDonna and Kamien have made explicit constructions of focal domains based on a variety of minimal surfaces (see figure 12 for an example based on the Schwartz P surface), complete with detailed predictions of their stability [19, 36].

5.2. Two-dimensional focal domains

In two dimensions, the situation with focal domains is slightly simpler. Here, the layers are curves and there is only one intrinsic curvature, the geodesic curvature. However, only straight lines have no geodesic curvature. Hence there are also no nontrivial curves with vanishing curvature.

We can set up a two-dimensional analogue of the type of frustration that occurs in three dimensions by considering smectics-A on a curved substrate. One advantage of this approach is that it can be arranged experimentally, for example by a monolayer

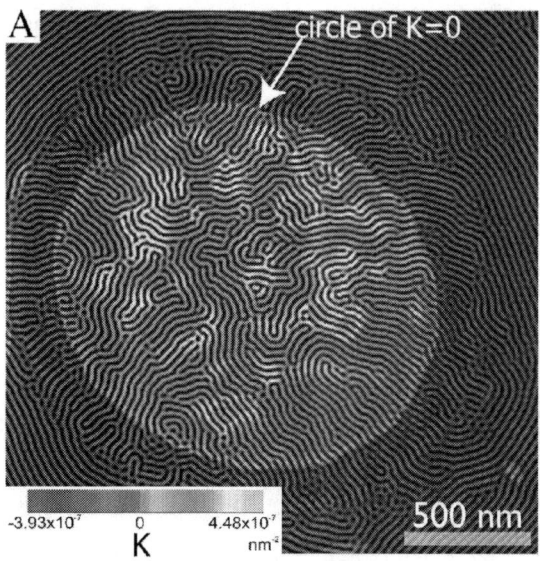

FIGURE 13. Block copolymer layers on a bump as viewed through an atomic force microscope (AFM). The color indicates the magnitude of Gaussian curvature of the surface. Only the top layer can be imaged experimentally, though the film is several layers thick. Reproduced from reference [17].

of diblock copolymer cylinders on a curved substrate. Hexemer and Kramer used the cylindrical phase of a triblock copolymer on a corrugated (bumpy) surface to study the properties of curved smectics [37]. A typical experimental image from a single bump is shown in figure 13.

For comparison, a typical configuration on a flat surface displays characteristic regions of tens of layers which are otherwise randomly oriented throughout the sample. These regions are joined by defects. The most striking feature of the layer configuration on a single bump is that there is a region which exhibits greatly enhanced order, with the layer normal pointing away (or alternatively, toward) the center of the bump. The top of the bump is still disordered, as it would be on a flat substrate. This pattern suggests a hypothesis: there is a mechanism selecting an orientation for the layers around the outside (but not top) of a bump, and the origin of this selection mechanism is geometrical.

5.2.1. Geometry in two dimensions

What remains then is to understand how the geometry of the surface impacts the smectic configuration. To simplify the problem, we will focus only on focal domains. In fact, for the block copolymer used to generate figure 13, λ is one tenth the layer spacing, suggesting that compression strain is far more important than bending energy in determining the layer configurations.

143

On a flat substrate, equal spacing implies $(\hat{n} \cdot \nabla)\hat{n} = 0$, implying the normals lie on straight lines. On an arbitrarily curved surface, this becomes [17]

$$\hat{n}^i D_i \hat{n}^j = 0, \tag{87}$$

where we assume repeated indices are summed and D_i is the covariant derivative on the substrate. Perhaps unsurprisingly, this is just the equation for geodesics on a curved surface. We conclude from this that there is no obstruction to finding equally spaced layers on a curved surface since there is none for finding geodesics. Moreover, there are geodesics in any direction at any point on the surface so there is no obstruction to finding layers with normals in any particular direction.

For focal domains, these layer normals are analogous to the rays in geometrical optics. In this analogy, the Gaussian curvature plays the role of the index of refraction: in regions of negative Gaussian curvature the rays diverge, in regions of positive Gaussian curvature they converge. A better analogy may be to gravitational lensing, where the curvature of space-time itself causes light rays to bend. In geometrical optics, there are places where light rays converge into singularities, known as caustics. If we take this analogy seriously, we will need to identify the analogue of caustics in smectic liquid crystals. These will turn out to be singularities of layer curvature. As a first step, we turn our attention to the layer bending energy in focal domains on curved surfaces.

5.2.2. Bending energy

From where, then, might a preferred orientation arise? Clearly the answer is with the layer bending energy (there is no other energy left, after all!). The layer bending energy is properly understood within the context of the Frenet-Serret formulas for curves in space [18]. Parametrizing each layer by the arc length s and denoting $\hat{t}$ as the unit tangent vector we define the unit normal vector $\hat{n}'$ to be proportional to $d\hat{t}/ds$. The prime is to distinguish this from the unit normal of the layers lying on the surface, denoted $\hat{n}$, which is actually a tangent vector of the surface. We will also find occasion to use $\hat{N}$ as the unit normal vector of the surface itself. The unit binormal vector is $\hat{b} = \hat{t} \times \hat{n}'$.

The Frenet-Serret formulas can be written in the convenient form

$$\frac{d}{ds} \begin{pmatrix} \hat{t} \\ \hat{n}' \\ \hat{b} \end{pmatrix} = \begin{pmatrix} 0 & \kappa & 0 \\ -\kappa & 0 & \tau \\ 0 & -\tau & 0 \end{pmatrix} \begin{pmatrix} \hat{t} \\ \hat{n}' \\ \hat{b} \end{pmatrix}, \tag{88}$$

where κ is the curvature and τ is the torsion. The matrix is a generator of the rotation group $SO(3)$. For curves lying on a surface, the curvature can be further decomposed into two pieces. The geodesic curvature is given by $\kappa_g = \hat{n} \cdot d\hat{t}/ds$ and the normal curvature by $\hat{N} \cdot d\hat{t}/ds$. Note that the component of $d\hat{t}/ds$ that lies on the surface must lie in the direction $\hat{n}$ since $\hat{t}$ is a unit vector, implying that $\hat{t} \cdot d\hat{t}/ds = d/ds(\hat{t}^2)/2 = d/ds(1/2) = 0$. These definitions imply that all three curvatures are related by

$$\kappa^2 = \kappa_g^2 + \kappa_n^2. \tag{89}$$

144

The geodesic curvature is zero when the layers are geodesics and is related to the analogue of the nematic splay energy on a curved surface, $D_i \hat{n}^i$. Thus, one might have generalized the smectic bending energy density to be just κ_g^2 in analogy with smectics in three dimensions. Of course, even on a two dimensional surface, real columns bend in three dimensional space and κ_n^2 should presumably play an important role. Thus, we might also consider a generalized bending energy given by

$$F_B = \int dA \ \left(K_g \kappa_g^2 + K_n \kappa_n^2 \right), \qquad (90)$$

in which the two coefficients K_g and K_n depend on the precise nature of the film itself. This generalization takes into account that anisotropic nature of the columnar film – the direction along the layer is distinguished from the direction normal to the surface by the presence of the surrounding columns and material.

5.2.3. Geometrical frustration in smectics

We return now to the mechanism of frustration of smectics on curved surfaces. We consider the case $K_n = 0$ first, though this is unrealistic. Suppose we consider a single, geodesic layer (having $\kappa_g = 0$). The celebrated Gauss-Bonnet theorem is given by

$$\int_R dA \ K = 2\pi - \sum_i \Delta\theta_i - \sum_i \int_{\partial R_i} ds \ \kappa_g. \qquad (91)$$

The first term is the integral of the Gaussian curvature of the surface in a region with piecewise smooth boundary R on the surface. $\Delta\theta_i$ is the turning angle of the ith vertex of the boundary. In other words, if an ant were traversing the perimeter of the region R, it would turn by an angle $\Delta\theta_i$ when it reached the ith vertex. The last term is the integral of the geodesic curvature along the smooth boundary between vertex i and $i+1$.

Now we ask, is it possible to build a set of equally spaced layers which are also geodesics (hence, with no bending energy)? To see why this is impossible, consider the square in figure 14. Two of the sides are made from layers and the remaining two are the layer normals connecting two adjacent layers. The turning angles are all $\pi/2$ and the layer normals are geodesics if the layers are equally spaced. Suppose one of the layers is also a geodesic, then we apply the Gauss-Bonnet formula to the square to find [17]

$$\int_R dA \ K = -\int_{\partial R_4} ds \ \kappa_g. \qquad (92)$$

Since the left hand side of this equation is nonzero on a curved surface, the remaining layer cannot also be a geodesic. Hence, substrate curvature implies either layer curvature or nonequal spacing between the layers.

In fact, for equally spaced layers, the geodesic curvature grows according to the equation

$$\partial_n \kappa_g = -\kappa_g^2 - K. \qquad (93)$$

145

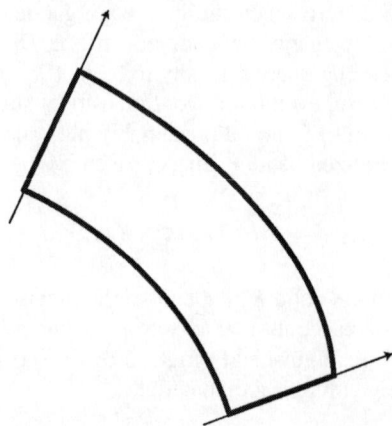

FIGURE 14. The unit normals (represented by thin curves with arrows) and two adjacent layers (thick lines) are always perpendicular if equally spaced. Therefore they form a square on which the Gauss-Bonnet theorem can be applied. If any three of these curves are geodesics and the square contains Gaussian curvature, then the remaining curve cannot also be a geodesic.

Interestingly, this equation possesses no more content than the Gauss-Bonnet theorem though there are a few steps involved in seeing this. Additionally, this form is very similar to the case of three-dimensional focal domains, where we found a similar equation for the evolution of the mean curvature. There, the Gaussian curvature of the layers frustrated our ability to find minimal, equally-spaced layers; here, it the background Gaussian curvature plays an almost identical role. The hope is that this simpler analogue will yield insights into the more delicate case in three dimensions.

A possible mechanism for selecting a direction is apparent in equation (93). A set of layers will preferentially lie in the direction that the geodesic curvature grows the least. However, this is an extremely weak mechanism since the curvature growth depends on the number of layers involved. We can get a sense for how this might work by considering a particular arrangement of layers. Suppose the surface contains a single, circularly symmetric bump with height $h(r)$, where r is the distance from the center of the bump. Starting with a single flat layer infinitely far from the center of the bump, we construct the remaining layers using the equal spacing condition, $e = 0$. Projected onto a flat plane, this construction gives the configuration of layers in figure 15. The ring is the ring of zero Gaussian curvature on the bump, the solid lines the layers and the dashed lines the normals.

On the side opposite where we started the construction, we see that the layers develop cusp-like singularities along a line. This structure, known as a grain boundary, occurs because the geodesic normals on one side of the bump cross the grain boundary at a fixed angle. It is this convergence of geodesics that is equivalent to the formation of caustics in geometrical optics. In other words, caustics lead to the formation of defects in smectics. These defects, however, can form arbitrarily far from the initial layer and depend on the precise shape of the surface as well as initial layer shape.

By considering the triangle outlined in thick solid lines in figure 15, we can compute

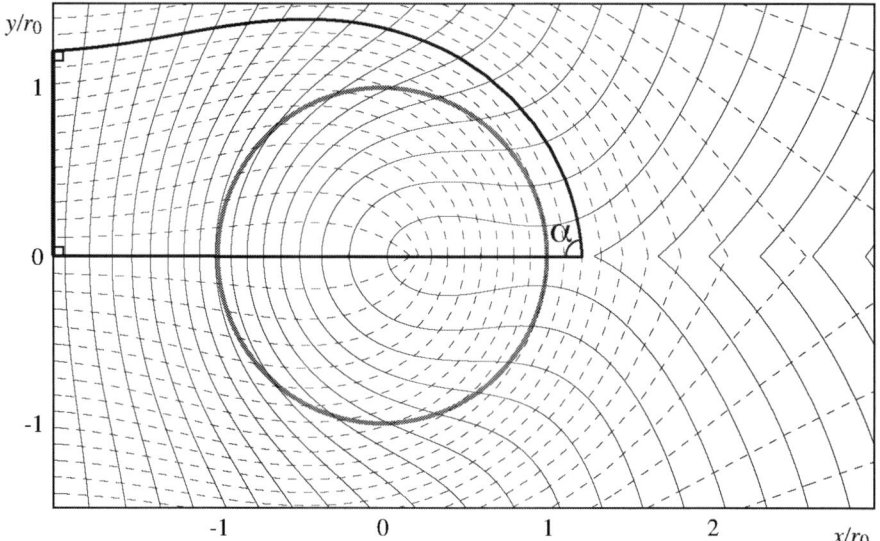

FIGURE 15. Theoretical layers (solid lines) on a curved surface with height profile $h(x, y) = He^{-x^2/(2\sigma^2)}$ for ratio $H/\sigma = 4$. The layers are shown in projection. The ring of zero Gaussian curvature is indicated by the circle. The layer normals are shown as dashed lines. Notice the cusp-like grain boundary that forms on one side of the origin. Reproduced from reference [17].

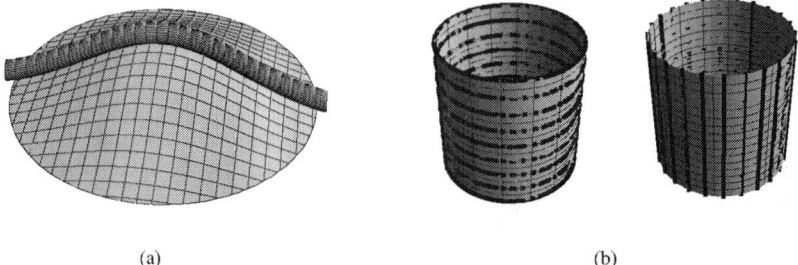

(a)   (b)

FIGURE 16. (a) A cylinder bent to lie on a curved surface still has curvature even if it lies along a surface geodesic (b) There is a preferred direction for layers on a cylinder endowed by the normal curvature. The layers have a lower curvature energy when lying straight along the axis (right image).

the angle of the grain boundary, α, using the Gauss-Bonnet theorem. This triangle is constructing to have one side along a layer infinitely far away from the bump (which must then be straight), a normal line which is a geodesic, and the line passing through the center of the bump along the symmetry axis, which is also a geodesic. Hence, the angle $\alpha = -\int dA\, K$ directly [17]. The triangle contains the positive part of the Gaussian curvature, and by making it larger it contains more and more of the negative region until $\alpha \to 0$. Hence, the grain boundary, though it remains singular, slowly unkinks.

Since geodesic curvature grows so slowly, it seems unlikely that this geodesic curvature mechanism can adequately explain the large, ordered regions seen experimentally. In fact, there is a more local mechanism at work which results directly in a preferred layer direction. That is, the energy of a three dimensional column curving in three dimensional space is nonzero even if that column follows a surface geodesic. This energy arises from the normal curvature κ_n. Suppose that the bump is a surface of revolution, that κ_r is the principle curvature radial to the center of the bump and κ_θ the principle curvature in the azimuthal direction. Then

$$\kappa_n = \cos^2(\theta)\kappa_r + \sin^2(\theta)\kappa_\theta, \tag{94}$$

where θ is the angle the layer makes with the radial direction. Along the ring of zero Gaussian curvature, $\kappa_r = 0$ (but not, in general, κ_θ. When the Gaussian curvature is negative, equation (94) can be solved for $\kappa_n = 0$ for the angle θ, resulting in [17]

$$\tan^2(\theta) = -\frac{\kappa_r}{\kappa_\theta}. \tag{95}$$

Hence there is a direction in which the normal curvature always vanishes. Furthermore, near the ring of zero Gaussian curvature, $\kappa_r = 0$ so the layers will prefer to lie along the radial direction ($\theta = 0$) in order to minimize their normal curvature. This reflects a simple fact about cylinders (for which $K = 0$ as well): geodesics winding around the cylinder have curvature in the three dimensional ambient space whereas geodesics lying along the cylinder axis do not (figure 16). This tipping toward radial of the layers can be seen in figure 13 near the circle of $K = 0$.

6. CONCLUSION

We have only touched on a smattering of the connections between liquid crystals and geometry. There are a number of connections to topology in the form of topological defects which I have barely mentioned at all. Nevertheless, it should be clear that this is a developing field. Though geometry has played a role in liquid crystal research since liquid crystals were first discovered, there has been significant recent progress in applying geometry to understand how frustration in resolved in liquid crystalline systems.

We have seen the important role of nonlinearities throughout these lectures. In the typical case, nonlinear theories are studied by computer, since there is no guarantee that any analytical progress can ever be made. In that light, it is remarkable that so much progress has been made in studying these nonlinear theories. However, these nonlinearities are the direct consequence of the geometry of the liquid crystal phases. Our ability to handle them is related to the fact that geometry itself imposes additional structure on the nonlinearities that can arise that can sometimes be exploited.

It is my hope that, by developing a proper nonlinear theory of defects in smectic liquid crystals that treats them as the fundamental objects, there will be further progress on the types of minimization problems where the boundary conditions (in the form of defects) can themselves change. There is a great deal of hidden structure in the free energy functionals of smectic liquid crystals that still remains to be exploited to any great degree.

ACKNOWLEDGMENTS

I would like to thank the organizing committee for the invitation to give these lectures and for their hospitality in Santiago de Compostela, Spain. Most of the work on smectics described in these lectures was done in collaboration with R.D. Kamien.

REFERENCES

1. P. G. de Gennes and J. Prost, *The Physics of Liquid Crystals*, Oxford University Press, 1995.
2. N. D. Mermin, *Reviews of Modern Physics* **51**, 591 – 648 (1979).
3. M. Kléman and J. Friedel, *Reviews of Modern Physics* **80**, 61 – 115 (2008); arXiv:0704.3055v1.
4. M. Nagasawa and R. Brandenberger, *Phys. Lett. B* **467**, 205 – 210 (1999).
5. S. Meiboom, J. Sethna, P. W. Anderson and W. F. Brinkman, *Physical Review Letters* **46**, 1216 (1981).
6. J. P. Sethna, *Physical Review B* **31**, 6278 (1985).
7. O. Lehman, *Z. Phys. Chem.* **56**, 750 (1906).
8. J. B. Becker and P. J. Collings, *Mol. Cryst. Liq. Cryst.* **265**, 163 (1995).
9. H. S. Kitzerow and P. P. Crooker, *Phys. Rev. Lett.* **67**, 2151 (1991).
10. H. M. Hornreich, *Phys. Rev. Lett.* **67**, 2155 (1991).
11. J. P. Sethna, D. C. Wright and N. D. Mermin, *51*, 467 – 470 (1983).
12. J.-F. Sadoc and R. Mosseri, *Geometrical Frustration*, Cambridge University Press, Cambridge, UK, 1999.
13. J. R. Schrieffer, *Theory of Superconductivity*, Addison-Wesley, Reading, MA, 1983.
14. S.R. Renn and T. Lubensky, *Physical Review A* **38**, 4392 (1990).
15. C. D. Santangelo and R. D. Kamien, *Proc. R. Soc. A* **461**, 2911–2921 (2005).
16. R. D. Kamien and C. D. Santangelo, *Geometriae Dedicata* **120**, 229 (2006).
17. C. D. Santangelo, V. Vitelli, R. D. Kamien and D. R. Nelson, *Phys. Rev. Lett.* **99**, 017801 (2007); arXiV:cond-mat/0703206.
18. R. D. Kamien, *Rev. Mod. Phys.* **74**, 953 (2002).
19. B. A. DiDonna and R. D. Kamien, *Phys. Rev. Lett.* **89**, 215504 (2002).
20. X. Xing, arXiV:0708.3182 (2007).
21. R. Osserman, *A Survey of Minimal Surfaces*, Dover, Mineola, NY, 1986.
22. R .D. Kamien and T. C. Lubensky, *Phys. Rev. Lett.* **82**, 2892 (1999).
23. C. D. Santangelo and R. D. Kamien, *Phys. Rev. Lett.* **96**, 137801 (2006).
24. R. D. Kamien, *Appl. Math. Lett.* **14**, 797 (2001).
25. H. Ogawa and N. Uchida, *Phys. Rev. E* **73**, 060701(R) (2006).
26. C. D. Santangelo and R. D. Kamien, *Phys. Rev. E* **75**, 011702 (2007).
27. Abramowitz and Stegun (Eds.), *Handbook of Mathematical Functions with Formulas, Graphs, and Mathematical Tables*, Dover, New York, 1972.
28. J. Fernsler, L. Hough, R.-F. Shao, J. E. Maclennan, L. Navailles, M. Brunet, N. V. Madhusudana, O. Mondain-Monval, C. Boyer, J. Zasadzinski, J. A. Rego, D. M. Walba and N. A. Clark, *Proc. Natl. Acad. Sci. USA* **102**, 14191 (2005).
29. C. D. Santangelo and R. D. Kamien, *Phys. Rev. Lett.* **91**, 045506 (2003).
30. G. Grinstein and R. Pelcovits, *Phys. Rev. Lett.* **47**, 856 (1981).
31. G. Grinstein and R. Pelcovits, *Phys. Rev. A* **26**, 915 (1982).
32. H. Minkowski, *Math. Ann.* **57**, 447 – 495 (1903).
33. Ph. Boltenhagen, O. D. Lavrentovich, and M. Kléman, *Phys. Rev. A* **46**, R1743 – R1746 (1992).
34. M. Kléman and O. D. Lavrentovich, *Phys. Rev. E* **61**, 1574 – 1578 (2000).
35. M. Kléman, O. D. Lavrentovich, *Eur. Phys. J. E* **2**, 47 – 57 (2000).

36. B. A. DiDonna and R. D. Kamien, *Phys. Rev. E* **68**, 041703 (2003).
37. A. Hexemer, Ph.D. Thesis, U.C. Santa Barbara (2007).

Semiflexible Polymers and Filaments: From Variational Problems to Fluctuations

Jan Kierfeld[*,†], Krzysztof Baczynski[†], Petra Gutjahr[†] and Reinhard Lipowsky[†]

[*]*Physics Department, Dortmund University of Technology, 44221 Dortmund, Germany*
[†]*Max Planck Institute of Colloids and Interfaces, Science Park Golm, 14424 Potsdam, Germany*

Abstract. We discuss shapes and shape fluctuations of semiflexible polymers or filaments, which are polymers with an appreciable bending rigidity. The physical properties of semiflexible polymers are governed by their persistence length. On length scales smaller than the persistence length thermal fluctuations can be neglected and polymer shapes are obtained by bending energy minimization. On length scales larger than the persistence length, however, thermal shape fluctuations play an important role and cannot be neglected in general. After a general definition of the persistence length based on the bending rigidity renormalization we will review some problems related to single semiflexible polymers where both variational problems of energy minimization and thermal fluctuations play an important role. We will discuss the buckling instability of semiflexible polymers, their force-induced desorption, and the shapes of adsorbed semiflexible polymers on structured substrates.

Keywords: Semiflexible Polymers, Filaments, Persistence length, Buckling, Ring polymers, Adsorption, Desorption
PACS: 02.30.Xx, 05.10.Cc, 46.32.+x, 82.35.Gh, 87.16.Dg, 87.16.Ka

INTRODUCTION

Many polymers in chemical and biological physics behave as flexible chains with segments that can freely rotate against each other. Such flexible polymers are governed by an entropic tension, which tends to minimize the end-to-end distance in order to maximize the number of possible chain conformations. Typical examples of flexible polymers are synthetic polymers with a carbon backbone, such as polyethylene, where the carbon-carbon bonds along the backbone can easily rotate against each other. By now, there is a rather complete theoretical description of flexible polymers [1, 2, 3], which includes both statics and dynamics, effects from self-avoidance, and the cooperative behavior of flexible polymers in solutions or gels.

Apart from flexible polymers, there is another important class of more rigid polymers, which are governed by their bending energy rather than their entropic tension over a wide range of length scales. The competition between thermal energy and the bending energy of the polymer sets a characteristic length scale, the *persistence length*,

$$L_p \equiv \frac{2}{d-1}\frac{\kappa}{T} \tag{1}$$

where κ is the bending rigidity of the polymer and T the temperature (in the following we will use energy units, i.e., the Boltzmann constant k_B is contained in the symbol T) and d the number of spatial dimensions of the embedding space. Physically relevant

CP1002, *Curvature and Variational Modeling in Physics and Biophysics*
edited by O. J. Garay, E. García-Río, and R. Vázquez-Lorenzo
© 2008 American Institute of Physics 978-0-7354-0521-9/08/$23.00

cases are polymers in three spatial dimensions ($d = 3$) and polymers, which are adsorbed on a planar substrate and, thus, confined to two dimensions ($d = 2$). We will give a detailed derivation of the persistence length (1) below. In short, the persistence length can be characterized as the length scale above which tangent-tangent correlations decay exponentially, see eq. (3) below.

On length scales much larger than L_p, any polymer behaves as a flexible chain with a segment size of the order of L_p, i.e., the polymer decays into effectively independent segments of size L_p. For large values of the persistence length, the polymer attains the limit of a rigid rod, deformations of which are described by classical mechanics [4], i.e., by minimization of the bending energy of the polymer. We will be interested in *semiflexible polymers*, which can be defined as polymers with a persistence length that is not infinite, but large enough that it becomes comparable to other important length scales governing the polymer's behavior. For a single free polymer, for example, the only other length scale which is relevant is its contour length, and we call the polymer semiflexible if the persistence length becomes of the order of the contour length. In this review, we will focus both on the regime where the persistence length is very large and the effects of thermal fluctuations can be neglected, such that we are dealing with the variational problem of minimizing the bending energy and on the semiflexible regime, where thermal fluctuations start to become relevant.

In biological and chemical physics one finds many examples of semiflexible polymers or "nanorods". Typically, these polymers are supramolecular assemblies with a relatively large diameter, which can vary between about one nanometer and tens of nanometers, i.e., they have a high molecular weight per monomer and are much "thicker" than a flexible synthetic polymer with a carbon backbone. Such polymers are generically semiflexible because their large diameter leads to stronger entropic or enthalpic interactions along their backbones, which increases the bending rigidity. Often these thick supramolecular structures are the result of a rather complicated assembly process. Recent examples from chemical physics are provided by dendronized polymers [5], polyisocyanides (in particular, polyisocanidepeptides [6]), and many supramolecular polymers, such as polyelectrolyte complexes [7]. Another prominent example are carbon nanotubes, which can be viewed as a two-dimensional graphene sheet rolled into a quasi-one-dimensional tubular structure hold together by many carbon bonds. Single-walled carbon nanotubes have a diameter $D \simeq 1 - 2$nm and a persistence length $L_p \simeq 1.5 \mu$m [8] (for $d = 3$). In biological physics an important example for a semiflexible polymer is double-stranded DNA, the carrier of the genetic code, with a mechanical persistence length of 50nm [9] (for $d = 3$). Other important examples are cytoskeletal filaments, such as filamentous (F-) actin and microtubules, or other protein fibers such as sickle hemoglobin (HbS) fibers. Cytoskeletal filaments, such as filamentous (F-) actin and microtubules, are supramolecular structures, which assemble spontaneously from globular protein monomers [10, 11, 12]. F-actin assembles from globular (G-) actin monomers, microtubules from tubulin monomers. These monomers bind by weak non-covalent bonds, typically each monomer binds by several hydrogen bonds. Therefore, cytoskeletal filaments represent self-assembled rod-like structures, sometimes called association colloids, rather than macromolecules that are connected by covalent bonds. [1] The persistence lengths of cytoskeletal filaments range from 15μm for F-actin [13, 14, 15] (with a diameter of $D \simeq 7$nm and in $d = 3$) to the mm-range for microtubules [14] (which have a diameter of $D \simeq 25$nm) and is comparable to typical contour lengths of these polymers.

It is instructive to estimate the bending rigidity of large supramolecular filaments by modeling them in a simplified manner as an isotropic elastic rod made from an isotropic elastic material with Young's modulus E. For such rods, elasticity theory [4] gives a bending rigidity $\kappa = EI$, where I is the moment of inertia of the rod's cross section S_{cross}, $I \equiv \int_{S_{\text{cross}}} r^2 dS$, which is given by the surface integral over the square of the radius r. Therefore, $I \propto D^4$, where D is the filament diameter, and the persistence length $L_p \sim EI/T \propto D^4$ strongly increases with the filament diameter, i.e., thick filaments are much more rigid. For a filament diameter from 5 to 25nm and an elastic modulus $E \sim 1$GPa, persistence lengths $L_p \sim EI/T$ from 10 to 600μm are estimated. This simple model thus explains why thick filaments, such as microtubules with a diameter of 25nm, are much stiffer than slender filaments like F-actin with a diameter of 8nm. The elastic description works well for cytoskeletal filaments because they assemble from relatively large protein monomers; globular (G-) actin, for example, has a molecular weight of 43kD.

Large supramolecular assemblies often form helices to optimize enthalpic interactions, which then also increases the rigidity. Filamentous actin, for example, assembles into a two-stranded helical structure. Another prominent example of a semiflexible helical polymer is DNA. The DNA helix forms as a result of the stacking interactions between planar nucleotide base pairs, which in turn arise from the hydrophobicity of the base pairs and electronic interactions. The helical structure allows base pairs to move closer and expel water from the space between bases. Both in F-actin and DNA the helical structure increases attractive interactions between monomers and increases the bending rigidity. The bending rigidity of a polymer is also increased by repulsive electrostatic interactions between unscreened charges along the polymer backbone [16]. The additional electrostatic contribution to the persistence length leads to a stiffening of charged polymers at low salt concentrations. At physiological salt concentrations, the Debye-Hückel screening length is $l_{DH} \simeq 1$nm so that we can usually neglect electrostatic interactions in modeling cytoskeletal filaments.

Typically, we are interested in fluctuations of semiflexible polymers on the scale of the persistence length such that molecular details are not relevant and a continuous description is justified. Moreover, many semiflexible polymers of interest, like F-actin, microtubules, or DNA, are inextensible to a good approximation. Then they can be modeled by the so-called *worm-like chain (WLC) model* introduced by Kratky and Porod [17] to interpret X-ray scattering experiments on solutions of "filamentous" polymers, e.g., cellulose. Their WLC model describes an inextensible continuous polymer governed by

---

[1] It is interesting to note that, at the beginning of the 20th century, all polymers were viewed as association colloids, whereas, nowadays, most scientists follow the terminology of Hermann Staudinger and use the term "polymer" as a synonym for a chain-like macromolecule.

its bending energy

$$\mathcal{H}_{\mathrm{WLC}}[\mathbf{t}(s)] = \int_0^L ds \frac{\kappa}{2} (\partial_s \mathbf{t})^2 \quad \text{with } \mathbf{t}^2(s) = 1. \tag{2}$$

Here, L is the contour length of the polymer, which is parameterized by its arc length s, and the polymer configuration is described by the unit tangent vectors $\mathbf{t}(s)$ so that $(\partial_s \mathbf{t})^2$ is the square of the local curvature. In the corresponding partition sum of the worm-like chain model we have to sum over all configurations of unit tangent vectors. The unit length constraint makes the statistical physics of the worm-like chain model non-trivial and equivalent to the one-dimensional non-linear σ-model. Tangent correlations decay exponentially, and in d spatial dimensions one finds [17, 18, 19]

$$\langle \mathbf{t}(s) \cdot \mathbf{t}(s') \rangle = \exp\left(-\frac{|s-s'|}{L_p} \right). \tag{3}$$

For $L \gg L_p$, the worm-like chain can then be described as an effective Gaussian chain with $N = L/L_p$ segments with Kuhn length $b \sim L_p$. Internal interaction potentials, such as, e.g., the excluded volume interaction between different parts of the chain, are neglected in the worm-like chain model. This approximation can be justified as long as the bending rigidity is sufficiently large such that the persistence length L_p is of the same order as the contour length L of the polymer. Then the bending energy is effectively preventing intra-polymer contacts.

During the past decade single-molecule techniques, such as atomic force microscopy (AFM) [20], optical [21] and magnetic tweezers [22], have become available which allow to measure mechanical properties of *individual* molecules and polymers. These techniques give a force resolution in the piconewton range and a spatial resolution in the nanometer regime. Experiments on individual molecules allow to measure distribution functions of observables independent of the spatial averaging, which is always present in usual bulk measurements. Other observables are not directly accessible in a bulk measurement, for example, the extension of a single polymer chain can only be deduced indirectly from scattering experiments in the bulk. Single-molecule techniques are also most suited to study dynamical fluctuations of individual molecules in or out of equilibrium. Applied to polymers, these techniques permit quantitative experimental studies of single polymer deformations and, thus, provide the basis for a quantitative understanding of the mechanical properties of more complex polymer assemblies, such as polymer solutions, gels, or the cytoskeleton of a living cell.

In this review we discuss a number of properties of single semiflexible polymers, which are accessible to such single molecule experiments. The buckling instability of a semiflexible polymer is one example. An elastic rod undergoes a buckling instability if the compressional force F exceeds a certain threshold value. Such buckling instabilities also play a role in biological systems, whenever rigid filaments or semiflexible polymers, such as cytoskeletal filaments or DNA, are under a compressive load. In a single molecule experiment it has been shown that polymerization forces are sufficient to buckle microtubules of micrometer length [23], and the shape of buckled microtubules growing against a hard obstacle has been analyzed to measure microtubule polymerization forces, which were found to lie in the piconewton range.

Novel types of single molecule manipulation experiments become possible with *adsorbed* semiflexible polymers because both visualization and manipulation are easier for adsorbed polymers [24, 25]. In this review we will discuss two single polymer manipulation experiments for adsorbed semiflexible polymers, the force-induced desorption and the forced sliding over an adhesive substrate. Force-induced desorption has been realized experimentally by attaching single polymers to AFM tips, which allows to measure the force exerted by the polymer as a function of the distance from the adsorbing substrate [26]. The force-induced desorption is similar to another type of single molecule manipulation experiment, the unzipping of two semiflexible polymers. Unzipping of polymers has first been studied for the unzipping of the two rather flexible strands of DNA [27] but has recently also been realized for much stiffer protein fibers [28]. Single polymers or other molecules on surfaces can be imaged using scanning tunneling microscopy (STM) [29] on metal or semiconductor surfaces or atomic force microscopy (AFM) [20]. These techniques do not only permit imaging on the surface but the microscopy tips can also be used to manipulate and position individual molecules [30] or individual semiflexible polymers such as DNA [31] on the substrate. In addition, the adsorbing substrate can be modified chemically or lithographically to present a patterned surface structure to the adsorbing polymer, which typically leads to contrasts in adsorption strength. The spatial extension of these regular structures ranges from the micrometer scale down to nanometers. The interaction with a structured substrate surface can be exploited for the immobilization and controlled manipulation of semiflexible polymers which is an important requirement for applications in bionanotechnology.

In comparison to flexible polymers, semiflexible polymers exhibit a more diverse and versatile behavior. First, they exhibit a variety of interesting shapes as obtained from the first variation of the bending energy. From the viewpoint of statistical physics, these shapes represent the states of the polymers at zero temperature. Furthermore, at nonzero temperature, these polymers undergo thermally excited shape fluctuations that "renormalize" their properties and can induce transitions between these different shapes. All shapes of minimal bending energy – the shapes of buckled filaments, the shapes of semiflexible polymer rings adhered to a striped substrate, or the shape of a desorbing polymer – are built from segments of Euler Elastica. These non-trivial minimal energy shapes are subject to thermal fluctuations which can modify the results strongly if the persistence length of the semiflexible polymer becomes comparable to other relevant length scales of the problem.

This article is organized as follows. In the following section "Persistence Length" we introduce a generalized definition of the persistence length based on a renormalization group analysis. In the section "Buckling of Semiflexible Polymers", we consider the buckling instability, in particular the effects of thermal fluctuations on the classical mechanical buckling problem. Then we consider problems related to adsorbed semiflexible polymers. In the section "Adsorption and Desorption", we study force-induced desorption both at zero temperature and taking into account the combined effects of force and temperature. In the section "Semiflexible Polymers on Structured Substrates", we investigate morphologies of adsorbed semiflexible polymers, in particular semiflexible rings, on substrates containing stripe surface structures before we end with conclusions and an outlook.

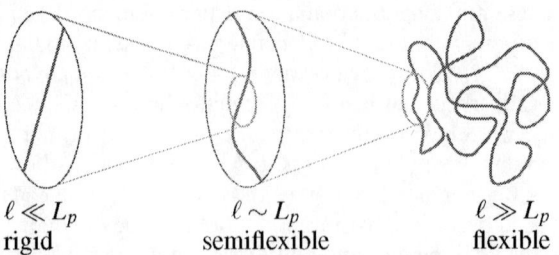

$$\ell \ll L_p \qquad\qquad \ell \sim L_p \qquad\qquad \ell \gg L_p$$
rigid   semiflexible   flexible

FIGURE 1. Schematic view of a thermally fluctuating semiflexible polymer on different length scales ℓ. On very short length scales $\ell \ll L_p$ the thermal energy is not sufficient to introduce a bend in the contour of the semiflexible polymer (left), whereas on large scales $\ell \gg L_p$ the conformational entropy dominates, so that the orientational order is completely destroyed (right). The intermediate regime $\ell \sim L_p$ is characterized by a subtle competition between thermal fluctuations and the stiffness of the semiflexible polymer (middle).

PERSISTENCE LENGTH

Thermal fluctuations of one-dimensional semiflexible polymers or filaments are governed by their bending energy and can be characterized using the concept of a *persistence length* L_p, which is illustrated in Fig. 1. In the absence of thermal fluctuations at zero temperature, semiflexible polymers are straight because of their bending rigidity. Sufficiently large and thermally fluctuating membranes or semiflexible polymers lose their planar or straight conformation. Only subsystems of size $\ell \ll L_p$ appear rigid and maintain an average planar or straight conformation with a preferred normal or tangent direction, respectively. Membrane patches or polymer segments of sizes $\ell \gg L_p$, on the other hand, appear flexible. In the "semiflexible" regime $\ell \sim L_p$ the statistical mechanics is governed by the competition of the thermal energy T and the bending rigidity κ.

The persistence length L_p of semiflexible polymers is usually defined by the characteristic length scale for the exponential decay of the two-point correlation function between unit tangent vectors **t** along the polymer, see eq. (3). The exponential decay of the tangent correlations can also be interpreted as the result of a softening of a semiflexible polymer on large length scales, which is caused by a coupling between bending modes of different wave lengths. This effect can be quantified using renormalization group (RG) methods. Here, we want to present our results for the persistence length from this more general approach, which employs an exact real-space RG scheme for the bending rigidity of a semiflexible polymer to define the persistence length as the characteristic decay length of the renormalized bending rigidity [32].

This problem is not only of interest in the context of one-dimensional semiflexible polymers or filaments. Fluid membranes are objects with two internal dimensions and are also governed by their bending rigidity. For fluid membranes, the quantity which is analogous to the tangent correlations function (3) is the correlation function of normal vectors. An approximate result has been given in Ref. [33], but a rigorous treatment is missing because of the more involved differential geometry. Surfaces cannot be fully determined by specifying an arbitrary set of normal vectors, but have to fulfill additional compatibility conditions in terms of the metric and curvature tensors, the equations of

Gauss, Mainardi and Codazzi, which ensure their continuity. Implementations of these constraints lead to a considerably more complicated field theory than (2) describing a two-dimensional fluid membrane in terms of its normal and tangent vector fields [34].

For two-dimensional fluid membranes, an alternative definition of the persistence length L_p has been given, which is linked to the effect of bending rigidity renormalization. The mode coupling between thermal shape fluctuations of different wave lengths modifies the large scale bending behavior, which can be described by an effective or *renormalized* bending rigidity κ. The renormalized κ has been calculated using different perturbative renormalization group (RG) approaches [35, 36, 37, 38, 39, 40]. The results are still controversial: Several authors [35, 36, 37, 38] find a thermal softening of the membrane with increasing length scales, but differing prefactors, whereas Pinnow and Helfrich [40] obtained the opposite result. Furthermore, different definitions of the persistence length are considered in these approaches: In Refs. [36, 37, 38], L_p is identified with the length scale, where the renormalized bending rigidity κ vanishes, while Helfrich and Pinnow defined L_p via the averaged absorbed area [35, 40]. Whereas the correct bending rigidity renormalization of two-dimensional fluid membranes is still a matter of controversy, we want to present a real-space RG scheme for the bending rigidity renormalization for the one-dimensional case of a semiflexible polymer or a one-dimensional fluid membrane, which allows us to define the persistence length as the characteristic decay length of the renormalized bending rigidity.

In principle, a perturbative result for the effective κ can be deduced from the RG analysis of the one-dimensional nonlinear σ-model, which is equivalent to the WLC Hamiltonian (2). After a Wilson-type momentum-shell RG analysis, one obtains a renormalized reduced effective rigidity $\kappa = \kappa(\ell)$, which depends on the length scale ℓ up to which fluctuations have been integrated in the partition sum. The persistence length can be defined by the condition $\kappa(L_p) \equiv 0$ leading to $L_p \simeq \pi^2 \kappa/(d-2)T$. For the case of the polymer in the plane ($d = 2$), the Hamiltonian simplifies to a free or Gaussian field theory such that κ is unrenormalized to all orders and, thus, the resulting L_p would become infinitely large.

A similar perturbative momentum-shell RG procedure is possible in the so-called Monge parametrization of a weakly bent semiflexible polymer, analogous to the RG analysis for two-dimensional membranes [36]. Then the polymer is parametrized by its projected length x with $0 < x < L_x$, where L_x is the fixed *projected* length of the semiflexible polymer, while its contour length becomes a fluctuating quantity. The renormalized $\kappa = \kappa(\ell_x)$ becomes a function of the projected length scale ℓ_x. Using the analogous criterion $\kappa(L_p) = 0$ we obtain $L_p \simeq 2\pi^2 \kappa/(3d-1)T$ within this approach.

A comparison of the RG results from the non-linear σ-model and the one obtained in the Monge parametrization with eq. (3) for the tangent correlations shows that the RG results for the persistence length L_p are incompatible with each other and with the result (1) based on the tangent correlation function (3). In order to resolve these discrepancies we use a discrete description for semiflexible polymers, which is equivalent to the one-dimensional classical Heisenberg model. This model has the advantage that the κ-renormalization as well as the tangent correlation function are exactly computable in arbitrary dimensions d. Consequently a direct comparison of the persistence length determined via κ-renormalization and via the tangent correlation function is possible.

A discretization of the WLC Hamiltonian (2) should preserve its local inextensibility.

In addition, we want to use a discretized Hamiltonian, which is locally invariant with respect to full rotations of single tangents $\mathbf{t}_i$ – in addition to the global rotational symmetry of the polymer as a whole. A suitable discrete model is an inextensible semiflexible chain model as given by [41]

$$\mathcal{H}\{\mathbf{t}_i\} = \frac{\kappa_0}{b_0} \sum_{i=1}^{M} (1 - \mathbf{t}_i \cdot \mathbf{t}_{i-1}), \quad \text{with } \mathbf{t}_i^2 = 1, \tag{4}$$

with M bonds or chain segments of fixed length b_0. In the following we denote the "bare" parameters before renormalization by κ_0 and b_0. The semiflexible chain model is equivalent to the one-dimensional classical Heisenberg model (except for the first term, which represents a constant energy term) describing a one-dimensional chain of classical spins.

The partition sum reads

$$Z_M = \left(\prod_{j=0}^{M} \int d\mathbf{t}_j \right) \exp[-\mathcal{H}\{\mathbf{t}_j\}/T] = \left(\prod_{j=0}^{M} \int d\mathbf{t}_j \right) \prod_{i=1}^{M} T_{i,i-1}, \tag{5}$$

where we have introduced the transfer matrix

$$T_{i,i-1} = \exp[-K_0(1 - \mathbf{t}_i \cdot \mathbf{t}_{i-1})], \quad \text{with } K_0 \equiv \kappa_0/b_0 T. \tag{6}$$

We can parametrize the scalar product of unit tangent vectors using the azimuthal angle difference $\Delta\theta_{i,i-1}$ as $\mathbf{t}_i \cdot \mathbf{t}_{i-1} = \cos(\Delta\theta_{i,i-1})$. Then the transfer matrix can be expanded as

$$T_{i,i-1} = \sum_{m=-\infty}^{\infty} \lambda_m^{(0)} e^{im\Delta\theta_{i,i-1}}, \quad \lambda_m^{(0)}(K_0) = e^{-K_0} I_m(K_0) \tag{7}$$

in two dimensions and

$$T_{i,i-1} = \sum_{l=0}^{\infty} (2l+1)\lambda_l^{(0)} P_l(\cos\Delta\theta_{i,i-1}), \quad \lambda_l^{(0)}(K_0) = \sqrt{\frac{\pi}{2K_0}} e^{-K_0} I_{l+1/2}(K_0) \tag{8}$$

in three dimensions, where $I_k(x)$ denotes the modified Bessel function of the first kind and $P_l(x)$ the Legendre polynomials [42]. In the following, the sums $\sum_{m=-\infty}^{\infty}$ for $d = 2$ and $\sum_{l=0}^{\infty}(2l+1)$ for $d = 3$ are abbreviated by $\sum_n^{(d)}$.

For simplicity, we restricted our analysis to $d = 2$ and $d = 3$ spatial dimensions, but our results can easily be generalized to arbitrary dimensions d: The transfer matrix is then expanded in Gegenbauer polynomials and the eigenvalues $\lambda_l^{(0)}$ are proportional to modified Bessel functions $I_{l+d/2-1}(K_0)$.

The real-space functional RG analysis for the semiflexible chain (4) proceeds in close analogy to the one-dimensional Heisenberg model [43] and similarly to the Ising-like case where the $\mathbf{t}_i$'s are confined to discrete values [44]. Similar real-space functional RG methods have also been used to study wetting transitions or the unbinding transitions of strings [45, 46]. In each RG step, every second tangent degree of freedom is eliminated. We introduce a general transfer matrix

$$T_{i,i-1} = \exp[h(\mathbf{t}_i \cdot \mathbf{t}_{i-1}, K)], \tag{9}$$

where $h = h(u,K)$ defines an *arbitrary* interaction function depending on the scalar product of adjacent tangents $u = \mathbf{t}_i \cdot \mathbf{t}_{i-1}$ and the parameter K. We start the RG procedure with an initial value $K = K_0$ and an initial interaction function $h(u,K) = -K(1-u)$, see eq. (6). Also for an arbitrary interaction function $h(u,K)$ we can expand the transfer matrix in the same sets of functions as in (7) and (8), which defines eigenvalues $\lambda_m = \lambda_m(K)$ in two dimensions and $\lambda_l = \lambda_l(K)$ in three dimensions. Initially, these eigenvalues are given by $\lambda_m(K) = \lambda_m^{(0)}(K)$ and $\lambda_m(K) = \lambda_l^{(0)}(K)$, see (7), (8).

Integration over one intermediate tangent $\mathbf{t}'$ between $\mathbf{t}$ and $\mathbf{t}''$ defines a recursion formula resulting in a new interaction function $h' = h'(u,K)$ and an energy shift g' by

$$\exp[h'(\mathbf{t} \cdot \mathbf{t}'',K) + g'(K)] = \int d\mathbf{t}' \, \exp[h(\mathbf{t} \cdot \mathbf{t}',K) + h(\mathbf{t}' \cdot \mathbf{t}'',K)], \qquad (10)$$

where the energy shift g' is determined by the condition that $h'(1,K) = h(1,K) = 0$, i.e., the energy is shifted in such a way that the interaction term is zero for a straight polymer. This leads to

$$\exp[g'(K)] = \int d\mathbf{t} \, \exp[2\,h(\mathbf{t} \cdot \mathbf{t}',K)]. \qquad (11)$$

The recursions (10) and (11) are exact and can be used to obtain an exact RG relation for the eigenvalues $\lambda_k^{(N)}$ after N RG recursions,

$$\lambda_k^{(N+1)} = \left[\lambda_k^{(N)}\right]^2 \bigg/ \left\{ \sum_n^{(d)} \left[\lambda_n^{(N)}\right]^2 \right\} = \left[\lambda_k^{(0)}\right]^{2^N} \bigg/ \left\{ \sum_n^{(d)} \left[\lambda_n^{(0)}\right]^{2^N} \right\}. \qquad (12)$$

In general, the new and old interactions $h'(u,K)$ and $h(u,K)$ will differ in their functional structure. Thus the renormalization of the parameter K cannot be carried out in an exact and simple manner as for one-dimensional Ising-like models with discrete spin orientation [44]. The only fixed point function of the recursion (10) is independent of u, i.e., $h^*(u,K) = 0$ because of $h^*(1,K) = 0$. This result, together with the condition $h'(1,K) = 0$, which is imposed at every RG step, suggests that the function $h'(u,K)$ can be approximated by a *linear* function

$$h'(u,K) \simeq -K'(K)(1-u) \quad \text{for } u = \mathbf{t} \cdot \mathbf{t}'' \simeq 1, \qquad (13)$$

as long as the scalar product $u = \mathbf{t} \cdot \mathbf{t}''$ is close to one, i.e., sufficiently close to the straight configuration. This approximation should improve when the whole function $h'(u,K)$ becomes small upon approaching the fixed point $h^*(u,K) = 0$ after many iterations, i.e., on large length scales. Using the approximation (13), $K'(K)$ is defined by the slope of $h'(u,K)$ at $u = 1$,

$$K'(K) \equiv \frac{dh'(u,K)}{du}\bigg|_{u=1} = \frac{d}{du}\exp\left[h'(u,K)\right]\bigg|_{u=1}, \quad \text{with } u = \mathbf{t} \cdot \mathbf{t}''. \qquad (14)$$

Equivalently, one could expand the explicit expression for $h'(x,K)$ given by (10) and the right hand side of (13) for small tangent angles and compare the coefficients. In order to extract the renormalized bending rigidity κ' from the result for K', one has to take into

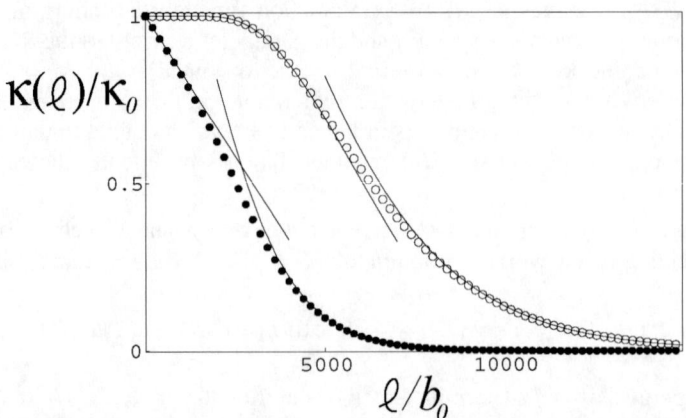

FIGURE 2. Renormalized bending rigidity $\kappa(\ell)/\kappa_0$ as a function of the length scale $\ell/b_0 = 2^N$ for small-scale bending rigidity $\kappa_0 = 1000 b_0 T$ ($K_0 = 1000$) in spatial dimension $d = 2$ ($\circ$) and $d = 3$ ($\bullet$) according to the recursion relation (16). The lines show the asymptotic behavior for $\ell \gg \kappa_0/T$ and $\ell \ll \kappa_0/T$ according to eqs. (17) and (19), respectively. Note that, in two dimensions, the renormalized bending rigidity remains essentially unchanged up to $\ell/b_0 \simeq \kappa_0/b_0 T$, see eq. (19).

account that K' also contains the new bond length $b' = 2b$, which increases by a factor of 2 at each decimation step. Therefore,

$$\kappa'(K) = 2bT\,K'(K). \tag{15}$$

Using this procedure we can calculate the renormalized bending rigidity κ_N after N RG recursions in two and three spatial dimensions starting from the exact RG recursions (12) for the eigenvalues. Inserting the renormalized eigenvalues into (7) or (8), taking the derivative according to (14) and applying the rescaling (15) finally yields the result

$$\frac{\kappa_N}{\kappa_0} = \frac{2^N}{K_0}\left\{\sum_n^{(d)}\left[\lambda_n^{(0)}(K_0)\right]^{2^N}A_n^{(d)}\right\}\Big/\left\{\sum_n^{(d)}\left[\lambda_n^{(0)}(K_0)\right]^{2^N}\right\}, \tag{16}$$

with $A_n^{(2)} \equiv n^2$ and $A_n^{(3)} \equiv \frac{1}{2}n(n+1)$. In the following we will interpret κ_N as a continuous function $\kappa(\ell)$ of the length scale ℓ by replacing the rescaling factor $2^N = b_N/b_0$ by the continuous parameter ℓ/b_0.

The sums in the expressions for the effective bending rigidity (16) can be computed numerically. Fig. 2 displays the results for $\kappa(\ell)/\kappa_0$ as a function of ℓ/b_0 for $K_0 = 1000$ and in two and three spatial dimensions. The value $K_0 = 1000$ is appropriate for a semiflexible polymer with $\kappa_0/T = 10\,\mu m$ and a bond length $b_0 = 10\,nm$, which is close to experimental values for F-actin [14, 15]. For DNA, appropriate values are $\kappa_0/T \simeq 50\,nm$ and $b_0 \simeq 0.3\,nm$ and, thus, $K_0 \simeq 150$.

As long as ℓ is small, κ decays almost linearly in $d = 3$, which is also in qualitative agreement with the result from the RG of the nonlinear σ-model. For $d = 2$ the decay is much slower at small length scales, but, in contrast to the non-linear σ-model where κ is *not* renormalized. This qualitative difference is due to the following important difference between the Heisenberg and the nonlinear σ-model: Parametrizing the WLC model (2) via tangent angles leaves only quadratic terms $\propto (\Delta\theta_{i,i-1})^2$, whereas the discrete semiflexible chain (4) gives terms $\propto 1 - \cos(\Delta\theta_{i,i-1})$, which represent the full expansion of the cosine and obey the local invariance under full rotations.

As ℓ increases $\kappa(\ell)$ approaches zero only asymptotically. Therefore, the definition of the persistence length as length scale where the renormalized κ vanishes, $\kappa(L_p) = 0$ – which is usually used for fluid membranes – would always give an *infinite* result. We propose not to ask at which length scale the renormalized κ reaches zero, but rather *how* it reaches zero. For $\ell \geq b_0 K_0 = \kappa_0/T$ the sums in (16) converge fast and one has to include only the first few terms for accurate results. In fact, one can replace the Bessel functions contained in the eigenvalues $\lambda_k^{(0)}$, see (7) and (8), by their asymptotic form $I_k(x) \approx (x/2\pi)^{-1/2} \exp[x - (k^2 - 1/4)(2x)^{-1}]$ for large x [42]. This is justified for sufficiently large $K_0 \gtrsim 100$, which is fulfilled by semiflexible polymers like F-actin ($K_0 \simeq 1000$) or DNA ($K_0 \simeq 150$). Using this asymptotic we find $(\lambda_m^{(0)}(K_0))^{\ell/b_0} \sim e^{-m^2\ell/2b_0K_0}$ for $d = 2$ and $(\lambda_l^{(0)}(K_0))^{\ell/b_0} \sim e^{-l(l+1)\ell/2b_0K_0}$ for $d = 3$. Moreover, we may expand (16) as a power series in $e^{-\ell T/\kappa_0}$ and obtain

$$\kappa(\ell)/\kappa_0 \approx (\ell T/\kappa_0)\left(2e^{-\ell T/2\kappa_0} - 4e^{-\ell T/\kappa_0} + 8e^{-3\ell T/2\kappa_0} - \ldots\right) \quad \text{for} \quad d = 2,$$
$$\kappa(\ell)/\kappa_0 \approx (\ell T/\kappa_0)\left(3e^{-\ell T/\kappa_0} - 9e^{-2\ell T/\kappa_0} + 42e^{-3\ell T/\kappa_0} - \ldots\right) \quad \text{for} \quad d = 3. \tag{17}$$

The characteristic length scales in the expansions are $2\kappa_0/T$ in $d = 2$ and κ_0/T in $d = 3$, which are, therefore, a natural definition for the persistence length L_p. For general dimensionality d, the exponent of the first term is determined by the order of the Bessel function appearing in the eigenvalue. Thus the RG calculation leads to a persistence length

$$L_p = \frac{2\kappa_0}{T(d-1)}, \tag{18}$$

which agrees exactly with the result (3) based on the tangent correlation function.

Our definition based on the large-scale asymptotics of the exact RG flow is qualitatively different from the definition used in perturbative RG calculations. While the result from the nonlinear σ-model is only valid for small length scales $\ell \ll \kappa_0/T$, where $\kappa(\ell)$ is close to κ_0, the expansions (17) describe the region $\ell \gg \kappa_0/T$. Indeed, taking the expansion of (16) for $\ell \ll \kappa_0/T$, that is

$$\kappa(\ell)/\kappa_0 \approx 1 - (8\pi^2\kappa_0/\ell T)e^{-2\pi^2\kappa_0/\ell T} + \mathcal{O}(\ell e^{-4\pi^2\kappa_0/\ell T}) \quad \text{for} \quad d = 2,$$
$$\kappa(\ell)/\kappa_0 \approx 1 - (\ell T/6\kappa_0) - \mathcal{O}(\ell^2 T^2/\kappa_0^2) \quad \text{for} \quad d = 3, \tag{19}$$

and defining L_p by the exponential decay length in two dimensions, respectively, by the linear term in three dimensions leads to a persistence length, which is considerably bigger than the value (18) found above. The slow exponential decay in the expansion (19) for $d = 2$ is reminiscent of the non-renormalization of κ in the non-linear σ-model and leads to a "plateau" in the numerical result for $\kappa(\ell)/\kappa_0$ for $\ell \ll \kappa_0/T$ in Fig. 2.

In conclusion we have presented a definition of the persistence length L_p of a semi-flexible polymer based on the large scale behavior of the RG flow of the bending rigidity κ, as obtained from a functional real-space RG calculation. Our result (18) for L_p generalizes the conventional definition based on the exponential decay of a particular two-point tangent correlation function and gives identical results for L_p, thus justifying past experimental and theoretical work based on this conventional definition. The RG flows (16) or (17) allow us to follow the behavior of a semiflexible polymer from a stiff polymer on short length scales to an effectively flexible polymer on large length scales quantitatively as a function of the length scale. On large length scales, our functional RG gives qualitatively different results from perturbative RG techniques, which have been used for the closely related problem of fluid membranes [35, 36, 37, 38, 39, 40] and which we also applied to the one-dimensional semiflexible polymer. The generalization of our renormalization approach to two-dimensional fluid membranes is complicated by the more involved differential geometry of these two-dimensional objects and remains an open issue for future investigation.

BUCKLING OF SEMIFLEXIBLE POLYMERS

Buckling of elastic rods is a ubiquitous mechanical problem, which is relevant in elasticity theory and mechanical engineering [4]. An elastic rod undergoes a buckling instability if the compressional force F exceeds a certain threshold value, the critical force F_c, for constant rod length or if the rod length L exceeds a certain critical length L_c for constant force. Such buckling instabilities also play a role in biological systems, whenever rigid filaments or semiflexible polymers, such as cytoskeletal filaments or DNA, are under a compressive load. In a living cell compressive loads can be generated by the polymerization of filaments or by molecular motors, both of which are driven by the hydrolysis of adenine triphosphate (ATP) [12]. Both processes can generate forces in the piconewton range. On the other hand, biological filaments also show pronounced thermal shape fluctuations, which should influence their buckling behavior.

It has been shown experimentally that polymerization forces are sufficient to buckle microtubules of micrometer length [23]. In Ref. [23], the shape of buckled microtubules growing against a hard obstacle has been analyzed to measure microtubule polymerization forces, which were found to lie in the piconewton range. Forces in the piconewton range can also be generated by motor proteins, and it has also been demonstrated experimentally that molecular motors can buckle microtubules of micrometer length [48]. Experiments on microtubules growing inside lipid vesicles demonstrate that microtubules also buckle under the compressive forces exerted by a lipid bilayer under tension [49].

All these experiments show that small forces in the piconewton range are sufficient to buckle cytoskeletal filaments. Such small buckling forces suggest that additional thermal forces, which also generate piconewton forces on a nanometer scale, could modify the buckling instability considerably. In this section we review our results on the influence of thermal fluctuations on the classical buckling instability in two spatial dimensions [47], which can be realized experimentally in confined geometries, i.e., for filaments adsorbed or confined to a planar substrate. We use a systematic expansion in the ratio L/L_p of contour length to persistence length, and integrate out small scale fluctuations to

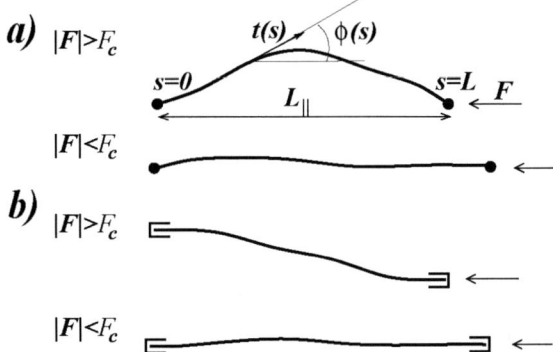

FIGURE 3. Thermally fluctuating filament under a compressive force **F** for (a) free and (b) clamped boundary conditions at both ends. If the absolute value |**F**| of the force exceeds the critical value F_c, the filament buckles; for |**F**| < F_c it remains unbuckled. The filament has contour length L, unit tangent vector $\mathbf{t}(s)$, and tangent angle $\phi(s)$ at arc length s. The parameter $L_\parallel$ is the projected length in the force direction.

obtain an effective theory governing the buckling instability in the presence of thermal fluctuations. This leads to a shift of the buckling force in the presence of thermal fluctuations, and we find that the buckling force *increases* in two dimensions. We also calculate the mean projected length as a function of the applied force (at fixed contour length) and as a function of the contour length (at fixed applied force) in the presence of thermal fluctuations, and our results show that thermal fluctuations lead to a *stretching* of buckled filaments, whereas they compress unbuckled filaments.

To describe a semiflexible polymer under a compressive force in two spatial dimensions we use the worm-like chain Hamiltonian (2) and fulfill the constraint $|\mathbf{t}(s)| = 1$ explicitly by using a parametrization in terms of the tangent angle $\phi(s)$, i.e., $\mathbf{t}(s) = (\cos\phi(s), \sin\phi(s))$. The Hamiltonian also contains an additional energy contribution from the compressive force and becomes

$$\mathcal{H} = \int_0^L ds \left[\frac{\kappa}{2}(\partial_s\phi)^2 + F\cos\phi(s) \right], \tag{20}$$

where $F \equiv |\mathbf{F}|$ is the absolute value of the compressive force. We consider the buckling instability of the straight state $\phi(s) = 0$ and the compressive force is acting in the direction $\phi = \pi$. An important quantity, which can serve as an order parameter for the buckling instability, is the projected length $L_\parallel$, which is given by

$$L_\parallel = \int_0^L ds\cos\phi(s). \tag{21}$$

Buckling at zero temperature

The classical buckling instability is obtained by minimizing the total energy (20) with respect to the angle configuration $\phi(s)$. This minimization leads to the beam equation

$$\kappa \partial_s^2 \phi + F \sin \phi(s) = 0 \tag{22}$$

which has to be solved for appropriate boundary conditions. Boundary conditions at each end of the rod can be classified as free or clamped, where "free" means that the tangent at the end point can freely adapt to the compressional force and "clamped" means that it is constrained to a certain direction, which is usually parallel to the applied force. In the following, we will focus on boundary conditions with two clamped or two free ends. In two dimensions, as considered here, we use either clamped boundary conditions $\phi(0) = \phi(L) = 0$ with both tangent vectors (anti-)parallel to the applied force or free boundary conditions, which correspond to $\partial_s\phi(0) = \partial_s\phi(0) = 0$, i.e., a vanishing curvature and thus a vanishing torque at the filament ends.

Solving the beam equation (22) one finds that a non-zero buckled solution exists at zero temperature above a critical buckling force $F_{c,0}$, which is given by

$$F_{c,0} = \pi^2 \kappa / L^2 \tag{23}$$

both for free and clamped ends and fixed contour length L. Alternatively, if the filament polymerizes against a fixed compressive load F, it will buckle above a critical contour length

$$L_{c,0} = \pi \left(\kappa/F\right)^{1/2} \tag{24}$$

at zero temperature.

Energy minimization gives the contour length L as well as the projected length $L_\parallel$ as a function of the maximal buckling angle ϕ^*, which is attained at $s = 0$ or $s = L$ for two free ends and for $s = L/2$ for two clamped ends,

$$\frac{L}{L_{c,0}} = \sqrt{\frac{F}{F_{c,0}}} = \frac{\mathscr{I}_1(\phi^*)}{\mathscr{I}_1(0)}, \tag{25}$$

$$\frac{L_\parallel}{L_{c,0}} = \frac{\mathscr{I}_1(\phi^*) - \mathscr{I}_2(\phi^*)}{\mathscr{I}_1(0)} \tag{26}$$

with the two integrals

$$\mathscr{I}_1(y) \equiv \int_0^y dx \frac{1}{\sqrt{2(\cos x - \cos y)}} \quad \text{and} \quad \mathscr{I}_2(y) \equiv \int_0^y dx \frac{1 - \cos x}{\sqrt{2(\cos x - \cos y)}}. \tag{27}$$

As y goes to zero, the first integral has the finite limit $\mathscr{I}_1(0) = \pi/2$ whereas $\mathscr{I}_2(0) = 0$. Relations (25) and (26) can also be used as implicit equations to determine the buckling angle ϕ^* for given contour length L or projected length $L_\parallel$, respectively.

Using the Eqs. (25) and (26), one can obtain parametric representations of the reduced projected length $L_\parallel/L$ or $L_\parallel/L_{c,0}$ as a function of the reduced force or the reduced contour length

$$\bar{F} \equiv F/F_{c,0} \quad \text{and} \quad \bar{L} \equiv L/L_{c,0}, \tag{28}$$

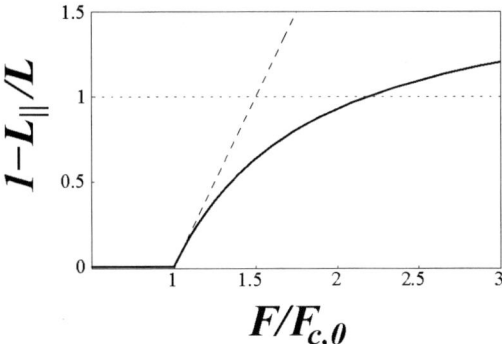

FIGURE 4. Reduced and shifter projected length $1 - L_\parallel/L$ as a function of the reduced force $F/F_{c,0}$ with the critical buckling force $F_{c,0}$ as given by (23). For $F < F_{c,0}$, the filament is straight with $L_\parallel = L$ and $1 - L_\parallel/L = 0$. The buckled solution appears for $F > F_{c,0}$. The solid curve is obtained numerically from relations (25) and (26) by a parametric plot using the buckling angle ϕ^* as the curve parameter. The dashed line is the linear approximation (29). For $F/F_{c,0} > 2.183$, the projected length $L_\parallel$ becomes negative.

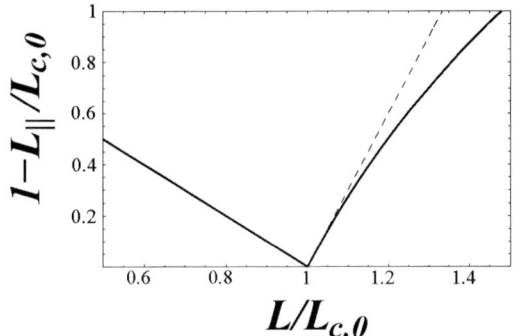

FIGURE 5. Reduced and shifted projected length $L_\parallel/L_{c,0}$ as a function of the reduced contour length $L/L_{c,0}$ with the critical contour length $L_{c,0}$ as given by (24). For $L < L_{c,0}$, the filament is straight with $L_\parallel = L$ which corresponds to the left part of the diagram with $L/L_{c,0} < 1$. The buckled solution appears for $L > L_{c,0}$. The solid curve is obtained numerically by a parametric plot using the buckling angle ϕ^* as the curve parameter in eqs. (25) and (26). The dashed line is the linear approximation (29). For $L/L_{c,0} > 1.478$, the projected length $L_\parallel$ becomes negative.

in the buckled state with $F > F_{c,0}$ or $L > L_{c,0}$, see Figs. 4 and 5, where we use the buckling angle ϕ^* as a curve parameter. Close to the buckling instability we find the asymptotic behavior

$$1 - L_\parallel/L \approx 2(\bar{F} - 1) \text{ for small } \bar{F} - 1 > 0,$$
$$1 - L_\parallel/L_{c,0} \approx 3(\bar{L} - 1) \text{ for small } \bar{L} - 1 > 0.$$

$$(29)$$

165

For $L < L_{c,0}$, on the other hand, the filament is unbuckled which implies that the projected length $L_\parallel$ is identical with the contour length L and

$$1 - L_\parallel/L_{c,0} = 1 - \bar{L} \text{ for } \bar{L} - 1 < 0. \tag{30}$$

Combining the two results for $L > L_{c,0}$ and $L < L_{c,0}$, we see that the relation between projected and contour length exhibits a cusp at the buckling point with $L = L_{c,0}$ [50], as shown in Fig. 5. The cusp could be used to detect the buckling threshold in experiments on growing filaments under a fixed compressive load, which could be generated, for example, by optical traps. The parametric representations shown in Figs. 4 and 5 and thus the asymptotic behavior (29) just above the buckling threshold are valid both for two free and two clamped ends.

Buckling in the presence of thermal fluctuations

In order to consider the effects of thermal fluctuations on the buckling instability, several approaches are possible. We can expand around the "classical" configuration obtained in the previous section and integrate out fluctuations up to quadratic (or higher) order. This approach, however, does not allow us to calculate a fluctuation-induced shift of the threshold force for buckling. Therefore we employ a renormalization-like procedure where we integrate out short wavelength fluctuations in order to obtain an effective theory governing the long wavelength buckling instability. We focus on the regime close to the buckling instability where we can expand the Hamiltonian (20) in tangent angles up to quartic order, and obtain

$$\mathcal{H} = \int_0^L ds \left[\frac{\kappa}{2} (\partial_s \phi)^2 + F \left(1 - \frac{1}{2} \phi^2(s) + \frac{1}{24} \phi^4(s) \right) \right]. \tag{31}$$

For free and clamped boundary conditions, Fourier expansion of $\phi(s)$ leads to

$$\phi(s) = \sum_{n=1}^N \tilde{\phi}_n \cos(n\pi s/L) \quad \text{(free)}, \tag{32}$$

$$\phi(s) = \sum_{n=1}^N \tilde{\phi}_n \sin(n\pi s/L) \quad \text{(clamped)}, \tag{33}$$

respectively, with Fourier coefficients $\tilde{\phi}_n$. The maximal wave number N is given by the number of degrees of freedom, $N = L/a$, where a is a microscopic cutoff, which is set by the monomer size or the filament diameter. The $n = 0$ mode is absent for free boundary conditions because we apply the additional constraint $z(L) - z(0) = \int_0^L ds \sin \phi(s) = 0$ that the end points have the same height coordinate (perpendicular to the force direction). This constraint is automatically fulfilled by the zero temperature solution but has to be imposed separately in the presence of thermal fluctuations. The condition $\tilde{\phi}_0 = 0$ satisfies this constraint up to terms of order $\mathcal{O}(\tilde{\phi}_n^3)$.

In order to investigate the effect of the anharmonic quartic terms, we write the Hamiltonian (31) as

$$\mathcal{H} = \mathcal{H}_2 + \mathcal{H}_4, \tag{34}$$

where $\mathcal{H}_2$ contains all terms up to quadratic order and $\mathcal{H}_4$ the remaining terms up to quartic order. Using the Fourier expansions (32) or (33), the quadratic part can be rewritten as

$$\mathcal{H}_2\{\tilde{\phi}_n\} = FL + \sum_{n\geq 1} \frac{F_{c,0}L}{4} \left(n^2 - \bar{F}\right) \tilde{\phi}_n^2. \tag{35}$$

This representation in Fourier modes shows that buckling is an instability of the $n = 1$ mode for $\bar{F} > 1$, which attains a non-zero equilibrium value in this regime at zero temperature. Higher modes $n > 1$ remain stable up to higher order buckling forces, i.e., for $\bar{F} < n^2$. In the following we focus on the regime $\bar{F} \ll 4$, where only the $n = 1$ mode can become unstable and large. Expectation values for higher modes $n, m \geq 2$,

$$\langle \tilde{\phi}_n \tilde{\phi}_m \rangle = \delta_{nm} \frac{2T}{F_{c,0}L} \frac{1}{n^2 - \bar{F}}, \tag{36}$$

as calculated with the Hamiltonian (35) are of the order of

$$\frac{T}{F_{c,0}L} = \frac{1}{\pi^2} \frac{L}{L_p} \equiv t. \tag{37}$$

The dimensionless parameter t is a reduced temperature, which is small for semiflexible filaments with $L \lesssim L_p$. Expectation values $\langle \tilde{\phi}_n^2 \rangle \sim t$ of higher modes are thus small as well. The parameter t will be used in the following as an expansion parameter for the systematic treatment of fluctuations. This parameter is small in the limit of small temperature, large bending rigidity, or small contour length. A typical value for a microtubule of contour length $L = 10\mu$m and $L_p = 1$mm is $t \simeq 10^{-3}$, whereas an actin filament of contour length $L = 10\mu$m and $L_p = 15\mu$m has a much larger value $t \simeq 6.7 \times 10^{-2}$.

This motivates our treatment of the quartic Hamiltonian $\mathcal{H}_4$. Because fluctuations of higher Fourier modes $n \geq 2$ will remain small at the buckling transition, we neglect terms of cubic and quartic order in the Fourier modes $n \geq 2$. The corresponding terms for the unstable $n = 1$ mode have to be retained, and we obtain

$$\mathcal{H}_4\{\tilde{\phi}_n\}/T = \frac{\bar{F}}{64t} \tilde{\phi}_1^4 \pm \frac{\bar{F}}{48t} \tilde{\phi}_1^3 \tilde{\phi}_3 + \sum_{n\geq 2} \frac{\bar{F}}{16t} \left(\tilde{\phi}_1^2 \tilde{\phi}_n^2 \pm \tilde{\phi}_1^2 \tilde{\phi}_n \tilde{\phi}_{n+2}\right). \tag{38}$$

The upper and lower signs in Eq. (38) are for free and clamped boundary conditions, respectively.

We first trace over all higher order modes $n \geq 2$ in order to obtain an effective Hamiltonian for the single mode $n = 1$, which is the relevant mode for the buckling instability.

$$e^{-\mathcal{H}_{\text{eff}}\{\tilde{\phi}_1\}/T} = \left(\prod_{n\geq 2} \int_{-\infty}^{\infty} d\tilde{\phi}_n\right) e^{-\mathcal{H}_2\{\tilde{\phi}_n\}/T - \mathcal{H}_4\{\tilde{\phi}_n\}/T}. \tag{39}$$

The Hamiltonian $\mathcal{H}_2 + \mathcal{H}_4$, as given by Eqs. (35) and (38), is quadratic in the higher order modes and the Gaussian integrals in Eq. (39) can be performed to obtain

$$\mathcal{H}_{\text{eff}}\{\tilde{\phi}_1\}/T \;=\; \bar{F}/t + \alpha \tilde{\phi}_1^2 + \beta \tilde{\phi}_1^4 \tag{40}$$

with

$$\alpha \;\equiv\; \frac{1}{4}\left(\frac{1-\bar{F}}{t} + \frac{1}{2}h(\bar{F})\right), \tag{41}$$

$$h(\bar{F}) \;\equiv\; \sum_{n\geq 2}\frac{\bar{F}}{n^2 - \bar{F}}, \tag{42}$$

$$\beta \;\equiv\; \frac{1}{64}\frac{\bar{F}}{t} \tag{43}$$

to leading order in the small parameter t. We point out that up to this order there is no difference between clamped and free boundary conditions. Therefore, our results regarding the critical force and the mean projected length will be identical for both types of boundary conditions also in the presence of thermal fluctuations. The function $h(\bar{F})$ can be approximated by $h(\bar{F}) \simeq \sqrt{\bar{F}}\,\text{arccoth}\left(2/\sqrt{\bar{F}}\right)$ by converting the sum into an integral. Close to the buckling threshold around $\bar{F} = 1$ we can also find an exact expression for the Taylor expansion $h(\bar{F}) \approx 3/4 + (1-\bar{F})(\pi^2/12 + 1/16)$. For $t \ll 1$ we can therefore use

$$\alpha \approx \frac{1}{4}\left(\frac{3}{8} + \frac{1-\bar{F}}{t}\right) \tag{44}$$

to a good approximation.

The resulting effective theory (40) for the single mode $\tilde{\phi}_1$ is a fourth order Ginzburg-Landau-type theory. The buckling instability occurs if the coefficient $\alpha(\bar{F})$ of the quadratic term changes sign. This determines the critical force F_c in the presence of thermal fluctuations,

$$F_c = F_{c,0}\left(1 + \frac{t}{2}h(\bar{F}_c)\right) \approx F_{c,0}\left(1 + \frac{3t}{8}\right) \tag{45}$$

where the last approximation is to leading order in the reduced temperature t such that $h(\bar{F}_c) \approx h(1) = 3/4$. Using the relation $\bar{F} = \bar{L}^2$, we obtain the corresponding result for the critical contour length L_c in the presence of thermal fluctuations,

$$\bar{L}_c = \sqrt{\bar{F}_c} \approx 1 + \frac{3t}{16} \tag{46}$$

to leading order in t. It is remarkable that, in two dimensions as considered here, the critical buckling force *increases* because of fluctuation effects as described by eq. (45). In the special case of two dimensions, the short wavelength fluctuations always *weaken* the effect of the applied force on a larger scale because the fourth order contribution to the force term in the Hamiltonian (31) has a sign opposite to the leading quadratic contribution. On the other hand, it is well known that short wavelength fluctuations do *not* affect the bending rigidity on a larger scale in two dimensions because there is no bending rigidity renormalization in two dimensions for the continuous worm-like chain model (20) according to the results of the previous section. The combination of these two effects leads to the increase of the critical buckling force in the presence of thermal fluctuations.

For arbitrary spatial dimensions d, a more general argument based on the RG flow of the nonlinear σ model is given in Ref. [47] and suggests that thermal fluctuations increase the buckling force for dimensions $d < 3$, whereas they decrease the buckling force for dimensions $d > 3$.

The partition sum Z is obtained by performing the one-dimensional integral over the remaining Fourier amplitude mode $\tilde{\phi}_1$,

$$Z = \int_{-\infty}^{\infty} d\tilde{\phi}_1 e^{-\mathcal{H}_{\text{eff}}\{\tilde{\phi}_1\}/T}. \tag{47}$$

From the force dependence of the partition sum the mean value of the projected filament length $L_{\parallel}$ from eq. (21) can be determined from the relation

$$\langle L_{\parallel} \rangle = -T \partial_F \ln Z(F). \tag{48}$$

Performing this calculation we finally obtain

$$1 - \frac{\langle L_{\parallel} \rangle}{L} = -\frac{t}{4\bar{F}} - \mathscr{F}_1\left(\frac{\alpha}{\beta^{1/2}}\right)\frac{t^{1/2}}{\bar{F}^{3/2}}(\bar{F}_c + \bar{F}) \tag{49}$$

with

$$\frac{\alpha}{\beta^{1/2}} \approx \frac{2}{t^{1/2}\bar{F}^{1/2}}(\bar{F}_c - \bar{F}), \tag{50}$$

where

$$\mathscr{F}_1(y) \equiv \begin{cases} \dfrac{\mathscr{F}'(y)}{\mathscr{F}(y)} = \dfrac{y}{4}\left(1 - \dfrac{K_{3/4}(y^2/8)}{K_{1/4}(y^2/8)}\right) & \text{for } y > 0, \\[3mm] \dfrac{y}{4}\left(1 + \dfrac{I_{3/4}(y^2/8) + I_{-3/4}(y^2/8)}{I_{1/4}(y^2/8) + I_{-1/4}(y^2/8)}\right) & \text{for } y < 0 \end{cases} \tag{51}$$

is a monotonously increasing, negative function. The solid curves in Fig. 6a show the result (49) for $1 - \langle L_{\parallel} \rangle/L$ as a function of the reduced force $\bar{F}$ for different values of the parameter t.

Further analysis of the result (49) in the vicinity of the buckling threshold reveals an interesting behavior: Thermal fluctuations as described by the small parameter t *decrease* the mean projected length $\langle L_{\parallel} \rangle$ below its zero temperature value $L_{\parallel} = L$ for $\bar{F} < \bar{F}_c$ whereas they *increase* the mean projected length above the zero temperature value $L_{\parallel} = L\bar{F}^{-2}$ in the buckled state for $\bar{F} > \bar{F}_c$. We thus conclude that thermal fluctuations lead to a stretching of buckled filaments, whereas they compress unbuckled ones. This implies that two curves for the mean projected length $\langle L_{\parallel} \rangle$ as a function of force, which are taken at different temperatures t, should *intersect* in the vicinity of the

169

buckling force. This characteristic behavior is clearly confirmed in Figs. 6a, where the full analytical result (49) is shown at different temperatures.

A characteristic feature of the buckling instability at zero temperature is the cusp in the relation between projected and contour length at the critical contour length $L_{c,0}$, see Fig. 5. For $L < L_{c,0}$ in the unbuckled state, the projected length is given by $L_\| = L$ and grows with the contour length. The projected length becomes maximal at the critical length $L = L_{c,0}$, where the filament buckles. If the filament grows further after buckling, $L > L_{c,0}$, the projected length decreases and $L_\| < L_{c,0}$. In the presence of thermal fluctuations, the cusp becomes modified, and we obtain the reduced mean projected length $1 - \langle L_\| \rangle / L_{c,0}$ as a function of the reduced contour length $\bar{L}$ by applying the relations $\bar{F} = \bar{L}^2$ and

$$1 - \frac{\langle L_\| \rangle}{L_{c,0}} = \left(1 - \frac{\langle L_\| \rangle}{L} \right) \bar{L} + (1 - \bar{L}) \tag{52}$$

to our previous result (49). This gives

$$1 - \frac{\langle L_\| \rangle}{L_{c,0}} = 1 - \bar{L} - \frac{t}{4\bar{L}} - \mathscr{F}_1 \left(\frac{\alpha}{\beta^{1/2}} \right) \frac{t^{1/2}}{\bar{L}^2} \left[\bar{L}_c^2 + \bar{L}^2 \right] \tag{53}$$

with

$$\frac{\alpha}{\beta^{1/2}} \approx \frac{2}{t^{1/2}\bar{L}} \left[\bar{L}_c^2 - \bar{L}^2 \right]. \tag{54}$$

The solid curves in Fig. 6b represent the expression $1 - \langle L_\| \rangle / L_{c,0}$ as a function of $\bar{L}$ according to eq. (53). Thermal fluctuations lead to a rounding of the zero temperature cusp to a pronounced minimum and to a shift of the location $\bar{L}_m$ of this minimum. Because thermal fluctuations lead to a stretching of buckled filaments, whereas they compress unbuckled filaments, curves for different temperatures t *intersect* in Fig. 6b, In principle, the contour length $\bar{L}_m$, where the mean projected length $\langle L_\| \rangle$ is maximal, could be determined experimentally by observing filaments growing against an obstacle as in Ref. [23].

In order to check our analytical predictions, we perform Monte Carlo simulations of buckling filaments in two dimensions in the presence of thermal fluctuations. We simulate a discretized version of the Hamiltonian (20) and employ clamped boundary conditions. In order to equilibrate the filament, we use two kinds of Monte Carlo (MC) moves: (i) a local move in real space, which changes the angles of two neighboring segments in opposite directions and thus induces a displacement of the point connecting both segments in the direction perpendicular to the local filament orientation; (ii) a collective move in Fourier space, which changes the amplitude $\tilde{\phi}_n$ of Fourier mode n by a random amount. For the simulation results shown in Figs. 6, we used a discretization into $N = 200$ segments and performed 8×10^6 MC sweeps alternating local moves and moves in Fourier space.

The simulation results for the reduced projected length $\langle L_\| \rangle / L$ as a function of the reduced force $\bar{F}$ in Fig. 6a are in good agreement with our analytical result (49). Deviations become appreciable for the largest values of the reduced temperature $t \simeq 10^{-1}$ for which we performed simulations. For these values it becomes necessary to

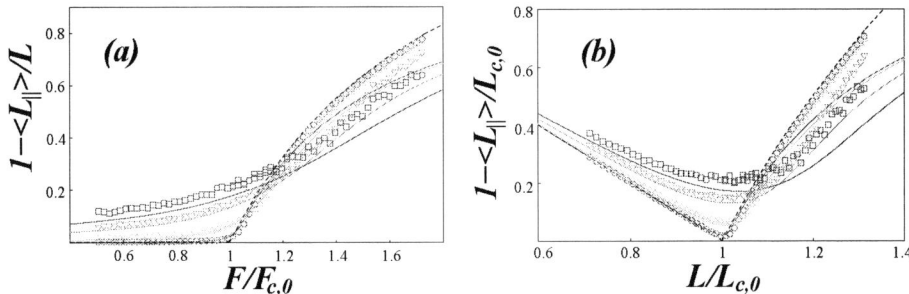

FIGURE 6. Monte Carlo simulation data for the reduced and shifted projected length $1 - \langle L_\parallel \rangle / L_{c,0}$ as a function of (a) the reduced force $F/F_{c,0}$ and (b) the reduced contour length $L/L_{c,0}$. The two ends of the filament are clamped using the model (20) for reduced persistence lengths $L_p/L = 100$ (red, ○), 10 (green, △), 2 (light blue, ▽), and 1 (blue, □) corresponding to reduced temperatures $t \simeq 10^{-3}, 10^{-2}, 5 \times 10^{-2}$, and 10^{-1} with t as defined in (37). The solid lines show the analytic results (49) and (53). The analytical zero temperature solutions from Figs. 4 and 5 are shown as a dashed line.

include higher order terms in the expansion in t underlying the analytical result (49). In particular, also the MC simulations confirm that curves for the mean projected length $\langle L_\parallel \rangle$ as a function of force, taken at different temperatures t, *intersect* in the vicinity of the buckling force. Also the MC results for the reduced projected length $\langle L_\parallel \rangle / L_{c,0}$ as a function of the reduced contour length $\bar{L}$ in Fig. 6b are in good agreement with the analytical result (53). The existence of a cusp rounded by thermal fluctuations close to the critical length L_c is clearly confirmed.

In conclusion, we presented in this section a systematic study of the buckling instability in the presence of thermal fluctuations in two spatial dimensions. By integrating over all short wavelength modes we derived an effective theory governing the buckling instability of the Fourier mode with the longest wavelength given by the filament length. We find that thermal fluctuations *increase* the critical force for buckling in two spatial dimensions. The increase in the critical buckling force is closely related to our main result that curves for the mean projected length $\langle L_\parallel \rangle$ measuring the end-to-end extension of the filament as a function of the applied compressive force, which are taken at different temperatures, *intersect* in the vicinity of the buckling force. This leads to the conclusion that, in two spatial dimensions, thermal fluctuations lead to a stretching of buckled filaments, whereas they compress unbuckled filaments. Before buckling a filament is governed by entropy and an increasing temperature leads to a shortening of the filament in order to maximize its configurational entropy, similar to the well-known elastic behavior of a flexible polymer, which gives rise to classical rubber elasticity [51]. A buckled filament, on the other hand, is governed by its bending energy and for increasing temperature also the bending energy decreases in favor of the entropy, which gives rise to the observed effect of stretching by thermal fluctuations.

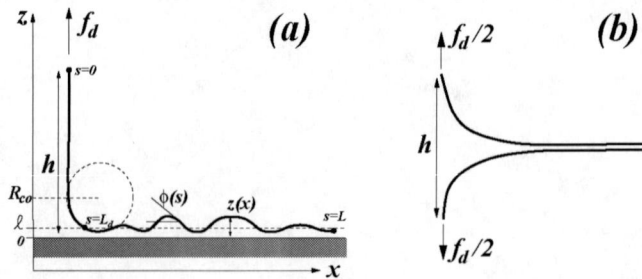

FIGURE 7. Force-induced desorption (a) and unzipping (b) of semiflexible polymers; f_d is the desorbing or unzipping force and h the height or separation of the end points.

ADSORPTION AND DESORPTION

In this section we want to consider the problem of peeling a single adhesive semiflexible polymer or filament from a surface, which is closely related to the problem of separating two adhesive filaments [52]. Over the past decade, experimental force spectroscopy techniques such as atomic force microscopy (AFM), optical or magnetic tweezers have been developed, which allow to perform manipulation experiments on individual polymers with spatial resolution in the nanometer range and force resolution in the piconewton range, for example, the mechanical unfolding of single proteins [53], the stretching of single DNA [54, 55]. Particularly suited for such single polymer manipulation experiments are semiflexible polymers with a high molecular weight per monomer and a large diameter. The quantitative analysis of force spectroscopy on semiflexible polymers requires theoretical models that take into account the combined effects of external force, temperature, and polymer bending energy. In this section we discuss results from theory and simulation for the force-induced desorption of semiflexible polymers. We will first discuss force-induced desorption at zero temperature, which is a variational or mechanical problem. Then we point out corrections from thermal fluctuations. Force-induced desorption experiments with single semiflexible polymers have been realized experimentally by attaching adsorbed polyelectrolytes to an AFM tip [56, 57, 58, 26, 59, 60]. The most recent experiments [26, 59, 60] give access to the single polymer force-distance curve. Force-induced desorption is assisted by thermal fluctuations and, thus, also gives additional insight into the fundamental problem of the adsorption transition of semiflexible polymers, which has been studied intensively both analytically [61, 62, 63, 64, 65, 66, 67, 68, 69, 70] and by simulations [71, 72].

Force-induced desorption at zero temperature

In the absence of thermal fluctuations ($T = 0$), a semiflexible polymer is only governed by its bending energy, and we recover a classical mechanics problem similar to fracture. At $T = 0$, polymer excursions parallel to the adhesive surface are suppressed, and the configuration of a polymer segment of contour length L can be parameterized by tangent angles $\phi(s)$ with respect to the adhesive surface, where s is the arc length ($0 < s < L$), see Fig. 7. The bending energy is given by the Hamiltonian $E_b = (\kappa/2) \int_0^L ds (\partial_s \phi)^2$, which is the same representation as in eq. (20). The adsorption energy is

$$E_a = \int_0^{L_c} ds V(z(s)), \tag{55}$$

where $z(s)$ is the distance of polymer segments from the adsorbing surface at $z = 0$ and $V(z)$ is a generic square well adhesion potential of small range ℓ with $V(z) = W < 0$ for $z < \ell$, $V(z) = 0$ for $z > \ell$, and $V(z) = \infty$ for $z < 0$ due to the hard wall. For van der Waals forces or screened electrostatic interactions the potential range ℓ is comparable to the polymer thickness or the Debye-Hückel screening length, respectively. For the discussion at $T = 0$, we consider a contact potential, i.e., the limit of small ℓ. In the absence of a desorbing force the polymer lies flat on the adhesive surface [$\phi(s) = 0$ for all s] gaining an energy $-|W|L$. The semiflexible polymer is peeled from the adhesive surface by a localized desorbing force $\mathbf{f}_d$ that is applied in z-direction at the end point $s = 0$. Under the influence of the force, a polymer segment $0 < s < L_d$ desorbs, which costs a potential energy $|W|L_d$. In order to map out the energy landscape of the desorption process, we consider a *constrained* equilibrium and minimize the sum of bending and potential energy of the polymer, $E = |W|(L_d - L_c) + E_b$, under the constraint of a fixed height $h = \int_0^{L_d} ds \sin \phi(s)$ of the polymer end at $s = 0$. Minimizing with respect to L_d gives the transversality condition $\partial_s \phi(L_d) = -(2|W|/\kappa)^{1/2} \equiv 1/R_{co}$ which determines the contact curvature radius R_{co} [73]; the boundary conditions are $\phi(L_d) = 0$ and $\partial_s \phi(0) = 0$ corresponding to a free tangent. Solving the shape equation in the presence of the height constraint, we find the scaling form $\Delta E(h) \equiv E(h) - E(0) = (\kappa |W|)^{1/2} \mathscr{F}_E (h/R_{co})$ for the total energy with the two limits

$$\Delta E(h) \approx \begin{cases} 2^{7/4} 3^{-1/2} h^{1/2} \kappa^{1/4} |W|^{3/4} & \text{for } h \ll R_{co} \\ |W|[h + 4(\sqrt{2} - 1)R_{co}] & \text{for } h \gg R_{co} \end{cases}. \tag{56}$$

For the desorbed polymer length we obtain the scaling result $L_d(h) = R_{co} \mathscr{F}_L (h/R_{co})$ with the limits

$$L_d \approx \begin{cases} \sqrt{3} h^{1/2} R_{co}^{1/2} & \text{for } h \ll R_{co} \\ h + 2(\sqrt{2} - 1)R_{co} & \text{for } h \gg R_{co} \end{cases}. \tag{57}$$

The results (56) and (57) can be corroborated by a scaling argument starting from the estimate $\Delta E(h, L_d) \sim \kappa h^2 / L_d^3 + |W|L_d$ of the energy cost to desorb a segment of length L_d. For $h \ll R_{co}$, energy minimization with respect to the desorbed length L_d gives $L_d \sim h^{1/2} \kappa^{1/4} |W|^{-1/4}$ and an energy cost $\propto h^{1/2}$ as in (56). For $h \gg R_{co}$ essentially the whole desorbed length L_d is lifted straight and perpendicular to the substrate except for a curved segment of length $\sim R_{co}$ around the contact point, which leads to $L_d \approx h + \mathcal{O}(R_{co})$ and an energy cost $\propto h$ as in (56).

Including the energy gain for a constant desorbing force f_d, we obtain the energy landscape $\Delta G(h) \equiv \Delta E(h) - f_d h$ at $T = 0$ as a function the height h, see Fig. 8b. The equilibrium height minimizes $\Delta G(h)$, and we find a first order desorption transition from $h = 0$ to infinite h above the critical force $f_{d,c} = |W|$. For *all* force values $f_d > |W|$,

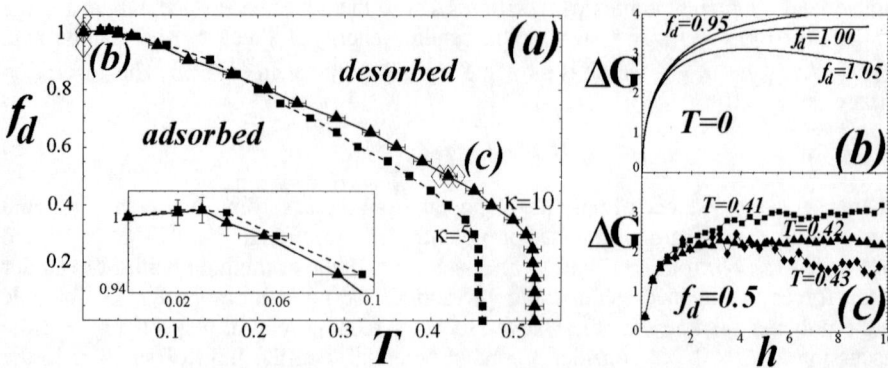

FIGURE 8. (a) Phase diagrams in the plane of the desorbing force f_d and temperature T from Monte Carlo simulations of a discretized semiflexible chain (with segments of length Δs) and bending rigidities $\kappa = 10$ (triangles) and $\bar{\kappa} = 5$ (squares), adsorption potential range $\ell = 0.1$, and contour length $L_c = 100$ (lengths ℓ and L_c in units of Δs, temperature T in units of the adhesion energy $|W|\Delta s$, desorbing force f_d in units of $|W|$, and bending rigidities κ in units of $|W|\Delta s^2$). Lines are guides to the eye. The inset shows the reentrance region at low temperatures. (b),(c) Free energy landscapes $\Delta G(h)$ as a function of the height h for $\kappa = 10$ and forces and temperatures as indicated by diamonds in the phase diagram b. (b) at $T = 0$ according to the analytical result, see eq. (56). (c) for $f_d = 0.5$ from Monte Carlo simulations in agreement with eqs. (60) and (61). In the desorbed phase $\Delta G(h)$ exhibits an energy barrier.

there remains a local minimum at $h = 0$ corresponding to the firmly adsorbed state, which is separated by an *energy barrier* ΔG_b from the desorbed equilibrium state. The energy barrier is given by $\Delta G_b \approx 2^{3/2}\kappa^{1/2}|W|^{3/2}/3f_d$ for all $f_d > |W|$, i.e., it decays as $1/f_d$ and scales with $\kappa^{1/2}$ and, hence, is a consequence of the bending rigidity of the polymer. The scaling behavior of ΔG_b also follows from equating the energy cost $\Delta E(h) \propto h^{1/2}\kappa^{1/4}|W|^{3/4}$ and the gain $f_d h$. Due to the energy barrier, force-induced desorption requires thermal activation or an h-dependent force $f_d(h) = \partial_h \Delta E(h)$, which diverges as $h^{-1/2}$ for small h.

Force-induced desorption in the presence of thermal fluctuations

In the absence of a desorbing force ($f_d = 0$), a semiflexible polymer can undergo thermal desorption [69, 70]. In order to discuss the influence of thermal fluctuations onto the force-induced desorption we first give some basic results on the thermal desorption transition in the absence of an additional force. Using a model connecting length scales below and above the persistence length L_p. At the desorption transition, the correlation length $\xi_\parallel$ diverges. For $\xi_\parallel < L_p$, we apply results for a weakly bent *semiflexible* polymer [69], whereas we use standard results for the adsorption of *flexible* Gaussian polymers in combination with an effective adsorption potential for $\xi_\parallel > L_p$. In the latter flexible regime, desorbed segments of typical length $\xi_\parallel$ decay into uncorrelated persistent Kuhn segments of length $\sim L_p$. The strength of the effective renormalized adsorption potential for these Kuhn segments is given by the free energy of adsorption $f_{W,\mathrm{SF}}$

of a semiflexible segment. This construction connects the semiflexible and the flexible regime. Using this approach we find the for critical potential strength for desorption, $W_c \approx -\frac{\sqrt{3}\pi}{2^{4/3}}(T/\ell^{2/3}L_p^{1/3})[1 + \frac{\pi^{3/2}}{2^{8/3}3^{3/2}}(\ell/L_p)^{2/3}]$, and a free energy of adsorption

$$|f_W| \approx \begin{cases} 3(W - W_c)^2 L_p/2T & \text{for } |W - W_c| \ll T/L_p \\ |f_{W,\text{SF}}| \sim |W - W_c| & \text{for } |W - W_c| \gtrsim T/L_p \end{cases}, \tag{58}$$

which is related to the correlation length by $|f_W| = T/\xi_\parallel$. The first line in (58) is the free energy of adsorption in the flexible regime; the second line is the free energy of adsorption in the semiflexible regime, which holds for $|f_W| \gtrsim T/L_p$ or outside a window of adhesion strengths of width T/L_p around the critical value W_c.

In the presence of thermal fluctuations the free energy of adsorption f_W replaces the bare potential strength W and the free energy per length $g(f_d)$ of a thermally fluctuating, stretched semiflexible polymer replaces the force $-f_d$. For small stretching forces $f_d \ll T/L_p$, the polymer is effectively flexible and entropic elasticity gives $g(f_d) \approx -f_d^2 L_p/6T$, whereas for strong stretching $f_d \gg T/L_p$, we have $g(f_d) \approx -f_d + (2Tf_d/L_p)^{1/2}$, where the square root contribution is typical for semiflexible behavior [74, 41]. The polymer desorbs if the stretching free energy $g(f_d)$ compensates for the free energy cost of desorption, i.e., for $|g(f_d)| > |f_W|$. This gives a first order force-induced desorption transition (similar to DNA unzipping [75, 76], where the single strands are flexible polymers), at a critical force

$$f_{d,c} \approx \begin{cases} (6T|f_W|/L_p)^{1/2} & \text{for } |f_W| \ll 2T/L_p \\ |f_W| + (2T|f_W|/L_p)^{1/2} & \text{for } |f_W| \gg 2T/L_p \end{cases} \tag{59}$$

and, thus, the phase boundary of the adsorbed phase in the f_d-$|W|$ or f_d-T plane. The line of first order force-induced desorption transitions ends in the critical point of thermal desorption at zero force. The results for the phase diagram were confirmed by Monte Carlo (MC) simulations of a discretized semiflexible polymer, see Fig. 7a. The discretized semiflexible chain consists of $N = L_c/\Delta s$ beads with heights z_i (i.e., $h = z_N$) and $N - 1$ connecting segments of length Δs with unit tangent vectors $\mathbf{t}_i$ using the Hamiltonian $\mathscr{H} = E_b + \sum_{i=1}^{N} \Delta s V(z_i) - f_d h$, where $E_b = (\kappa/2)\sum_{i=1}^{N-1}(\mathbf{t}_{i+1} - \mathbf{t}_i)^2/\Delta s$ is the bending energy. The analytical result (59) correctly describes three main features of the simulation results: (i) A characteristic square-root dependence $f_{d,c} \sim |f_{W,\text{SF}}|^{1/2} \sim |T - T_c|^{1/2}$ close to the thermal desorption transition typical for flexible behavior. (ii) A broad *linear* regime $f_{d,c} \approx |f_W| \sim |T - T_c|$ at lower temperatures, which is absent for flexible polymers and due to the bending rigidity effects. (iii) At low temperatures $T < |W|\ell^{2/3}L_p^{1/3}$, we find $f_{d,c} \sim |W| - T^{4/3}/\ell^{2/3}\kappa^{1/3} + T|W|^{1/2}/\kappa^{1/2}$, which gives a small *reentrant* region of the desorbed phase because thermal fluctuations weaken the adhesion strength less than the pulling force. Such "cold desorption/unzipping" has been reported previously for flexible polymers like DNA [76].

The energy landscape of the desorption process can be mapped by calculating the *constrained* free energy $\Delta F(h) = -T \ln[Z(h)/Z(0)]$ where $Z(h)$ is the restricted partition sum over all polymer configurations with a given height h of the end point. The transfer matrix treatment of the weakly bent semiflexible polymer [69] gives the constrained free energy

$$\Delta F(h) = -\frac{T}{2}\ln\left(\frac{h}{L_p}\right) + \frac{2^{7/4}}{3^{1/2}}h^{1/2}\kappa^{1/4}|f_W|^{3/4} \tag{60}$$

for the semiflexible regime $|f_W| \gtrsim T/L_p$. This is the exact generalization of the $T = 0$ result (56) for small h to finite temperatures, where the free energy of adsorption of a semiflexible polymer f_W replaces the bare contact potential W and a logarithmic entropic repulsion from the hard wall occurs. For large h, the free energy cost (60) is *always* exceeded by the linear energy gain $-f_d h$, which suggests the absence of a phase transition and a desorption instability even for small forces f_d [68].

However, the weak bending approximation breaks down upon increasing h if typical tangent angles $h/L_d > 1$ become large for $h > R_{co} \equiv (\kappa/2|f_W|)^{1/2}$. Then the whole desorbed tail of length L_d becomes lifted perpendicular to the substrate except for a curved segment of length $\sim R_{co}$, i.e., $L_d \approx h + \mathcal{O}(R_{co})$. In this limit the full free energy $\Delta G(h) = \Delta F(h) - h f_d$ in the presence of the desorbing force can be written as

$$\Delta G(h) \approx h[|f_W| + g(f_d)] + c R_{co}|f_W|, \tag{61}$$

where c is a numerical constant [$c = 4(\sqrt{2} - 1)$ at $T = 0$, see (56)]. Equation (61) is in accordance with our above free energy criterion $|g(f_d)| = |f_W|$ for the desorption transition.

Therefore, also for $T > 0$, the free energy landscape $\Delta G(h) = \Delta F(h) - h f_d$, as given by (60) for $h \lesssim R_{co}$ and (61) for $h \gg R_{co}$, exhibits a *barrier* for $f_d > f_{d,c}$, which arises from the bending rigidity although the microscopic adhesion potential is purely attractive. In the MC simulation, $\Delta G(h)$ can be calculated from the logarithm of the end point distribution function, which clearly confirms the existence of a barrier, see Fig. 7c. In the semiflexible regime for $|f_W| \gtrsim T/L_p$, we find an energy barrier $\Delta G_b \sim \kappa^{1/2}|f_W|^{3/2}/f_d$ for all forces $f_d > f_{d,c} \approx |f_W|$. The barrier scales with $\kappa^{1/2}$ and decreases as $1/f_d$ starting from $\Delta G_b \sim (\kappa|f_W|)^{1/2}$. The barrier is attained for a height $h \sim \Delta G_b/f_d$, which approaches $h \sim R_{co}$ for $f_d = |f_W|$. In the semiflexible regime $|f_W| \gtrsim T/L_p$, we have $\Delta G_b \gtrsim T$ and $R_{co} \lesssim L_p$. Upon entering the flexible regime the barrier becomes smaller than the thermal energy T and can thus be overcome quasi-spontaneously by thermal activation; the contact radius becomes larger than the Kuhn segment length L_p.

The existence of a barrier in the semiflexible regime has important consequences for single polymer desorption experiments. In equilibrium, the necessary desorption force $f_d(h) = \partial_h \Delta F(h) \propto h^{-1/2}$ *diverges* for small h, before a plateau of constant separation force $f_d = f_{d,c} \approx |f_W|$ is reached at large h. If the desorption experiment is performed out of equilibrium at constant desorption force larger than the threshold force, $f_d > f_{d,c}$, the energy barrier ΔG_b has to be overcome by thermal activation with an Arrhenius-type desorption rate which has a strong influence on the kinetics of *initial* desorption [52].

In summary, we presented the phase diagram for force-induced desorption of semi-flexible polymers and derived the existence of a characteristic energy barrier which is a consequence of the bending rigidity and absent for flexible polymers. Both results are confirmed by Monte Carlo simulations. The energy barrier gives rise to activated des-orption kinetics and leads to an enhanced dynamic stability of the bound state of stiff adhesive polymers or fibers under force. This effect plays a role for biological polymers under force, e.g., in DNA, protein, or filament unzipping and desorption as well as for numerous materials science applications ranging from the delamination of thin sheets to the peeling of adhesive hairs. The results can also shed new light on the zipping or adsorption dynamics of semiflexible filaments, which plays an important role in cytoskeletal networks [77].

SEMIFLEXIBLE POLYMERS ON STRUCTURED SUBSTRATES

Many applications in (bio-)nanotechnology, such as the construction of electric devices or sensors containing nanotubes, require immobilization, controlled shape manipulation, and positioning of semiflexible polymers. Adsorption on substrates is the simplest technique to immobilize single polymers and a first step towards further manipulation and visualization of structure details using, e.g. modern scanning probe techniques [24, 25]. Such scanning force techniques can be used to apply localized point forces to a polymer adsorbed on a substrate and force a lateral movement [31, 5]. Polymers that are strongly adsorbed onto crystalline substrates such as graphite or mica experience a spatially modulated adsorption potential reflecting the underlying crystal lattice structure and giving rise to preferred orientations of the adsorbed polymer. For such systems, the dynamics of the adsorbed polymer is governed by thermal activation over the potential barriers of the surface potential.

The thermally activated motion over a translationally invariant potential barrier has been studied in Refs. [78, 79, 80] for homogeneous and point driving forces. At low forces the activated dynamics is governed by the nucleation of localized kink-like excitations as shown in Fig. 9. The activated dynamics of semiflexible polymers is different from that of flexible polymers as kink properties are not governed by entropic elasticity of the polymer chain but rather by the bending energy of the semiflexible polymer. In principle, this allows to extract material parameters such as the persistence length or the height of substrate energy barriers from kink properties. The barrier crossing of the polymer requires kink nucleation, diffusion and recombination and time scales for barrier crossing and the mean velocity of the semiflexible polymer can be linked to the dynamic properties of kink excitations.

In this review we want to focus on another important aspect of structured adsorbing surfaces, which is the control of the shape of an adsorbed semiflexible polymer. The controlled adsorption of single carbon nanotubes at predefined positions, e.g. at electrodes, has been achieved by using chemically structured substrates [81, 82, 83]. The substrate patterns used in these experiments are composed of domains of arbitrary shape, which are characterized by a greater binding affinity compared to the surrounding substrate. Theoretically, this effect can be described by a laterally modulated adhesion potential, where, the polymer gains a constant adsorption energy $W < 0$ per unit length.

We will explore the possibility to gain such shape control for semiflexible polymer rings using simple striped surface structures. Examples of circular semiflexible polymers are provided by carbon nanotubes [84, 85], DNA minicircles [86], filamentous actin [87], and amyloid fibrils [88]. The substrate patterns studied in the following correspond to a single chemically striped domain, see Fig. 10a, and to a single topographical surface channel, see Fig. 10b. It turns out, that a shallow topographical step of a certain width

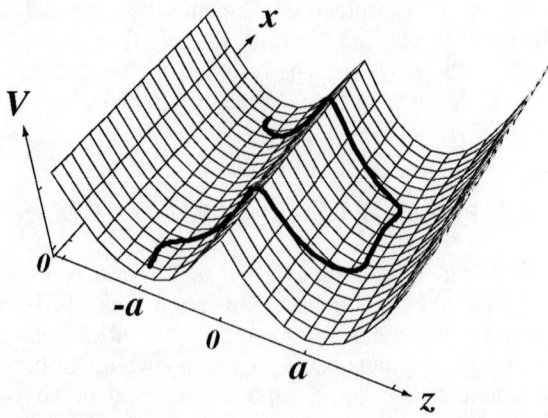

FIGURE 9. Typical conformation of a semiflexible polymer (thick line) with a kink-antikink pair in the xz-plane and in a translationally invariant double-well potential V, which depends on the coordinate z, is independent of the coordinate x, and has minima at $z = \pm a$.

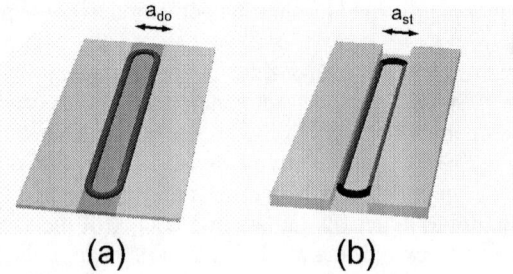

(a)   (b)

FIGURE 10. Adsorbed circular polymer on a striped surface containing (a) a chemically structured surface domain of width a_{do} and (b) two topographical surface steps forming a channel of width a_{st}.

has an effect very similar to a thin adhesive stripe, because the polymer is attracted to the surface steps. Such topographical surface steps have been employed in recent manipulation experiments on semiflexible polymer rings [89].

In the following, we will focus on the limit of persistence lengths much larger than the stripe width, where thermal fluctuations can be neglected. We find that the competition between the bending rigidity of the circular semiflexible polymer and its attraction to the striped domain allows a controlled switching between four distinct stable morphologies, see Fig. 11: Apart from a weakly bound almost circular shape (I) and a strongly bound confined shape (II_0), bulged intermediate shapes (II_1, II_2) become stable for large contour lengths. Our results are summarized in the full morphological diagrams for semiflexible ring shapes depending on the size of the ring compared to the size of the structure as well as the material parameters, namely, the bending rigidity κ and the adhesion energy gain W. This analysis can be used to (i) control the ring shape and (ii)

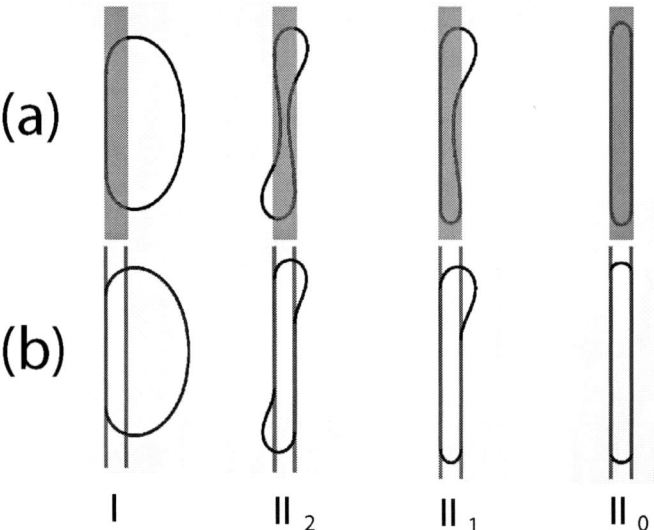

FIGURE 11. Top views of the four types of stable morphologies of a ring polymer adhering to a surface with a chemical domain (a) or topographical stripe (b) as obtained by energy minimization for contour lengths $L/a_{st} = 20$.

analyze material properties of the substrate or the semiflexible polymer ring experimentally. Flexible polymer rings, on the other hand, exhibit random coil configurations and such morphological transitions are absent.

We will first discuss topographical surface channels. We consider a planar substrate in the xy-plane that contains two parallel topographical surface steps at $x = \pm a_{st}/2$ forming an infinitely long surface channel of width a_{st}. A semiflexible polymer ring, characterized by the bending rigidity κ and the fixed contour length L, gains an overall adsorption energy $W_{st} < 0$ per polymer length by adsorbing to the substrate surface. At the edges of the surface channel the adsorption energy is doubled, since the semiflexible polymer can bind to two adjacent surfaces at the same time. We will refer to this extra energy contribution as the adsorption energy and omit the constant energy offset. The resulting effective lateral adsorption potential can be described by

$$V_{st}(x) = \begin{cases} W_{st} & \text{for } |x \pm a_{st}/2| < \ell/2, \\ 0 & \text{otherwise,} \end{cases} \tag{62}$$

where ℓ denotes the adhesive range of the steps, which is of the order of the polymer diameter and assumed to be small compared to the channel width, $\ell \ll a_{st}$. Furthermore, we neglect small energy corrections arising where the polymer crosses the surface steps, and kink configurations, similar to the one displayed in Fig. 9, will occur. The resulting confined shapes are therefore not a consequence of pure geometric constraints, but of the adsorption energy gain induced by the surface steps.

The bending energy of the polymer in the two-dimensional substrate plane is given by $E_b = (\kappa/2) \int_0^L ds (\partial_s \phi(s))^2$, where κ is the bending rigidity, and the contour is parameterized by the arc length s ($0 < s < L$) using the tangent angles $\phi(s)$, see (20). The adhered length L_{st} is given by the polymer length on the edges at $x = \pm a_{st}/2$, and the adhesion energy is $E_{ad} = -|W_{st}|L_{st}$. The polymer configuration is determined by minimizing the total energy $E_{tot} = E_{ad} + E_b$ under the constraints imposed by ring closure, i.e., $\int_0^L ds (\cos\phi(s), \sin\phi(s)) = (0,0)$. This yields a shape equation for $\phi(s)$ and an implicit equation for the Lagrange multiplier. Solving these equations, the polymer shape and the resulting energies can be calculated analytically. Instead of an analytical solution to the variational problem we want to focus here on results from the numerical minimization of the total energy $E_{tot} = E_{ad} + E_b$ using the dynamical discretization algorithm of the SURFACE EVOLVER 2.14 [90]. The contour of the polymer is discretized and represented by a set of vertices and directed edges, which connect neighboring vertices. The contour length of a shape segment is identified with the sum over the length of the corresponding edges and a discretized version of the bending energy is assigned to each vertex. The basic operations are the refinement of the discretization and energy minimization by a conjugate gradient algorithm. In order to determine all metastable states of the total energy $E_{tot} = E_b + E_{ad}$, we minimize the *constrained* energy, where we fix the adhered length L_{st} and, thus, the adhesion energy. In the presence of an additional constraint some metastable configurations can be stabilized.

Analyzing the total energy, one finds that two control parameters, which are combinations of the four parameters L, κ, a_{st} and W_{st}, are sufficient to characterize the system. In the following we measure lengths in units of the channel width and energies in units of the typical bending energy,

$$\bar{L} \equiv L/a_{st} \text{ and } \bar{E} \equiv E a_{st}/\kappa, \tag{63}$$

which leads to a *reduced adhesion strength*

$$|w_{st}| \equiv |W_{st}|a_{st}^2/\kappa \text{ or } |\tilde{w}_{st}| \equiv |W_{st}|L^2/\kappa = |w_{st}|\bar{L}^2. \tag{64}$$

The reduced parameters w_{st} and $\bar{L}$ are advantageous in discussing the control of shapes as a function of the contour length L, which changes the parameter $\bar{L}$ only. If the channel width a_{st} is varied, the reduced parameters $\tilde{w}_{st}$ and $\bar{L}$ are better suited because the channel width a_{st} then only changes the parameter $\bar{L}$ at constant $\tilde{w}_{st}$.

The numerical minimization gives the following results for the topographical channel. For small $|L_{st}|$, the ring will attach only to one step edge and adopt the rather round toroidal configuration I, see Fig. 10. For $L_{st} \gtrsim L/2$, conformations, where the ring binds to both step edges, will become relevant. These shapes may be classified by the number of bulges or segments outside the channel and are referred to as II_0, II_1, and II_2, accordingly, see Fig. 10. The energy landscape $\bar{E}_{tot}(\bar{L}_{st})$ exhibits up to four local minima, representing the four different (meta-)stable ring morphologies I, II_0, II_1, and II_2. The round configuration I is the state of minimal energy for small $|w_{st}|$, whereas the adhesion energy gain dominates for large $|w_{st}|$, and the elongated shape II_0 becomes the only stable conformation. The morphological transition between these two shapes is *discontinuous* with a jump in the adhered length L_{st}. In the vicinity of this transition also

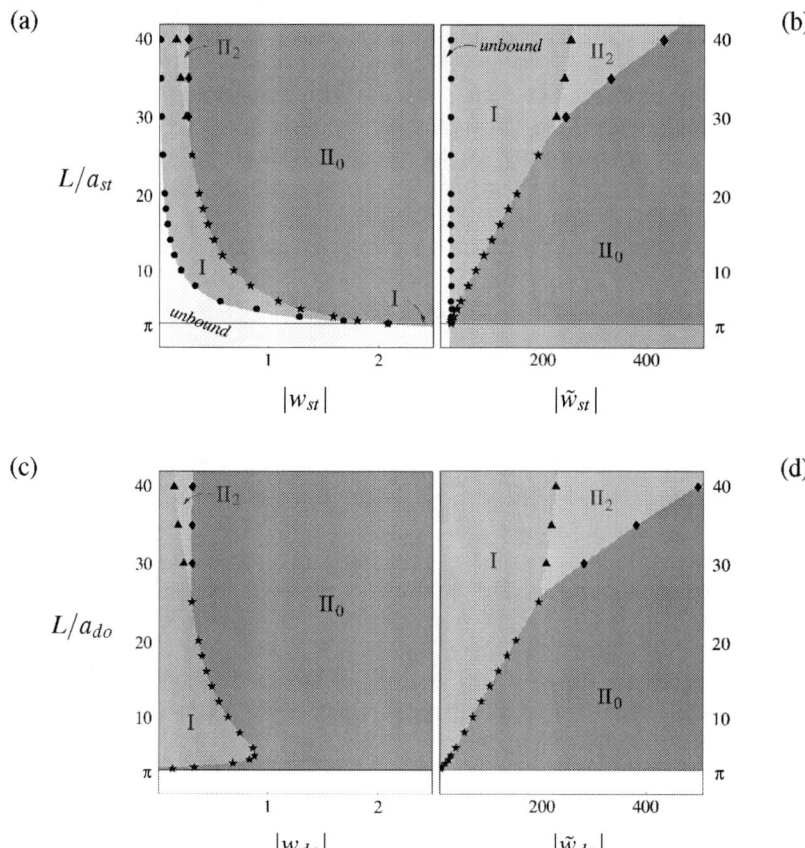

FIGURE 12. Morphology diagram for a ring polymer of length L adhering to the topographical surface channel of width a_{st} as a function of (a) L/a_{st} and $|w_{st}| \equiv |W_{st}| a_{st}^2/\kappa$ and (b) L/a_{st} and $|\tilde{w}_{st}| \equiv |W_{st}| L^2/\kappa$. If the topographical structure is replaced by a chemical domain of width a_{do} and adhesion strength $|W_{do}|$, the system is characterized by (c) L/a_{do} and $|w_{do}| \equiv |W_{do}| a_{do}^2/\kappa$ or (d) L/a_{do} and $|\tilde{w}_{do}| \equiv |W_{do}| L^2/\kappa$. The parameter choice in (b) and (d) is advantageous if the structure width a_{st} and a_{do} is varied, while the other system parameters are kept constant. Morphological transitions are obtained from numerical energy minimization and are represented by stars, triangles, diamonds and dots.

the shapes II_1 and II_2 can become stable or metastable, which develop from the elongated shape II_0 by the formation of one and two *bulges*, respectively. Shape I undergoes an additional unbinding transition from a single surface step, which is also known for vesicles adhering to a planar surface, where the interplay between adhesion and bending energy leads to an unbinding transition, which is not driven by thermal fluctuations [73].

These results can be summarized in morphology diagrams Figs. 12a and 12b. They show how the stability of the four shapes I, II_0, II_1 and II_2 and the unbound circle is controlled by the parameters $|w_{st}|$ and $\bar{L}$ (Fig. 12a) or $|\tilde{w}_{st}|$ and $\bar{L}$ (Fig. 12b). Morphology boundaries from the numerical minimization procedure are denoted by stars, triangles, dots and diamonds. The main feature of the morphology diagram is the discontinuous transition between morphologies I and II_0 (stars). This transition line terminates at $\bar{L} = \pi$ and $|w_{st}| = 2$ where it intersects the unbinding transition line (dots) because a short ring can bind at most to one surface step. At the vertical transition line between configurations II_0 and II_2 (diamonds), which occurs for large $\bar{L}$, it becomes energetically favorable to form bulges on top of the confined shape II_0.

Now we turn to the chemically striped domain of width a_{do}, which is modeled by an additional adhesion energy gain $W_{do} < 0$ per polymer length for $|x| \leq a_{do}/2$, which leads to a generic square well adsorption potential with

$$V_{do}(x) = \begin{cases} W_{do} & \text{for } |x| \leq a_{do}/2, \\ 0 & \text{for } |x| > a_{do}/2. \end{cases} \tag{65}$$

The adhered length L_{do} is given by the polymer length within the stripe $|x| \leq a_{do}/2$, and the adhesion energy is $E_{ad} = -|W_{do}|L_{do}$.

Performing a numerical energy minimization we find the same four types of morphologies I, II_0, II_1, and II_2 as for the topographical surface channel, see Fig. 10(b). Remarkably, we find that ring shapes minimizing the bending energy are almost identical as compared to a topographical surface channel of the same width $a_{st} = a_{do}$, see Fig. 10. In contrast to the channel, the stripe domain is also adhesive between its boundaries for $|x| < a_{do}/2$ such that the same ring shape has a larger adhered length. This turns out to be the major difference between both types of stripes and leads only to approximately constant shifts in the adhered lengths.

Therefore the corresponding morphology diagrams Figs. 12c and 12d of the chemically structured surface domain look very similar to the morphology diagrams for the topographical stripes. In particular, the discontinuous transition between shapes I and II_0 (stars) and the appearance of stable bulged states along the vertical line between configurations II_0 and II_2 (diamonds) are very similar. However, the unbinding transition of shape I is absent for the chemical stripe domain: It is always energetically favorable for the ring to adhere to the striped domain. Furthermore, the two phase diagrams differ in the behavior of small rings. Small rings can fully bind to the chemical stripe without deformation and shapes I and II_0 become equivalent, which leads to the re-entrance of shape II_0 close to $L/a_{do} = \pi$.

In summary, the morphology diagrams in Figs. 12 give a complete classification of the morphologies of an adsorbed semiflexible polymer on a substrate containing an adhesive stripe domain, which can be either topographically or chemically structured. Applied to experiments such morphology diagrams allow a control of the shape of the adsorbed ring. Both types of structures lead to very similar behavior with a *discontinuous morphological transition* between the two dominant shapes I and II_0 and with intermediate bulged shapes II_1 and II_2 for large contour lengths.

CONCLUSION AND OUTLOOK

In this review we presented several examples of non-trivial equilibrium or zero temperature shapes of semiflexible polymers and on thermal fluctuations of these shapes. The examples – buckling, force-induced desorption, and shapes of adsorbed polymers on structured substrates – are motivated by single polymer manipulation experiments. At zero temperature the shapes can be calculated by minimizing the bending energy using variational methods, which leads to shapes composed of segments of different Euler Elastica. Thermal fluctuations are also governed by bending energy and display a distinct behavior different from flexible polymers.

In biological systems, filaments are not only subject to thermal forces. They are also constantly rearranged and reorganized by small forces generated by motor proteins or polymerization forces [12]. Both of these forces are generated by the hydrolysis of adenine triphosphate (ATP). This "active" filament dynamics gives rise to more complex non-equilibrium phenomena. One important example relevant to the buckling instability discussed in section is the buckling of a single polymerizing filament [23]. In systems containing many filaments and molecular motors active filament dynamics can lead to much more complex phenomena such as the formation of various filament patterns [91, 92, 93, 94, 95]. Issues related to filament pattern formation in *in vitro* systems, filament organization in the cytoskeleton, or force generation in the cell are subject of current research. Also in these active systems the shapes of filaments and their corresponding bending energies will play an important role.

ACKNOWLEDGMENTS

We acknowledge financial support by the Deutsche Forschungsgemeinschaft via Sonderforschungsbereich 448.

REFERENCES

1. P. G. De Gennes, *Scaling Concepts in Polymer Physics*, Cornell University Press, Ithaca, 1979.
2. M. Doi and S. F. Edwards, *The Theory of Polymer Dynamics*, Clarendon, Oxford, 1986.
3. A. Y. Grosberg and A. R. Khokhlov, *Statistical Physics of Macromolecules*, American Institute of Physics Press, New York, 1994.
4. E. M. Lifshitz and L. D. Landau, *Theory of Elasticity*, Pergamon Press, New York, 1986.
5. C. Ecker, N. Severin, L. Shu, A. D. Schlüter, and J. P. Rabe, *Macromolecules* **37**, 2484 (2004).
6. P. Samori, C. Ecker, I. Gössl, P. A. J. de Witte, J. J. L. M. Cornelissen, G. A. Metselaar, M. B. J. Otten, A. E. Rowan, R. J. M. Nolte, and J. P. Rabe, *Macromolecules* **35**, 5290.
7. D. G. Kurth, N. Severin, and J. P. Rabe, *Angew. Chem.* **114**, 3833 (2002).
8. C. Dekker, *Physics Today* **52**, 22, (1999).
9. W. H. Taylor and P. J. Hagerman, *J. Mol. Biol.* **212**, 363 (1990).
10. B. Alberts, D. Bray, J. Lewis, M. Raff, K. Roberts, J. D. Watson, *Molecular Biology of the Cell*, Garland Publishing, New York, 1994.
11. H. Lodish, A. Berk, L. S. Zipursky, P. Matsudaira, D. Baltimore, and J. Darnell, *Molecular Cell Biology*, W.H. Freeman, New York, 2000.
12. J. Howard, *Mechanics of Motor Proteins and the Cytoskeleton*, Sinauer Associates, Inc., Sunderland, 2001.
13. A. Ott, M. Magnasco, A. Simon, and A. Libchaber, *Phys. Rev. E* **48**, R1642 (1993).

14. F. Gittes, B. Mickey, J. Nettleton, and J. Howard, *J. Cell Biol.* **120**, 923 (1993).
15. J. Käs, H. Strey, and E. Sackmann, *Nature* **368**, 226 (1994).
16. J.-L. Barrat and J.-F. Joanny, *Advances in Chemical Physics, Vol. XCIV*, edited by I. Prigogine and S. A. Rice, John Wiley & Sons, New York, 1996.
17. O. Kratky and G. Porod, *Recl. Trav. Chim. Pays-Bas* **68**, 1106 (1949).
18. E. M. Lifshitz and L. D. Landau, *Statistical Physics, Part 1*, Pergamon Press, New York, 1969.
19. H. Kleinert, *Path Integrals in Quantum Mechanics, Statistics, and Polymer Physics, and Financial Markets*, World Scientific, Singapore, 2004.
20. G. Binnig, C. F. Quate, and C. Gerber, *Phys. Rev. Lett.* **56**, 930 (1986).
21. A. Ashkin, *Science* **210**, 1081 (1980).
22. S. B. Smith, L. Finzi, and C. Bustamante, *Science* **258**, 1122 (1992).
23. M. Dogterom and B. Yurke, *Science* **278**, 856 (1997).
24. S. S. Sheiko and M. Möller, *Chem. Rev.* **101**, 4099 (2001).
25. P. Samori, *J. Mater. Chem.* **14**, 1353 (2004).
26. M. Seitz, C. Friedsam, W. Jöstl, T. Hugel, and H. E. Gaub, *Chem. Phys. Chem.* **4**, 986 (2003).
27. B. Essevaz-Roulet, U. Bockelmann, and F. Heslot, *Proc. Natl. Acad. Sci. USA* **94**, 11935 (1997).
28. C. W. Jones, J. C. Wang, R. W. Briehl, and M. S. Turner, *Biophys. J.* **88**, 2433 (2005).
29. G. Binning and H. Rohrer, *Rev. Mod. Phys.* **59**, 615 (1987).
30. T. A. Jung, R. R. Schlittler, J. K. Gimzewski, H. Tang, C. Joachim, *Science* **271**, 181 (1996).
31. N. Severin, J. Barner, A. A. Kalachev and J. P. Rabe, *Nano Lett.* **4**, 577 (2004).
32. P. Gutjahr, R. Lipowksy, and J. Kierfeld, *Europhys. Lett.* **76**, 994 (2006).
33. P. G. De Gennes and C. Taupin, *J. Phys. Chem.* **86**, 2294 (1982).
34. R. Capovilla and J. Guven, *J. Phys. A: Math. Gen.* **38**, 2593 (2005).
35. W. Helfrich, *J. Physique* **46**, 1263 (1985); *ibid.* **47**, 321 (1986); *ibid.* **48**, 285 (1987).
36. L. Peliti and S. Leibler, *Phys. Rev. Lett.* **54**, 1690 (1985).
37. D. Förster, *Phys. Lett.* **114A**, 115 (1986).
38. H. Kleinert, *Phys. Lett.* **114A**, 263 (1986).
39. W. Helfrich, *Eur. Phys. J. B* **1**, 481 (1998).
40. H. Pinnow and W. Helfrich, *Eur. Phys J. E* **3**, 149 (2000).
41. J. Kierfeld, O. Niamploy, V. Sa-yakanit, and R. Lipowsky, *Eur. Phys. J. E* **14**, 17 (2004).
42. M. Abramowitz and A. I. Stegun, *Handbook of Mathematical Functions*, National Bureau of Standards, Washington, 1965.
43. T. Niemeijer and T. W. Ruijgrok, *Physica* **81A**, 427 (1975).
44. M. Nauenberg, *J. of Math. Phys.* **16**, 703 (1975).
45. F. Jülicher, R. Lipowsky, and H. Müller-Krumbhaar *Europhys. Lett.* **11**, 657 (1990).
46. H. Spohn, *Europhys. Lett.* **14**, 689 (1991).
47. K. Baczynski, R. Lipowsky, and J. Kierfeld, *Phys. Rev. E* **76**, 061914 (2007).
48. F. Gittes, E. Meyhöfer, S. Baek, and J. Howard, *Biophys. J.* **70**, 418 (1996).
49. M. Elbaum, D. Kuchnir Fygenson, and A. Libchaber, *Phys. Rev. Lett.* **76**, 4078 (1996).
50. J. Kierfeld, P. Gutjahr, T. Kühne, P. Kraikivski, and R. Lipowsky, *J. Comput. Theor. Nanosci.* **3**, 898 (2006).
51. L. R. G. Treloar, *The Physics of Rubber Elasticity*, Clarendon Press, Oxford, 1975.
52. J. Kierfeld, *Phys. Rev. Lett* **97**, 058302 (2006).
53. M. Rief, M. Gautel, F. Oesterhelt, J. M. Fernandez, H. E. Gaub, *Science* **276**, 1109 (1997).
54. S. B. Smith, L. Finzi, C. Bustamante, *Science* **258**, 1122 (1992).
55. S. Smith, Y. Cui, and C. Bustamante, *Science* **271**, 795 (1996).
56. T. J. Senden, J.-M. di Meglio, and P. Auroy, *Eur. Phys. J. B* **3**, 211 (1998).
57. X. Chatellier, T. J. Senden, J.-F. Joanny, and J.-M. Di Meglio, *Europhys. Lett.* **41**, 303 (1998).
58. T. Hugel, M. Grosholz, H. C. Clausen-Schaumann, A. Pfau, H. E. Gaub, and M. Seitz, *Macromolecules* **34**, 1039 (2001).
59. S. Cui, C. Liu, and X. Zhang, *Nano Lett.* **3**, 245 (2003).
60. C. Friedsam, A. Del Campo Becares, U. Jonas, M. Seitz, and H. E. Gaub, *New J. Phys.* **6**, 9 (2004).
61. T. M. Birshtein, E. B. Zhulina, and A. M. Skvortsov, *Biopolymers* **18**, 1171 (1979).
62. A. C. Maggs, D. A. Huse, and S. Leibler, *Europhys. Lett.* **8**, 615 (1989).
63. G. Gompper and T. Burkhardt, *Phys. Rev. A* **40**, 6124 (1989).
64. C. C. van der Linden, F. A. M. Leermakers, and G. J. Fleer, *Macromolecules* **29**, 1172 (1996).

65. R. R. Netz and J.-F. Joanny, *Macromolecules* **32**, 9013 (1999).
66. S. Stepanow, *J. Chem. Phys.* **115**, 1565 (2001).
67. A. N. Semenov, *Eur. Phys. J. E* **9**, 353 (2002).
68. P. Benetatos and E. Frey, *Phys. Rev. E* **67**, 051108 (2003).
69. J. Kierfeld and R. Lipowsky, *Europhys. Lett.* **62**, 285 (2003).
70. J. Kierfeld and R. Lipowsky, *J. Phys. A: Math. Gen.* **38**, L155 (2005).
71. E. Y. Kramarenko, R. G. Winkler, P. G. Khalatur, A. R. Khoklov, and P. Reineker, *J. Chem. Phys.* **104**, 4806 (1996).
72. T. Sintes, K. Sumithra, and E. Straube, *Macromolecules* **34**, 1352 (2001).
73. U. Seifert and R. Lipowsky, *Phys. Rev. A* **42**, 4768 (1990).
74. J. F. Marko and E. D. Siggia, *Macromolecules* **28**, (1995) 8759.
75. D. K. Lubensky and D. R. Nelson, *Phys. Rev. Lett.* **85**, 1572 (2000); *Phys. Rev. E* **65**, 031917 (2002).
76. D. Marenduzzo, A. Trovato, and A. Maritan, *Phys. Rev. E* **64**, 031901 (2001).
77. W. H. Roos, A. Roth, J. Konle, H. Presting, E. Sackmann, and J. P. Spatz, *Chem. Phys. Chem.* **4**, 872 (2003).
78. P. Kraikivski, R. Lipowsky, and J. Kierfeld, Europhys. Lett. **66**, 763 (2004).
79. P. Kraikivski, R. Lipowsky, and J. Kierfeld, Eur. Phys. J. E **16**, 319 (2005).
80. P. Kraikivski, R. Lipowsky, and J. Kierfeld, Europhys. Lett. **71**, 138 (2005).
81. M. Burghard, G. Duesberg, G. Philipp, J. Muster, and S. Roth, *Adv. Mater.* **10**, 584 (1998).
82. J. Liu, M. J. Casavant, M. Cox, D. A. Walters, P. Boul, W. Lu, A. J. Rimberg, K. A. Smith, D. T. Colbert, and R. E. Smalley, *Chem. Phys. Lett.* **303**, 125 (1999).
83. Y. Wang, D. Maspoch, S. Zou, G. C. Schatz, R. E. Smalley, and C. A. Mirkin, *Proc. Nat. Acad. Sci. USA* **103**, 2026 (2006).
84. R. Martel, H. R. Shea, and P. Avouris, *Nature* **398**, 299 (1999).
85. M. Sano, A. Kamino, J. Okamura, and S. Shinkai, *Science* **293**, 1299 (2001).
86. A. Amzallag, C. Vaillant, M. Jacob1, M. Unser, J. Bednar, J. D. Kahn, J. Dubochet, A. Stasiak, and J. H. Maddocks, *Nucl. Acid Res.* **34**, e125 (2006).
87. J. X. Tang, J. A. Käs, J. V. Shah, and P. A. Janmey, *Eur. Biophys. J.* **30**, 477 (2001).
88. D. M. Hatters, C. A. MacRaild, R. Daniels, W. S. Gosal, N. H. Thomson, J. A. Jones, J. J. Davis, C. E. MacPhee, C. M. Dobson, and G. J. Howlett *Biophys. J.* **85**, 3979 (2003).
89. N. Severin, W. Zhuang, C. Ecker, A. A. Kalachev, I. M. Sokolov, and J. P. Rabe, *Nano Lett.* **6**, 2561 (2006).
90. K. A. Brakke, *Exp. Math.* **1**, 141 (1992).
91. T. B. Liverpool and M. C. Marchetti, *Phys. Rev. Lett.* **90**, 138102 (2003).
92. K. Kruse, J. F. Joanny, F. Jülicher, J. Prost, and K. Sekimoto, *Phys. Rev. Lett.* **92**, 078101 (2004).
93. I. S. Aranson and L. S. Tsimring, *Phys. Rev. E* **71**, 050901(R) (2005).
94. F. Ziebert and W. Zimmermann, *Phys. Rev. E* **70**, 022902 (2004).
95. P. Kraikivski, R. Lipowsky, and J. Kierfeld, *Phys. Rev. Lett.* **96**, 258103 (2006).

An experimental trip to the Calculus of Variations

Josu Arroyo

Departamento de Matematicas, Universidad del Pais Vasco, Bilbao, Spain.
E-mail:mtparolj@ehu.es

Abstract. This paper presents a collection of experiments in the Calculus of Variations. The implementation of the *Gradient Descent* algorithm built on *cubic-splines* acting as "numerically friendly" elementary functions, give us ways to solve variational problems by constructing the solution. It wins a pragmatic point of view: one gets solutions sometimes as fast as possible, sometimes as close as possible to the true solutions. The balance speed / precision is not always easy to achieve.

Starting from the most well-known, classic or historical formulation of a variational problem, section 2 describes briefly the bridge between theoretical and computational formulations. The next sections show the results of several kind of experiments; from the most basics, as those about geodesics, to the most complex, as those about vesicles.

Keywords: Calculus of Variations, Gradient Descent, Elastic Curves, Vesicles
PACS: 02.30.Xx,02.60.Pn, 02.40.Hw

1. CALCULUS OF VARIATIONS

Let us consider the general set-up for the classical calculus of variations as can be read in [17]: let $y(x)$ denote a function of x defined in $[a,b]$ with fixed endpoints $y(a) = y_a$ and $y(b) = y_b$.

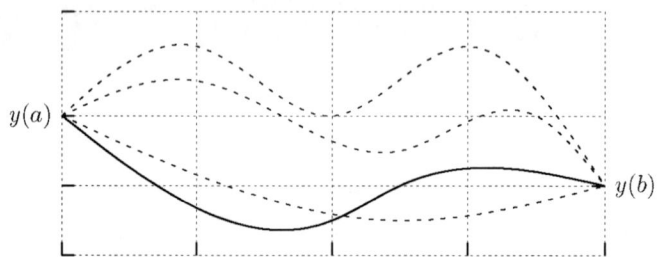

FIGURE 1. Functions (curves) with fixed endpoints.

Because of this picture, we often refer to $y(x)$ as a curve joining the endpoints. Now, there are many choices of curves $y(x)$ joining the given endpoints. It is only when we add some sort of condition to be satisfied that we can pick special $y(x)$ out of this collection. For example, the **Fixed Endpoint Problem**: find the curve $y(x)$ with $y(a) = y_a$ and

CP1002, *Curvature and Variational Modeling in Physics and Biophysics*
edited by O. J. Garay, E. García-Río, and R. Vázquez-Lorenzo
© 2008 American Institute of Physics 978-0-7354-0521-9/08/$23.00

$y(b) = y_b$ such that the following integral

$$\mathscr{F}[y] = \int_a^b f\left(t, y(t), \frac{dy}{dt}(t)\right) dt,$$

is minimized, where $f(x, y, y')$ is a function of x, y and $y' = dy/dx$, and the latter two are thought of as independent variables.

That is, we have a differentiable function $f \colon W \subset \mathbf{R}^3 \longrightarrow \mathbb{R}$, a space $\mathscr{H}$ of functions y, a functional $\mathscr{F} \colon \mathscr{H} \longrightarrow \mathbb{R}$, and we are looking for a $y^* \in \mathscr{H}$ such that $\mathscr{F}[y^*] \leq \mathscr{F}[y]$ for all $y \in \mathscr{H}$. A necessary condition for $y(x)$ to be a minimum is given by the Euler-Lagrange equation.

1.1. The Euler-Lagrange equation

To expose potential candidates for y^* the conventional approach utilizes the Euler-Lagrange equation (E-L). If $y^* = y^*(x)$ is a minimum for the fixed endpoint problem, then y^* satisfies the equation

$$\left[\frac{\partial f}{\partial y}(x, y, y') - \frac{d}{dx}\left(\frac{\partial f}{\partial y'}(x, y, y')\right)\right] = 0, \tag{1}$$

where $\frac{\partial f}{\partial y'}$ stands for the partial derivative of the function $f \colon \mathbf{R}^3 \longrightarrow \mathbb{R}$ with respect to the third variable.

Notice that we are not saying that a solution to the E-L equation is a solution to the fixed endpoint problem. The E-L equation is simply a first step towards solving the fixed endpoint problem. Nevertheless, because we deal with an unimaginably huge collection of possible solution functions (curves) for the fixed endpoint problem, the E-L equation is a powerful tool which is indispensable. Indeed, it is sometimes the case that only solutions to the E-L equations may be found with little or no other information to guide us to a solution of the fixed endpoint problem. For this reason solutions to the E-L equation are given the special name extremals and the fixed endpoint problem, for example, is often rephrased to say that a function $y(x)$ is desired which joins the given endpoints and extremizes the functional $\mathscr{F}$. In this case, solutions to the E-L equation solve the problem.

1.2. The most basic example

It is well known that the shortest distance between two points in the plane is attained by a straight line. If we use xy-coordinates, then the problem of determining the curve $y = y(x)$ of minimum arc-length is simply to minimize the functional

$$\mathscr{F}[y] = \int_a^b \sqrt{1 + (y'(x))^2}\, dx.$$

In this case $f(x,y,y') = \sqrt{1+(y'(x))^2}$. This is the most used example in the literature to introduce the calculus of variations. The E-L equation reduces to

$$y''(x) = 0.$$

As said before, it must be a segment of straight line.

1.3. Improving the original problem

The Variational problem, as written in the introduction to this section, can be immediately improved as the following options show:

1. The functional $\mathscr{F}$ defined in a space of functions $\mathscr{H}$ becomes now a functional defined in m spaces of functions $\mathscr{H}_1, \mathscr{H}_2, \ldots, \mathscr{H}_m$. That is to say, the integrand function f would be $f \colon \mathbf{R}^{2m+1} \longrightarrow \mathbb{R}$, with $m \geq 1$, and the functional $\mathscr{F}$ would look this way:

$$\mathscr{F}[y_1, \cdots, y_m] = \int_a^b f\left(x, y_1, y_1', y_2, y_2', \cdots, y_m, y_m'\right) dx. \tag{2}$$

2. There are more options with respect to the boundary conditions. It would be possible, for example, to find the functional $\mathscr{F}$ defined on three spaces of functions $\mathscr{H}_1$, $\mathscr{H}_2$ and $\mathscr{H}_3$ such that:
   - All the functions $y_1 \in \mathscr{H}_1$ have the same initial $y_1(a)$ and final $y_1(b)$ values (as in the Fixed End-points Problem -see figure 1-).
   - All the functions $y_2 \in \mathscr{H}$ have the same initial value $y_2(a)$ but not necessarily the final one $y_2(b)$, or vice versa;
   - There are no conditions on the initial and final values of the functions y_3 in $\mathscr{H}_3$.

This new options force to change the E-L Equation. It turns into a system of m equations, one for each function y_i in which is defined the functional given by equation (2). The absence of previous conditions over the initial and final values of the functions y_i is compensated by new equations added to the system of equations of E-L.

1.4. Isoperimetric constraints

The above endpoints constraints cause next to no harm since the corresponding domain of functions is at worst a translated linear space. In contrast, consider this new problem: among all the functions y of the space $\mathscr{H}$, with the endpoints $y(a)$ and $y(b)$ fixed or not, that satisfy the condition

$$\int_a^b g\left(x, y(x), y'(x)\right) dx = L = \text{constant}, \tag{3}$$

find the one that minimizes $\mathscr{F}$.

We have now a subspace $X \subset \mathscr{H}$ given by $X = \mathscr{G}^{-1}(L)$ for some functional $\mathscr{G}: \mathscr{H} \longrightarrow \mathbb{R}$ of the form

$$\mathscr{G}[y] = \int_a^b g\left(x, y(x), y'(x)\right) dx.$$

This new space X is not *flat*, which leads to considerable complications. In the classical theory, Lagrange multipliers are introduced to deal with this kind of *isoperimetric constraints*. Superficially it suffices to replace f by $f + \lambda g$ in equation (1) and then use the constraint as an additional condition to account for the unknown real number λ.

The most ancient of all these types of problems is the following: given a closed curve of fixed arc-length L, what is the shape the curve should assume to maximize the enclosed area A? See section 3.5.

1.5. In this world we will find

The original formulation, with the above mentioned improvements, would be enough to tackle all kinds of variational problems that appear as functionals defined on spaces of functions. For example, the following related ones are among them:

- Isoperimetric and isoareal
- Brachistochrones
- Geodesics
- (Hyper) Elasticae

- Minimal Surfaces
- Delaunay Surfaces
- Elastic Surfaces
    - Willmore
    - Membranes, Vesicles

2. 40% THEORY - 60% EXPERIMENTAL

There is a way that goes from the variational problem as introduced in section 1 to the experiments made in the following sections: the bridge from the theoretical to numerical worlds. In this section we briefly describe some of the steps that make this way.

1. Gradients in the space of functions $\mathscr{H}_i$ where the functionals are defined in.
2. An improved E-L equation.
3. The *gradient descent*, a way of finding the critical minima of these functionals.
4. The cubic splines, a way for representing functions from sets of points.

2.1. Gradients in the space of functions

Let $\mathscr{F}: \mathscr{H} \longrightarrow \mathbb{R}$ be a functional defined in a Hilbert space of functions $\mathscr{H}$ without boundary constraints (for the moment). The directional (Gâteaux) derivative in (the point) $y \in \mathcal{H}$, $D\mathcal{F}(y) : T_y\mathcal{H} \longrightarrow \mathbb{R}$, is defined for each $\vec{v} \in T_y\mathcal{H}$ by

$$D\mathcal{F}(y)\vec{v} = \lim_{t \to 0} \frac{\mathcal{F}(y + t\vec{v}) - \mathcal{F}(y)}{t}.$$

In this (simplest) space $\mathcal{H}$, the tangent space $T_y\mathcal{H} = \mathcal{H}$ itself. Suppose now that $\mathcal{H}$ is equipped with the inner product $\langle , \rangle$. Then the gradient vector field $\nabla\mathcal{F}$, such that for each $y \in \mathcal{H}$, $\nabla\mathcal{F}(y) \in T_y\mathcal{H} = \mathcal{H}$, is defined by the relationship

$$D\mathcal{F}(y)\vec{v} = \langle \nabla\mathcal{F}(y), \vec{v} \rangle,$$

for all $\vec{v} \in T_y\mathcal{H}$. The existence and uniqueness of $\nabla\mathcal{F}$ is the content of the Riesz-Fréchet Theorem. Different choices of metrics yield different gradient vector fields.

2.2. New Euler-Lagrange equation

A function $y^* \in \mathcal{H}$ is an extremal of the functional $\mathcal{F}$ in the sense pointed out in section 1, if and only if, $D\mathcal{F}(y^*)\vec{v} = 0$ for all vector $\vec{v} \in T_y\mathcal{H}$. In this way, the first necessary condition for $y^* \in \mathcal{H}$ to be a (local) minimum becomes

$$\nabla\mathcal{F}(y^*) = 0. \tag{4}$$

This statement incorporates the E-L equation as well as the necessary conditions given by the natural boundary conditions seen in 1.3. This equation is in general more selective that the E-L equation. Note that if the gradient is zero at a point with respect to a metric, then it is zero with respect to all metrics. More in [13, 14, 17].

2.3. Projected gradients

The cases of (one or both) fixed endpoints generate translated linear subspaces of functions and the need for an orthogonal projection to compute the gradient vector field. The tangent space to a translated linear subspace is the linear subspace itself.

For instance, consider $\mathcal{H}_{ab} \subset \mathcal{H}$ the space of functions $y \colon [a,b] \longrightarrow \mathbb{R}$ with fixed endpoints $y(a) = y_a$ and $y(b) = y_b$ in the sense of section 1. This new space $\mathcal{H}_{ab}$ is an example of a so-called translated linear space. Given the linear space of curves

$$\mathcal{H}_o = \{y \colon [a,b] \longrightarrow \mathbb{R}; y(a) = y(b) = 0\},$$

we can see that $\mathcal{H}_{ab} = y_o + \mathcal{H}_o$, where $y_o \in \mathcal{H}_{ab}$ is nothing but the function

$$y(x) = y_a + \frac{(x-a)}{(b-a)}(y_b - y_a),$$

and the tangent space of $T_y\mathcal{H}_{ab} = \mathcal{H}_o$, the same at every $y \in \mathcal{H}_{ab}$.

The restriction of the functional $\mathscr{F}$ to $\mathscr{H}_{ab}$ has a projected gradient $\nabla_{ab}\mathscr{F}$ that satisfies

$$\langle \nabla \mathscr{F} - \nabla_{ab}\mathscr{F}, \vec{v} \rangle = 0,$$

for all $\vec{v} \in \mathscr{H}_o$. This gradient vector field is the tangential part of the ambient gradient vector field $\nabla \mathscr{F}$. More in [13, 14].

2.4. Gradients with isoperimetric constraints

In the case of isoperimetric constraints, things are more complicated since the tangent space is no longer a single fixed subspace. Isoperimetric constraints $\mathscr{G}: \mathscr{H} \longrightarrow \mathbb{R}$ defined in a Hilbert space $\mathscr{H}$, as those of section 1.4, generate tangent spaces that vary from point to point. Nevertheless, there will be a projected gradient vector field on the tangent bundle of $\mathscr{H}_{\mathscr{G}} = \mathscr{G}^{-1}(L)$.

Consider the restriction of the functional $\mathscr{F}: \mathscr{H} \longrightarrow \mathbb{R}$ to the space $\mathscr{H}_{\mathscr{G}}$ of the functions $y \in \mathscr{H}$ such that $\mathscr{G}(y) = L = $ const. This means that the directional derivative satisfies $DG(y)\vec{v} = 0$ for all $\vec{v} \in T_y\mathscr{H}_{\mathscr{G}}$, so $\langle \nabla \mathscr{G}(y), \vec{v} \rangle = 0$ holds. The projected gradient we are looking for can be expressed in the form

$$\nabla_{\mathscr{G}}\mathscr{F}(y) = \nabla \mathscr{F}(y) - \lambda(y)\nabla \mathscr{G}(y),$$

where $\lambda(y)$ is a scalar field on $\mathscr{H}_{\mathscr{G}}$. The condition $\langle \nabla \mathscr{G}(y), \vec{v} \rangle = 0$ leads to

$$\lambda(y) = \frac{\langle \nabla \mathscr{G}(y), \nabla \mathscr{F}(y) \rangle}{\langle \nabla \mathscr{G}(y), \nabla \mathscr{G}(y) \rangle},$$

being necessary to assume the *regularity* condition given by $\nabla \mathscr{G}(y) \neq 0$ for all $y \in \mathscr{H}_{\mathscr{G}}$.

Suppose now a finite number of isoperimetric constraints $\mathscr{G}_i: \mathscr{H} \longrightarrow \mathbb{R}$ in the variational problem. The functional $\mathscr{F}: \mathscr{H} \longrightarrow \mathbb{R}$ is restricted to the space of functions $y \in \mathscr{H}$ such that $\mathscr{G}_i(y) = L_i = $ constant for $i \in \{1, \ldots, n\}$. This time the projected gradient is given by

$$\nabla_{\mathscr{G}}\mathscr{F}(y) = \nabla \mathscr{F}(y) - \sum_{i=1}^{n} \lambda_i(y)\nabla \mathscr{G}_i(y),$$

where each $\{\lambda_i\}$ is a scalar field. If $n > 1$ the following linear system must be solved:

$$\begin{pmatrix} \langle \nabla \mathscr{G}_1, \nabla \mathscr{G}_1 \rangle & \cdots & \langle \nabla \mathscr{G}_1, \nabla \mathscr{G}_n \rangle \\ \vdots & \ddots & \vdots \\ \langle \nabla \mathscr{G}_n, \nabla \mathscr{G}_1 \rangle & \cdots & \langle \nabla \mathscr{G}_n, \nabla \mathscr{G}_n \rangle \end{pmatrix} \begin{pmatrix} \lambda_1(y) \\ \vdots \\ \lambda_n(y) \end{pmatrix} = \begin{pmatrix} \langle \nabla \mathscr{G}_1, \nabla \mathscr{F} \rangle \\ \vdots \\ \langle \nabla \mathscr{G}_n, \nabla \mathscr{F} \rangle \end{pmatrix}.$$

More in [13, 14].

2.5. Inner product and gradient examples in $\mathcal{H}$

Let y_1, y_2 be two functions in $\mathcal{H}$. The following formula defines an inner product in $\mathcal{H}$, which is well known in the literature about this topic:

$$\langle y_1, y_2 \rangle = y_1(a)\, y_2(a) + \int_a^b y_1' y_2'\, dt. \tag{5}$$

Let $f: \mathbf{R}^3 \longrightarrow \mathbb{R}$ be a differentiable function and suppose the functional $\mathcal{F}$ is given by

$$\mathcal{F}[y] = \int_a^b f\left(x, y(x), y'(x)\right) dx.$$

Define the Euler operator and the Work integral by

$$E_y^f(x) = f_{y'}\left(x, y(x), y'(x)\right) - \int_a^x f_y\left(s, y(s), y'(s)\right) ds,$$

$$W_y^f = \int_a^b f_{y'}\left(x, x(x), y'(x)\right) dx,$$

where f_y and $f_{y'}$ are the partial derivatives respect to the second and third variables, respectively.

1. In the case of free endpoints, the gradient formula is given by

$$\nabla \mathcal{F}(y)(x) = \int_a^x E_y^f(s)\, ds + ((x-a)+1)\, W_y^f. \tag{6}$$

2. In the case of fixed endpoints,

$$\nabla \mathcal{F}(y)(x) = \int_a^x E_y^f(s)\, ds + \frac{(x-a)}{(b-a)} \int_a^b E_y^f(s)\, ds. \tag{7}$$

Observe that the gradient is a function of x that vanishes at both endpoints.

3. In the case of free right-hand endpoint,

$$\nabla \mathcal{F}(y)(x) = \int_a^x E_y^f(s)\, ds + (x-a)\, W_y^f. \tag{8}$$

The gradient vanishes at the left-hand endpoint, but no necessarily at the right-hand endpoint.

4. In the case of free left-hand endpoint,

$$\nabla \mathcal{F}(y)(x) = \int_a^x E_y^f(s)\, ds - \left\{ \int_a^b E_y^f(s)\, ds \right\} \left(\frac{(x-a)+1}{(b-a)+1} \right). \tag{9}$$

The gradient vanishes at the right-hand endpoint, but no necessarily at the left-hand endpoint.

More details in [13].

2.6. Gradient descent

Gradient descent is an optimization algorithm. To find a local minimum of a function using gradient descent, one takes steps proportional to the negative of the gradient (or the approximate gradient) of the function at the current point. If instead one takes steps proportional to the gradient, one approaches a local maximum of that function; the procedure is then known as *gradient ascent*.

Gradient descent is based on the observation that if the real-valued function F is defined and differentiable in a neighborhood of a point P, then F decreases fastest if one goes from P in the direction of the negative gradient of F at P, $-\nabla F(P)$. With this observation in mind, one starts with a guess $\mathbf{x}_o$ for a local minimum of F, and considers the sequence

$$\mathbf{x}_{n+1} = \mathbf{x}_n - h_n \nabla F(\mathbf{x}_n), n \geq 0.$$

We have

$$F(\mathbf{x}_o) \geq F(\mathbf{x}_1) \geq \ldots \geq F(\mathbf{x}_n) \geq \ldots,$$

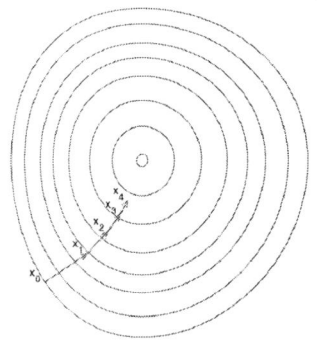

2: Descent towards the minimum.

so hopefully the sequence $\{\mathbf{x}_n\}$ converges to the desired local minimum as

$$\|\nabla F(\mathbf{x}_n) \to 0\|.$$

The value of the step size h_n is allowed to change at every iteration.

Gradient descent works in spaces of any number of dimensions, even in infinite dimensional ones. In the latter case the search space is typically a function space, and one calculates the Gâteaux derivative of the functional to be minimized to determine the descent direction.

Two weaknesses of gradient descent are:

1. The algorithm can take many iterations to converge towards a local minimum, if the curvature in different directions is very different.
2. Finding the optimal h_n per step can be time-consuming. Conversely, using a fixed h can yield poor results.

2.7. Cubic splines

The gradient descent is the flow along the the trajectories of the negative gradient vector field. In all but the simplest classes, this flow is not known explicitly.

All numerical algorithms will suffer from an irreversible loss of information at the start of the descent in the infinite dimensional case. The initial point is a curve that can only be represented by a finite number of points. The flow subsequently acts on the discrete representation and this severely limits the number of applicable numerical tools during the descent.

To deal with this difficulty we follow the cubic spline representation of functions, in such a way that we will see this space of splines as a closed set for the operations needed to implement the gradient descent.

The cubic spline representation is a numerical friendly way of getting a

- C^2 function from a tabulated list of values $\{x_i, y_i\}_{i=0}^N$,
- C^2 approximation of a differentiable function $y \in C^m[a,b]$,
- $C^{1(2)}$ approximation of the derivative function of y,
- C^3 approximation of the integral function of y,

More details in [14, 16].

2.8. How to get a cubic spline

Given a tabulated function $y_i = y(x_i)$, with $a = x_o < x_1, \ldots, < x_N = b$, focus attention on one particular interval, $[x_i, x_{i+1}]$, between y_i and y_{i+1}. Linear interpolation in that interval gives the known basic formula

$$y(x) = \left(\frac{x_{i+1} - x}{x_{i+1} - x_i} \right) y_i + \left(1 - \frac{x_{i+1} - x}{x_{i+1} - x_i} \right) y_{i+1}.$$

The advantage of the Cubic Spline Construction 1 is that it only needs two further values to get a (unique) $C^2[a,b]$ function $y(x)$, such that

- $y(x_i) = y_i$ for all $i = 0, \ldots, N$, and
- the function $y(x)$ is a cubic polynomial in each sub-interval $[x_i, x_{i+1}]$.

The mentioned two further values are typically taken as boundary conditions at (a, y_o) and (b, y_N). The most common ways of doing this are either

- set both of $y''(a)$ and $y''(b)$ equal to zero, giving the so-called *natural cubic spline*, or
- set both $y'(a)$ and $y'(b)$ to specified values.

More details in [16].

2.9. C^2 spline functions from points

Let us see some cubic splines with different boundary conditions obtained from a tabulated list of points. We don't really need many points to understand how they work. By now, take only $(x_i, y_i) = \{(-1,0), (0,2), (1,0)\}$. We see in figure 3 three different C^2 curves through these points, depending on the boundary conditions.

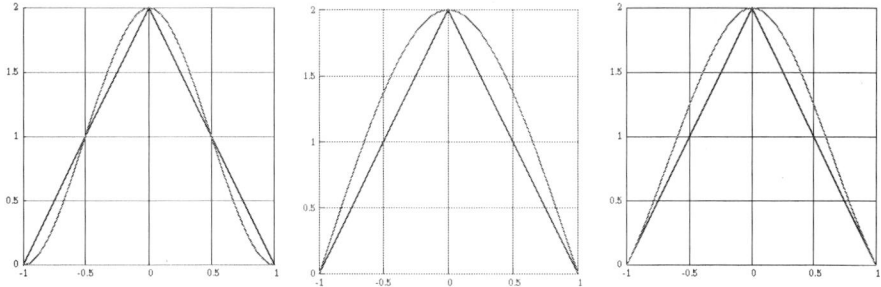

FIGURE 3. Cubic splines through three points. The border conditions are respectively (a) $y'(-1) = 0$, $y'(1) = -0$, (b) $y'(-1) = 2$, $y'(1) = -2$ and (c) $y''(-1) = 0$, $y''(1) = 0$.

2.10. C^2 spline functions from functions

Let $y(x)$ be a function defined in an interval $[a,b]$. Given a partition $a = x_0 < x_1 < \ldots < x_N = b$, we define the cubic spline $\tilde{y}(x) \in C^2[a,b]$, interpolator of y at the points $\{(x_j, y(x_j)); j = 0, \ldots, N\}$. We may say that $\tilde{y}$ "looks" close to function y.

- The derivative $\tilde{y}'(x)$ (as derivative of polynomials) of the spline $\tilde{y}(x)$ approximates the derivative $y'(x)$.
- Even more, we can integrate $\tilde{y}$ (integrating polynomials) to obtain a function that approximates the integral of y.

Choose the function $y(x) = \sin(10\pi x)$ with $x \in [0,1]$, take an equidistant partition $\{x_i\}$ of $N+1$ points in $[0,1]$ and get the cubic splines $\tilde{y}$ at the points

$$\{(x_j, y(x_j)); j = 0, \ldots, N\}.$$

We see in the table 1 how much the splines approach the function y, its derivative and its primitive, in L^2 metric, where ψ and $\tilde{\psi}$ are primitive functions of $y(x)$ and $\tilde{y}(x)$, respectively.

2.11. Algorithm at work

We must deal with a variational problem of the type 1, maybe improved following 1.3, and perhaps with isoperimetric constraints 1.4. The aim is to minimize the functional $\mathscr{F}$ by means of the gradient descent algorithm 2.6, defined in spaces of functions $\mathscr{H}_i$ that, in the presence of isoperimetric constraints, fails to be translated linear spaces. For notation simplicity, suppose for the moment the functional $\mathscr{F}$ defined only in one $\mathscr{H}$.

1. In order to construct the sequence described in 2.6, we start with a function $y_o \in \mathscr{H}$ that satisfies the boundary constraints. It is approximated by a spline $\tilde{y}_o$ defined on a partition $a = x_o < x_1 \cdots < x_N = b$ of the definition domain interval $[a,b]$, following

195

TABLE 1. Comparison between functions and cubic splines.

$y(t) = \sin(10\pi t)$, with $t \in [0,1]$				
points spline	$\|\tilde{y}-y\|_{L^2}$	$\|\tilde{y}'-y'\|_{L^2}$	$\|\tilde{\psi}(t)-\psi(t)\|_{L^2}$	
64	8×10^{-5}	1.8×10^{-2}	3.4×10^{-6}	
128	4×10^{-6}	2×10^{-3}	2.1×10^{-7}	
256	2×10^{-7}	2×10^{-4}	1.2×10^{-8}	
512	10^{-8}	3×10^{-5}	7.3×10^{-10}	
1024	10^{-9}	3×10^{-6}	5×10^{-11}	
2048	6×10^{-11}	4×10^{-7}	3×10^{-12}	

2.7; we have also $\tilde{y}_o \in \mathcal{H}$. This partition $\{x_i\}$ will be the same along the whole process.

2. Evaluate $\nabla \mathcal{F}(\tilde{y}_o)$; the gradient of the functional $\mathcal{F}$ at the (point) function $\tilde{y}_o$, according to 2.5, equations (6)-(9) and 2.4.

3. Put $\tilde{y}_{i+1} = \tilde{y}_i - h_i \nabla \mathcal{F}(\tilde{y}_i)$; this is a new spline function, for a small enough $h_i \in \mathbb{R}$ such that $\mathcal{F}[\tilde{y}_{i+1}] < \mathcal{F}[\tilde{y}_i]$.

4. If $\|\nabla \mathcal{F}(\tilde{y}_i)\| < \varepsilon$ admissible, the method breaks off.

5. Otherwise, go to 3 and repeat until 4 is fulfilled.

For a given spline function $\tilde{y}$ on the partition $\{x_i\}$, the functions $E_{\tilde{y}}^f$, $W_{\tilde{y}}^f$ and $\nabla \mathcal{F}(\tilde{y})$ from 2.5 will be also spline functions on the same partition. The linear combinations, products and integrals are closed operations in the space of splines.

Without isoperimetric constraints, the sequence of functions $\{\tilde{y}_i\}$ remains in $\mathcal{H}$. This is the best case. The worst, the sequence $\{\tilde{y}_i\}$ does not remain in $\mathcal{H}$ in presence of isoperimetric constraints. In this case we must try to maintain the sequence "close" to $\mathcal{H}$. That is to say, if $\mathcal{H} = \mathcal{G}^{-1}(L)$ as written in 2.4, the final minimum y^* must satisfy $|\mathcal{G}[y^*] - L| < \delta$, for a given small tolerance $\delta > 0$.

3. FIRST EXPERIMENTS

We may classify the experiments in four levels according to the complexity of the algorithm to deal with the functional.

1. At the first level, we find variational problems with functionals $\mathcal{F}$ defined in only one space of functions $\mathcal{H}$. These functions $y(x)$ satisfy the border conditions seen in 1.3. Known names among others are geodesics, brachistochrones, minimal surfaces of revolution.

2. The second level contains functionals defined on several spaces of functions $\mathcal{H}_i$, each other independent, as we wrote in 1.3.

3. The presence of isoperimetric constraints, as introduced in 1.4, extends greatly the playground. Think, for example, in (hyper) elasticae, elastic surfaces, membranes, vesicles, area and volume constraints, etc.

4. In the kind of functionals $\mathscr{F}$ we would deal with, as first introduced in 1, the integrand function f acts at most on the first derivative of given functions $y_i \in \mathscr{H}_i$. Nevertheless, there will be functionals that need more derivatives of the functions y_i. This is accomplished by introducing new functions together with isoperimetric constraints (as *bridges* between functions) to reduce the order of the above derivatives. For instance, given a functional $\mathscr{F}$ defined in a suitable space of functions $y \in \mathscr{H}$ that needs y, y' and y'', it could be written as a functional defined on $\mathscr{H} \times \mathscr{H}_2$ together with a new constraint given by $y' = z$ (the bridge) and $z \in \mathscr{H}_2$ (the new function). In this way, y'' becomes z' and $\mathscr{F}$ acts on y, y' and z'.

3.1. Path of shortest distance in $\mathbf{R}^2$

It is a well-known fact that the shortest distance between two given points in Euclidean ("flat") space is calculated along a straight line joining the two points. Let it be the problem of minimizing the integral

$$\mathscr{F}[y] = \int_0^1 \sqrt{1 + (y')^2}\, dx.$$

We saw in 1.2 that the function $y(x)$ that minimizes the length integral is the solution of the differential equation $y''(x) = 0$, subject to the boundary conditions $y(0) = 0$ and $y(1) = m$, i.e., $y(x) = mx$.

Given the boundary conditions $y(0) = 0$ and $y(1) = 1$, take a function $y_o(x)$ satisfying them; for example $y_o(x) = x^6$. We can see how the functional flow, by means of the Gradient Descent 2.6, creates a sequence $\{y_i(x)\}$ of spline-functions that converges to the straight segment that joins $(0,0)$ and $(1,1)$. Remember again that each y_i satisfies the same boundary constraints that y_o does. This is what happens when there are no isoperimetric constraints in the problem.

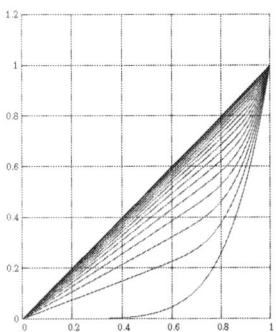

3.2. Geodesics in $\mathbf{H}^2(-1)$

4: The limit straight segment

Let $\mathbf{H}^2 = \{(x,y) \in \mathbf{R}^2 ; y > 0\}$ be the Poincare's Upper Half Plane model of the hyperbolic plane, with metric $g_{ii} = 1/y^2$. Let $\gamma(t) = (x(t), y(t))$ be a smooth curve joining the points $P = \gamma(a)$ and $Q = \gamma(b)$. The length of γ is

$$\mathscr{L}[\gamma] = \int_a^b \sqrt{(x')^2 + (y')^2}\, \frac{dt}{y}. \tag{10}$$

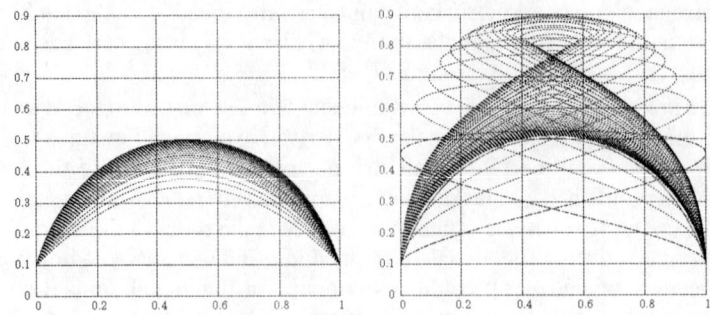

FIGURE 5. Looking for spline-geodesics in $\mathbf{H}^2$.

In this way we define the *curve length functional* in the space of curves that join P and Q. Without loss of generality, suppose that $x(t) = t$ so the curve is $\gamma(x) = (x, y(x))$ and the functional becomes

$$\mathscr{L}[\gamma] = \int_a^b \sqrt{1 + (y')^2}\, \frac{dx}{y}.$$

Now we have a functional $\mathscr{L} : \mathscr{H} \longrightarrow \mathbb{R}$ defined in the space of differentiable functions

$$\mathscr{H} = \left\{ y \in C^1[a,b] ; P = (a, y(a)), Q = (b, y(b)) \right\}.$$

Take for example the points $P = (0, 0.1)$, $Q = (1, 0.1)$ and the curve $\gamma_o(x) = (x, 0.1)$ joining them. This functional creates a curve flow, given by a sequence of functions $\{y_i\} \in \mathscr{H}$, searching for a length minimizing curve that joins those points. See figure 5-(a).

The same variational problem but starting from equation (10) leads us to a functional $\mathscr{L}$ defined in the two spaces of functions: $\mathscr{L} : \mathscr{H}_1 \times \mathscr{H}_2 \longrightarrow \mathbb{R}$. Now, we are not restricted to curves that are graphs of functions. For example, inspired in the plane curve $\gamma(t) = (\sin 2t, \sin t)$, we can take a pair $\{x_o, y_o\} \in \mathscr{H}_1 \times \mathscr{H}_2$ that starts a sequence of functions $\{x_i, y_i\}$ that defines the curves of the evolution seen in figure 5-(b).

Of course, the solution of the variational problem must be the same.

3.3. Geodesics on $\mathbf{T}^2$

Let $\mathbf{T}^2$ be a Torus of radii R and r, and take the standard parametrization

$$X(\varphi, \theta) = ((R + r\cos\theta)\cos\varphi, (R + r\cos\theta), r\sin\theta),$$

where $(\varphi, \theta) \in [0, 2\pi] \times [-\pi, \pi]$. Geodesic curves on $\mathbf{T}^2$ are expressed in terms of extremal curves of the length functional

$$\mathscr{L}[\gamma] = \int_x \left((\theta')^2 r^2 + (\varphi')^2 (R + r\cos\theta)^2 \right)^{\frac{1}{2}} dt.$$

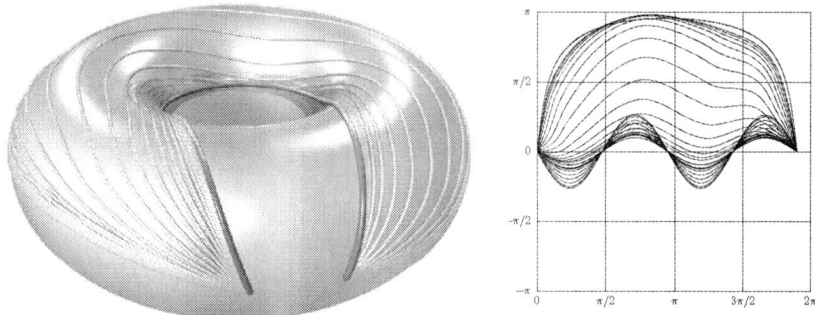

FIGURE 6. Looking for a geodesic in T^2.

There is no loss if we consider the angle $\varphi(\theta)$ as an arbitrary function of the angle θ and write the functional by

$$\mathscr{L}[\gamma] = \int_x \left(r^2 + (\varphi')^2 (R + r\cos\theta)^2\right)^{\frac{1}{2}} dt.$$

Take the points $P = X(0,0)$ and $Q = X(1.9\pi, 0)$ and the curve

$$\gamma_o(t) = X(1.9\pi t, \sin 4\pi t)$$

joining them. This functional creates a curve flow searching for a length minimizing curve that joins the points. The points P and Q are on the same geodesic equator, homotopical to the curve γ_o, but not minimal. Therefore, the curve evolution reaches the geodesic equator, as a locally length minimizing curve, but immediately jumps over it, looking for a globally length minimizing curve. See it in figure 6.

3.4. Least area surfaces of revolution [17]

What should the shape of the surface be so that it has the minimum possible area? Let the surface be generated by rotating the curve $y = y(x)$ around the x-axis. The boundary conditions are $y(a) = y_a$ and $y(b) = y_b$. Slicing the surface up into vertical rings, we see that the area is given by

$$\mathscr{A}[y] = 2\pi \int_a^b y\sqrt{1 + (y')^2}\, dx.$$

The goal is to find the function $y(x)$ that minimizes this integral, so we "simply" have to apply the E-L equation to this Lagrangian, a bit tedious calculation that gives

$$y(x) = \frac{1}{\lambda} \cosh(\lambda(x + \eta)),$$

199

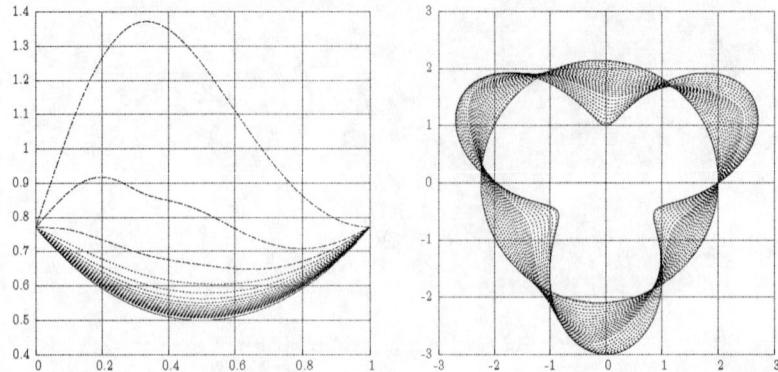

FIGURE 7. Catenary gives minimal surfaces of revolution. Circumference in the isoperimetric problem.

with the constants λ and η determined by the boundary conditions,

$$y_a = \frac{1}{\lambda}\cosh(\lambda a + \lambda \eta) \quad \text{and} \quad y_b = \frac{1}{\lambda}\cosh(\lambda b + \lambda \eta).$$

Take the interval $[0,1]$ and the boundary conditions

$$y(0) = y(1) = \frac{1}{2}\cosh(1).$$

The initial curve

$$y_o(t) = \frac{4}{5}\sin\left(\pi \sin\left(\frac{\pi}{2}t\right)\right) + \frac{1}{2}\cosh(1),$$

evolves towards the catenary

$$y(t) = \frac{1}{2}\cosh(2t-1),$$

as can be seen in figure 7-(a).

3.5. Isoperimetric problem

Isoperimetric problems represent some of the earliest applications of the variational approach to solving mathematical optimization problems. Pappus (ca. 290-350) was among the first geometers to recognize that among all the isoperimetric closed planar curves (i.e., closed curves that have the same perimeter length), the circle encloses the greatest area. The variational formulation of the isoperimetric problem requires that we maximize the area A while keeping the perimeter length L constant. Therefore, given a plane closed curve $\gamma(t) = (x(t), y(t))$ the functional to be maximized becomes

$$\mathscr{A}[\gamma] = \int_a^b \frac{1}{2}(xy' - yx')\,dt,$$

200

while keeping fixed the perimeter

$$\mathscr{L}[y] = \int_a^b \sqrt{(x')^2 + (y')^2}\, dt.$$

It also can be shown by the same variational methods that the circle solves the problem of finding the curve of shortest length which encloses a fixed area A. In this case the roles of $\mathscr{A}$ and $\mathscr{L}$ change.

Using a little algebra, let γ be a closed curve of fixed area A and length L. Since the circle solves the problem of finding a curve of fixed length having maximum area, we must have $A \leq L^2/4\pi$ (area of the circle of length L). Now, given a circle of the same area A, its length $\tilde{L}$ satisfies necessarily $\tilde{L} \leq L$, so the circle minimizes the problem.

Let $\gamma(t) = \rho(t)(\cos\theta(t), \sin\theta(t))$ be a plane closed simple curve in polar coordinates. Its area and length elements are given respectively by $ds = \rho^2\theta'$ and

$$dA = \sqrt{(\rho')^2 + \rho^2(\theta')^2},$$

respectively. Take γ_o as an initial curve defined by $\rho_o(t) = 2 + \sin(2\pi t)$ and $\theta_o(t) = 2\pi t$, with $t \in [0,1]$. The curve flow generated by this functional evolves this initial curve towards the circle of same area and least perimeter, as can be seen in the figure 7-(b). More details in [1, 17].

3.6. Least area surfaces of revolution with volume constraint

Let's go back again to the Least Area Surfaces of Revolution problem 3.4 and consider a surface of revolution **S** with parametrization

$$X(u,v) = (u, y(u)\cos v, y(u)\sin v).$$

We want to minimize the surface area

$$\mathscr{A}[y] = 2\pi \int y\sqrt{1 + (y')^2},$$

but now keeping the enclosed volume $\mathscr{V}[y] = \pi \int y^2(x)\, dx$ fixed. What are the resulting surfaces? The usual setup for such a constrained problem gives the functional to be minimized

$$\mathscr{F}[y] = \mathscr{A}[y] + \eta\mathscr{V}[y] = \pi \int_a^b 2y\sqrt{1 + (y')^2} + \eta y^2\, dx,$$

being $\eta \neq 0$ the Lagrange multiplier. The E-L equation 1 takes the form

$$\frac{d}{dx}\frac{\partial f}{\partial(y')} = \frac{d}{dx}\left[\frac{2yy'}{\sqrt{1 + (y')^2}}\right] = 2\sqrt{1 + (y')^2} + 2\eta y = \frac{\partial f}{\partial y}.$$

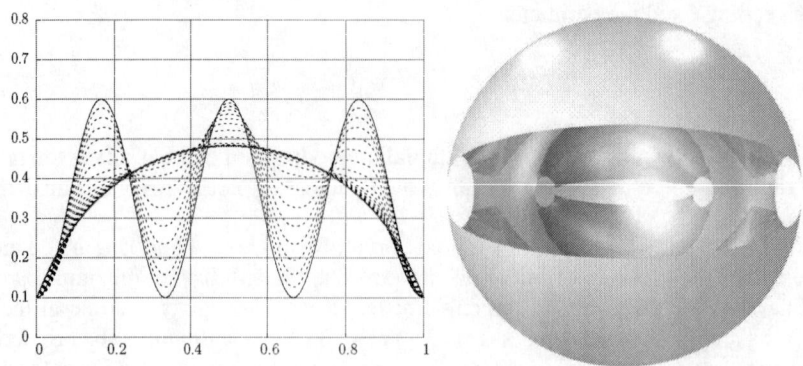

FIGURE 8.   The initial curve evolves towards the profile curve of a Delaunay surface.

By deriving and simplifying becomes

$$-1 + \left(y'\right)^2 - yy'' = -\eta y \left(1 + \left(y'\right)^2\right)^{\frac{3}{2}},$$

the same equation that characterizes the surfaces of Delaunay, surfaces of revolution of constant mean curvature, as can be read in [17].

For example, given the initial curve $\gamma_o(t) = (t, 0.1 + \frac{1}{2} \sin^2 3\pi t)$, with $t \in [0,1]$, the gradient flow created by this functional carries γ_o into the profile curve of a Delaunay surface, as can be seen in figure 8.

4. UNIT-SPEED PLANAR ELASTICAE

According to the Bernoulli-Euler theory of elastic rods the *bending energy* of a thin wire shaped as a curve γ is proportional to the total squared curvature of γ, $\mathscr{F}[\gamma] = \int_\gamma \kappa^2 ds$. It is not unreasonable to assume the shape the rod assumes at equilibrium is determined by minimizing this potential energy of the rod.

Let it consider the *bending energy* functional on plane curves. Given a unit-speed planar curve $\gamma(s) = (x(s), y(s))$ with curvature function $\kappa(s)$ and $s \in [a,b]$, the elastic energy stored in γ is proportional to

$$\mathscr{F}[\kappa] = \int_a^b \kappa^2(s) \, ds.$$

From the theory about plane curves, [17], the curvature of a plane curve is given by $\theta' = d\theta/ds$ where $\theta(s)$ is the angle the unit tangent $T(s)$ makes with the x-axis. In this way, the bending energy functional becomes

$$\mathscr{F}[\theta] = \int_a^b \left(\theta'(s)\right)^2 ds,$$

that is, the functional $\mathscr{F}: \mathscr{H} \longrightarrow \mathbb{R}$ is defined in the space $\mathscr{H}$ of functions $\theta(s)$. Remember here that one could recover (isometrically) the curve γ from θ as

$$\gamma(s) = \left(\int_a^s \cos\theta(u)\,du, \int_a^s \sin\theta(u)\,du \right). \tag{11}$$

4.1. Unit-speed elasticae and boundary constraints

1. The functional $\mathscr{F}$ is defined in spaces $\mathscr{H}$ of functions $\theta(s)$, s being the arc-length parameter defined in a fixed interval $[a,b]$. This means that all the curves are of the same length $L = b - a$. Therefore, this isoperimetric condition is assured by definition.

2. Suppose now, according to section 1.3, that one or both of the endpoints are fixed for all the functions $\theta \in \mathscr{H}$: $\theta(a) = \theta_a$ and/or $\theta(b) = \theta_b$. This means that the curves γ defined by the functions θ would be of the same initial and/or final tangent vector.

3. The *fixed endpoints* constraint on the curves γ, if needed, is accomplished by these isoperimetric constraints:

$$\mathscr{G}_1[\theta] = \int_a^b \cos\theta(s)\,ds, \qquad \mathscr{G}_2[\theta] = \int_a^b \sin\theta(s)\,ds.$$

   (a) By fixing $\mathscr{G}_1[\theta] = $ constant, the functions θ define curves γ with the same $x(b) - x(a)$.
   (b) Symmetrically, by fixing $\mathscr{G}_2[\theta] = $ constant the functions θ define curves γ with the same $y(b) - y(a)$.

Some of the experiments we have made in looking for planar spline-elasticae as minimizing solutions of variational problems are seen in figures 9, 11, 12 and 13. In this last one it is needed the *total turning* constraint, assured by the isoperimetric constraint

$$\mathscr{G}[\theta] = \int_a^b \theta(s)\,ds,$$

as can be read in [17].

5. ELASTICAE IN $\mathbb{R}^3$

Let $\gamma = (x(s), y(s), z(s))$ be a unit speed curve in $\mathbb{R}^3$. We write its tangent vector as

$$T(s) = (\cos\theta(s)\sin\varphi(s), \cos\theta(s)\sin\varphi(s), \sin\theta(s)).$$

By the Frenet-Serret formulae we have that

$$\kappa(s)N(s) = \frac{dT}{ds}(s),$$

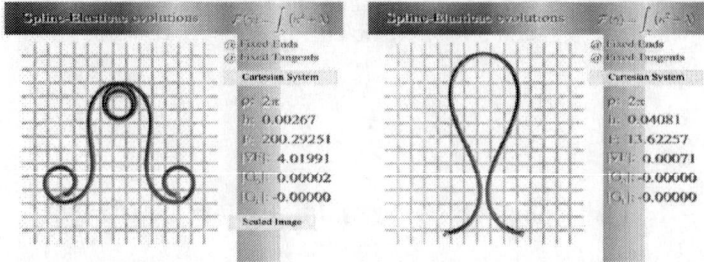

FIGURE 9. Starting from the same curve γ_o, the curve flow looks for the elastica in $\mathbf{R}^2$ under three border conditions: fixed boundary tangents, ...

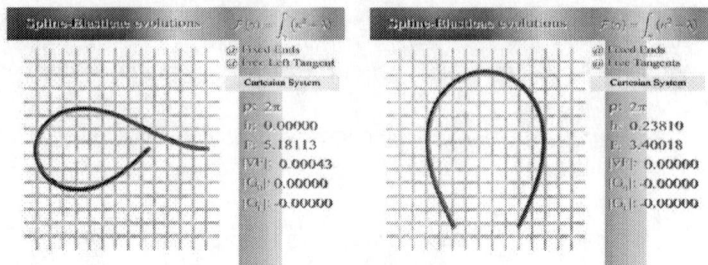

FIGURE 10. ...free left tangent, and border free tangents.

and so the squared curvature can be written by

$$\kappa^2\left(s\right) = \left(\theta'\right)^2 + \left(\varphi'\right)^2 \cos^2\theta\left(s\right).$$

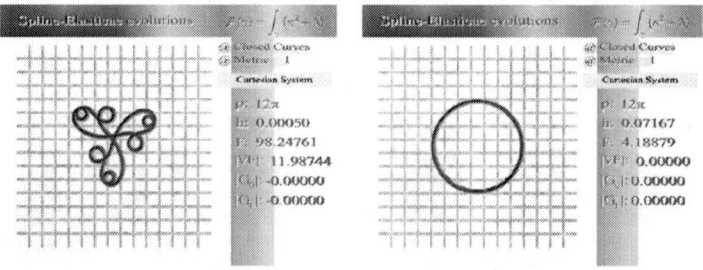

FIGURE 11. Starting from the left curve, rotation index > 0, the gradient flow evolves towards the known circle elastica.

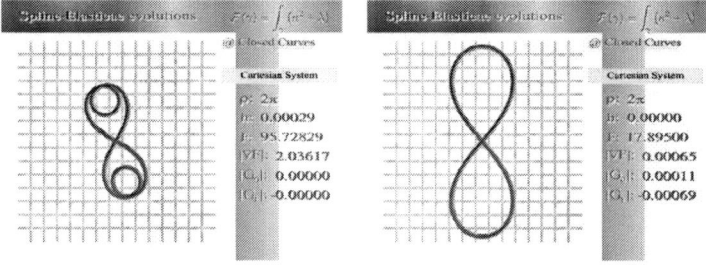

FIGURE 12. If the rotation index is 0, the final Elastica is the figure eight curve.

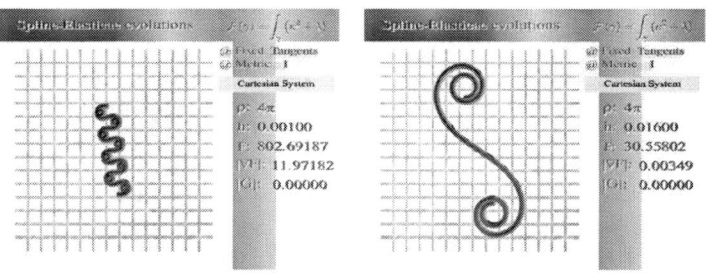

FIGURE 13. The clothoid is an Elastica, obtained under the *fixed total turning* constraint.

In this way, we can define the *elastic energy functional* in $\mathbf{R}^3$ as a functional defined in two spaces $\mathscr{H}_1$ and $\mathscr{H}_2$ of functions θ and φ respectively:

$$\mathscr{F}[\theta,\varphi] = \int_a^b \kappa^2(s)\,ds = \int_a^b \left\{ (\theta')^2 + (\varphi')^2 \cos^2\theta(s) \right\} ds.$$

There are of course more ways to do this. In this problem, the following integrals

$$\mathscr{G}_1[\theta,\varphi] = \int_a^b \cos\theta(s)\sin\varphi(s)\,ds,$$

$$\mathscr{G}_2[\theta,\varphi] = \int_a^b \cos\theta(s)\cos\varphi(s)\,ds,$$

$$\mathscr{G}_3[\theta,\varphi] = \int_a^b \sin\theta(s)\,ds,$$

would give us, if necessary, the constraints to keep the end-points $\gamma(a)$ and $\gamma(b)$ fixed.

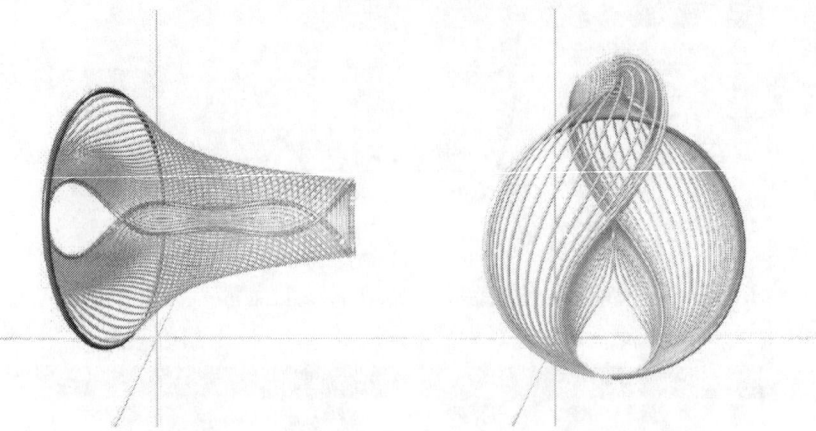

FIGURE 14. Curve flow in looking for an elastic circumference.

5.1. The circumference as absolute minimum

This is a well known fact; see [9] for instance. We may try an experiment about this fact. Let $\gamma_o(s)$ be the curve defined by integrating its tangent vector

$$\gamma_o'(s) = (\sin(4s)\cos(s), \cos(4s)\sin(s), \sin(s))$$

with $s \in [-\pi, \pi]$ under the usual boundary constraints: fixed tangents and points. The gradient flow evolves the curve γ_o towards the circumference of length 2π and so curvature $\kappa = 1$ and energy $\mathscr{F} = 2\pi$.

In the sequence described in 2.6, when the numerical algorithm breaks off in this experiment the deviation with respect to the true solution is less than 2×10^{-5}. See figure 14-(a).

The same happens in 14-(b) but starting from

$$\gamma_o'(s) = (\sin(3s)\cos(s), \cos(3s)\cos(s), \sin(s)).$$

5.2. Circular helical elasticae

It is also known that the circular helices are trivial elastic curves in $\mathbf{R}^3$. In the following two experiments the curves evolve towards elastic helices that verify the imposed constraints: fixed points in both cases, fixed tangents in the first one, figure 15-(a), and free tangents in 15-(b). Both start from $\theta_o(s) = \frac{\pi}{10}$ and $\varphi_o(s) = \frac{3}{2}s + \sin(3s)$ according to the model of section 5.

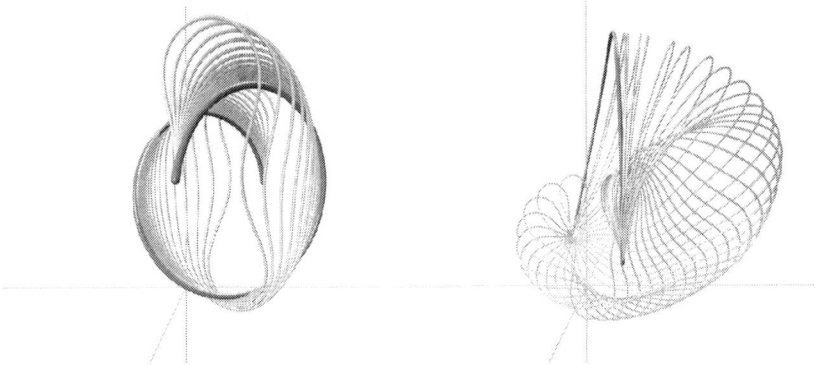

FIGURE 15. Looking for helical elasticae .

6. ELASTIC SURFACES

Elastic surfaces, such as those occurring in biological membranes, interfaces between polymers, or resilient metal plates, are of fundamental interest in science. A simple geometric model, proposed by Sophie Germain around 1810, sets the elastic energy $\mathscr{E}(S)$ of a surface S equal to the integral with respect to surface area of an even, symmetric function of the principal curvatures of S. The surface S may be embedded or immersed in three-space (typically $\mathbf{R}^3$, but possibly another 3-manifold of constant curvature, such as $\mathbf{S}^3$ or $\mathbf{H}^3$), perhaps with volume or boundary constraints. Assumed that the integrand is quadratic, it can be expressed in terms of the mean curvature H and Gauss curvature K as

$$\mathscr{E}[S] = \int_S \left(\lambda + \eta H^2 + \xi K\right) dA. \tag{12}$$

Physically, this formula is called Hooke's Law; η and ξ here are "bending" energies, while λ is a surface tension or "stretching" energy. When the two sides of the elastic surface are distinguished, as in a polymer interface, the assumption of evenness may not be satisfied. This case can be handled by replacing H with $H - H_o$ in Hooke's Law.

An equilibrium elastic surface S is critical for $\mathscr{E}[S]$, subject to the constraints, meaning that for any variation S_t of $S = S_o$, we have

$$\delta\mathscr{E} = \left.\frac{d}{dt}\right|_{t=0} \mathscr{E}[S_t] = 0.$$

Assuming that S is smooth enough, this implies that S satisfies the E-L equation

$$\Delta H + 2\left(H^2 - \left(K - \frac{\lambda}{\eta}\right)\right)H = p,$$

where Δ is the Laplace-Beltrami operator of S and p is a constant ("pressure"), which vanishes in the absence of a volume constraint. Observe that the coefficient ξ of K does

207

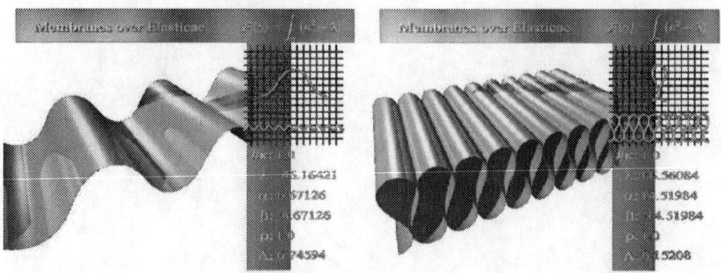

FIGURE 16. Membranes over elasticae.

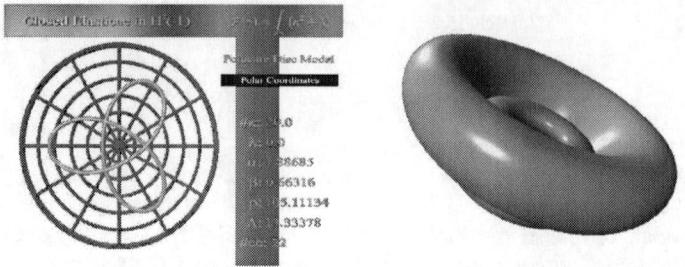

FIGURE 17. Vesicle over a closed hyperbolical elastic curve.

not enter into the E-L equation. Poisson noticed this (1815) decades before Gauss and Bonnet showed that the third term in $\mathscr{E}(S)$ is actually a topological constant:

$$\int_S K dA = 2\pi \chi(S),$$

where $\chi(S)$ is the Euler characteristic of S. Thus it suffices to consider energies of the form

$$\mathscr{E}[S] = \int_S \left(\lambda + \eta (H - H_o)^2 \right) dA.$$

Among many others, we could find here:

1. Minimal surfaces, 3.4;
2. Constant Mean Curvature Surfaces, 3.6;
3. Membranes and Vesicles, 6.1, 6.2;
4. Cell models, 6.3-6.6.

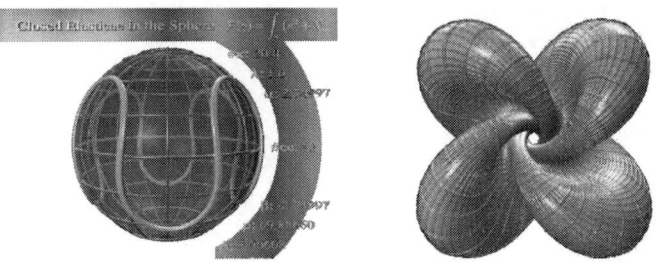

FIGURE 18. Vesicle over a closed spherical elastic curve.

FIGURE 19. Evolution of the profile curve looking for a Willmore Torus. .

FIGURE 20. Evolution of the profile curve looking for a Willmore Torus.

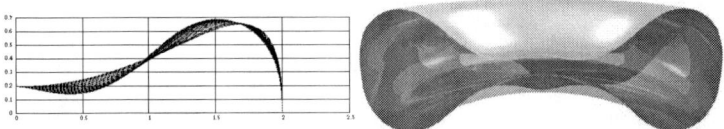

FIGURE 21. Evolution of the profile curve looking for a stable cell with area and volume constraints and fixed endpoints.

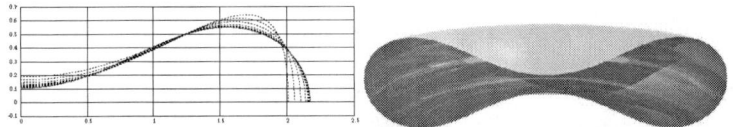

FIGURE 22. Evolution of the profile curve looking for a stable cell with area and volume constraints but free endpoints.

209

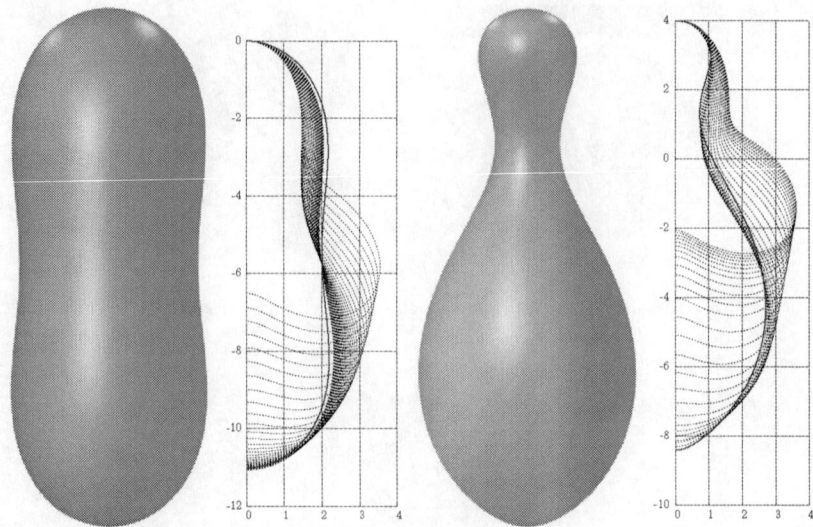

FIGURE 23. (a) Cell under area and volume constraints but free endpoints. (b) Cell with spontaneous curvature $H_o \neq 0$.

6.1. Membranes over elasticae

From [15], cylinders defined on curves, which rest vertically on parallel walls, are membranes (minimums for the elastic energy), if and only if, the profile curve is an planar elastic curve (for different types of constraints). The surfaces of figure 16 can be obtained by both ways, solving the problem as in [15] or by the experimental procedure described here (section 4).

6.2. Vesicles over elasticae

In [9, 18] there are described two ways to get vesicles in $\mathbf{R}^3$, created from closed elasticae in the hyperbolic plane $\mathbf{H}^2\,(-1)$ and the sphere $\mathbf{S}^2\,(1)$ respectively. The details are far from the scope of this lines, but two examples are given in the figures 17 and 18.

6.3. Rotational elastic surfaces

Let S be a rotational surface obtained by revolving the curve $\gamma(t) = (y(t), z(t))$ around the z axis, with $t \in [a, b]$. Take the standard parametrization of S given by

$$X(t, \theta) = (y(t)\cos\theta, y(t)\sin\theta, z(t)),$$

and get the mean curvature H, the area element dA and the volume element dV of the surface S. We write the functional of the *elastic energy*

$$\mathscr{E}[S] = \int_S H^2 dA,$$

in coordinates

$$\mathscr{E}[y,z] = \int_0^{2\pi} \int_a^b H^2(t) \sqrt{(y')^2 + (z')^2}\, dt\, d\theta \tag{13}$$

$$= 2\pi \int_a^b H^2(t) \sqrt{(y')^2 + (z')^2}\, dt, \tag{14}$$

with or without area and volume constraints $\mathscr{G}_1[S] = \int_S dA$ and $\mathscr{G}_2[S] = \int_S dV$.

6.4. Willmore tori with area and volume constraints

Consider the elastic energy functional (13) with the usual constraints. The descending flow of the gradient takes an initial curve given $\gamma_o(t) = (y(t), z(t))$ towards a limit curve such that the torus surface is a minimum for the energy. See, for instance, the figures 19 and 20 obtained by the descendant flow starting respectively from the curves

$$\gamma_o(t) = (5 + 2\cos(t), 3\sin(t)),$$
$$\gamma_o(t) = (9 + (3 + \cos(3t))\cos(t), (3 + \cos(3t))\sin(t)).$$

6.5. Biconcave and axis-symmetric

From [2, 5, 6, 7], an axis-symmetry surface is a closed embedded surface $S \subset \mathbf{R}^3$ with a rotation symmetry and reflection symmetry by the plane perpendicular to the rotation axis. It is biconcave if there are exactly two components of negative Gaussian curvature. Without loss of generality, the rotational axis is labeled as z-axis and the plane of reflexion is the xy-plane. Then the surface S can be obtained by revolving a radial curve about the z-axis on the upper half plane and reflecting it to the lower half.

A stable cell shape corresponds to a curve $\gamma(t) = (f(t), g(t))$, $t \in [a,b]$ (or maybe better $t \in [a, +\infty)$), which satisfies area and volume equations if these were constraints, and results in a minimum value for equation (13).

There are natural boundary conditions imposed by the rotation and reflection symmetries of the surface. The obvious ones are $g'(a) = 0$ and $g'(b) = 0$ (or better, $\lim_{t \to b} g'(t) = 0$).

6.6. Experimental axis-symmetric cells

Let z be the axis of symmetry for an axis-symmetric cell and $\gamma(t) = (f(t), g(t))$ be the parametrization for the cell shape in meridional section (zx-plane). In figures 21, 22 and 23 we take as initial curves

$$\gamma_0(t) = \left(2\cos\left(\frac{t}{4}\right), \frac{1}{16}\left(\pi + (2\pi - t)^2 \sin\left(\frac{t}{4}\right)\right)\right),$$

$$\gamma_0(t) = \left(\left(3 + \cos\left(\frac{3}{2}t\right)\right)\sin\left(\frac{t}{2}\right), \left(3 + \cos\left(\frac{3}{2}t\right)\right)\cos\left(\frac{t}{2}\right)\right),$$

where $t \in [0, 2\pi]$, and check under different constraints the evolution of these curves by means of the Willmore flow in searching for a stable cell of the same area and enclosed volume than the one generated by γ_0. In figure 23-(b) we can see what happens if the spontaneous curvature H_o is set $H_o \neq 0$.

REFERENCES

1. A. J. Brizard, *Introduction to Lagrangian and Hamiltonian Mechanics*, Department of Chemistry and Physics Saint Michaels College, Colchester, VT 05439.
2. P. B. Canham, *J. Theor. Biol.* **26**, 61-81 (1970).
3. M. P. do Carmo, *Differential Geometry of Curves and Surfaces*, Prentice-Hall, 1976.
4. L. Euler, *Methodus inveniendi lineas curves maximiminimive proprietate gaudentes, sive Solutio-problematis isoperimitrici latissimo sensu accepti*, Lausanne, Geneva, 1744.
5. W. Helfrich, *Z. Naturforsch. C* **28**, 693-703 (1973).
6. K. H. Adams, *Biophys. J.* **13**, 1049-1053 (1973).
7. Th. Kwok-keung Au and T. Yau-heng Wan, An Analysis on the Shape Equation for Biconcave Axisymmetric Vesicles, eprint arXiv:math/0001103, (2000).
8. J. Langer and D. A. Singer, *J. Diff. Geom.* **20**, 1-22 (1984).
9. J. Langer and D. A. Singer, *J. London Math. Soc.* **16**, 512-520 (1984).
10. J. Langer and D. A. Singer, *Bull. London Math. Soc.* **16**, 531-534 (1984).
11. A. Linnér, *Ann. Global Anal. Geom.* **16**, 445-475 (1998).
12. A. Linnér, *Trans. Amer. Math. Soc.* **350**, 3743-3765 (1998).
13. A. Linnér, *Gradients, preferred metrics and asymmetries*,
    http://www.math.niu.edu/~alinner/
14. A. Linnér, *Generating Optimal Curves via the C++ Standard Library*,
    http://www.math.niu.edu/~alinner/
15. J. C. C. Nitsche, *Quart. Appl. Math.* **LI**, 363-387 (1993).
16. W. H. Press, S. A. Teukolsky, W. T. Vetterling and B.P. Flannery, *Numerical Recipes in C*, Cambridge University Press, 2002.
17. J. Oprea, *Differential Geometry and its Applications*, Second edition. Classroom Resource Materials Series. Mathematical Association of America, Washington, DC, 2007.
18. U. Pinkall, *Invent. Math.* **81**, 379-386 (1985).
19. R. C. Veltkamp and W. Wesselink, *Eurographics* **14**, 97-109 (1995).

Computational methods in Mathematics

José Carlos Díaz Ramos

Department of Mathematics, University College Cork, Ireland

Abstract. These are the notes of an introductory course on the programming language of *Mathematica*. We use this tool to explore some elementary geometric concepts in Differential Geometry.

Keywords: *Mathematica*, Differential Geometry, Curves and Surfaces

GETTING STARTED WITH *MATHEMATICA*

It is important to point out here that these are the notes of course on *Mathematica* 5. Although, most of the commands explained can be run in any version of *Mathematica*, some of them could produce different behaviour. This is particularly true of the graphics section: some of the functionality of graphics has been changed in *Mathematica* 6. It is not so difficult to adapt things, however, and the code given in these notes should be general enough to figure out how to rewrite the commands on other versions.

This is just an introduction to *Mathematica*. A more complete description can be seen in [4].

The author has been supported by a Marie Curie Intra-European Fellowship and project PGIDIT06PXIB207054PR (Spain).

Kernel and Front End

Mathematica consists of two parts: the Kernel and the Front End.

The Kernel of *Mathematica* is responsible for doing the calculations and handling the data. It is the essential part of *Mathematica* and can be accessed as a stand-alone programme. However, interacting directly with the Kernel is usually tedious and it is not easy to take advantage of its full power. For that reason one almost never starts the Kernel directly and runs *Mathematica* through the Front End.

The Front End is essentially the Windows application that one runs to use *Mathematica*. The Front End is a kind of word processor that is capable of sending and retrieving information from the Kernel. It allows a convenient way of displaying results of calculations, graphics, text and other elements, presenting the results in a readable way.

Mathematica was designed to supply a powerful calculation tool for mathematicians and other scientists; this capability is accomplished by the Kernel. However, the Front End expanded *Mathematica* applications to other areas, such as education, because of its power to display and handle data in an interactive and visual way. When using *Mathematica* we should aim not only to perform the calculations we need but also to use the capabilities of the Front End to explain the problems that are tackled and the

CP1002, *Curvature and Variational Modeling in Physics and Biophysics*
edited by O. J. Garay, E. García-Río, and R. Vázquez-Lorenzo
© 2008 American Institute of Physics 978-0-7354-0521-9/08/$23.00

results that are obtained. By using this approach, one ensures that the data is understood in the future and that other people can take advantage of it.

Running *Mathematica*

After running the *Mathematica* Front End, take a while to get familiar with the appearance of the application. It looks pretty much like any other windows-like application: it has a title bar, a menu bar and a client window where we will type the text and commands and get the results from the Kernel. These client windows (or more precisely, the documents displayed in them) are called *notebooks*.

You open, save and print documents in the "File" menu; copy, cut and paste in the "Edit" menu; select fonts, colours and text effects in the "Format" menu; search and replace text in the "Find" menu...

Using the Help Browser

Go to "Help" and select "Help Browser...". The *Mathematica* Help Browser appears. Take some time to get familiar with it. By using it you get immediate access to thousands of pages of documentation and examples. No one remembers all the commands by heart, so you want to be efficient in finding the information you want.

Use the text field to enter search terms. Click "Go" to display the associated notebook. The highlighted items in the columns track your search path. The resulting help notebook is displayed below the columns. You can evaluate input, copy and paste from the help browser to another notebook, or print out information with the "Print" menu command.

Use the buttons for different search categories. The typical categories you will use are "Built-in functions" (information on built-in functions in the *Mathematica* kernel, including further examples), "The *Mathematica* Book" (the complete text of Stephen Wolfram's *The Mathematica Book* [4] which provides further explanations of the various *Mathematica* aspects) and "Add-ons & Links" (information on application packages installed with your *Mathematica* system). You may want to find out about menu commands in "Front End" or look up examples in "Demos", "Getting started" or "Tour".

To get information on a built-in function: with the cursor immediately after a function name, press F1; the Help Browser opens and displays the information from "Built-in functions".

Your first Mathematica calculations

You can use *Mathematica* just like a calculator. Type your input, press [SHIFT]-[RETURN] or [ENTER], and *Mathematica* returns the answer.

At this point, note how the *Mathematica* Kernel is called the first time you run a calculation: a new button appears in the task bar representing the fact that another applications is running. The first time you run a calculation with *Mathematica*, you will have to wait for a few seconds until the Kernel is loaded. The Kernel will run while the Front End is open or until you explicitly close it down. When you exit the Front End, the Kernel is automatically closed.

We are about to start our first Mathematica calculations. Type 3+5 and execute the command (by pressing [SHIFT]-[RETURN] or [ENTER]).

```
In[1]:= 3+5
Out[1]= 8
```

It is easy to get familiar with the basic arithmetic operations: 2+3, 2-3, 2/3, 2^3. These represent multiplication: a*b, a[SPACE]b, a(b+1); 2x means 2*x. Some usual functions are: Sin Cos Tan Sec Sinh ArcTan ArcSech Exp Log... Use the help browser to obtain further information on these functions and find related ones. The number π is represented by Pi and the Euler number e by E. All *Mathematica* built-in functions have their first character capitalised. Moreover, if a function takes parameters (such as the previous ones), then these parameters are enclosed by square brackets.

```
Out[1]= 3
```

```
In[2]:= (3 Sin[Pi/4]+1)^2
```
$$\text{Out[2]}= \left(1+\frac{3}{\sqrt{2}}\right)^2$$

Unless you explicitly say so, *Mathematica* always tries to give an exact result. If you want an approximation use N.

```
In[3]:= N[(3 Sin[Pi/4]+1)^2]
Out[3]= 9.74264
```

You can get any precision you want from an *exact* number.

```
In[4]:= N[(3 Sin[Pi/4]+1)^2,50]
Out[4]= 9.7426406871192851464050661726290942357090156261308
```

```
In[5]:= N[Pi,200]
Out[5]= 3.1415926535897932384626433832795028841971693993751058209749445923078164062862089986280348253421170679821480865132823066470938446095505822317253594081284811174502841027019385211055596446229489549303820
```

The imaginary unit is represented by I. You can perform calculations with complex numbers as well.

```
In[6]:= E^(-I Pi)
Out[6]= -1
```

```
In[7]:= N[Log[3+I],10]
Out[7]= 1.1512925465 + 0.3217505544 I
```

Stopping calculations

Although *Mathematica* is a powerful calculating tool, it has its limits. Sometimes it will happen that the calculations you tell *Mathematica* to do are too complicated, or maybe the output produced is too large. In these cases, *Mathematica* could be calculating for too long to get an output so you might want to stop these calculations. It is important to know how to do this. *To abort a calculation*: go to "Kernel" and select "Abort evaluation" or press [ALT]-[.].

It can take long to abort a calculation. If the computer does not respond an alternative is to close down the Kernel. By doing this you do not lose the data displayed in your notebooks but you do lose all the results obtained so far from the Kernel, so in case your are running a series of calculations, you would have to start again. *To close down the Kernel*: go to "Kernel" and select "Quit Kernel" and then "Local".

Closing down the Kernel is not a practise that is done only when you want to stop a calculation. Sometimes, when you have been using *Mathematica* for a long time you forget about the definitions and calculations that you have done before (maybe you have defined values for variables or functions, for example). Those definitions can clash with the calculations you are doing, so you might want to close down the Kernel and start your new calculations from scratch. In general, it is a good idea to close down the Kernel after you have finished with a series of calculations, so that when you move to a different problem your new calculations do not interfere with the previous ones.

Introduction to notebooks

Before continuing with the computational capabilities of *Mathematica*, we discuss a bit further how notebooks work. The purpose of this is to present a convenient way of documenting and formatting your code in a more mathematical way. We will content ourselves with the basic capabilities.

Cells and styles

Each Mathematica notebook consists of *cells*. A cell is essentially each of the paragraphs that you see in a notebook. They can be identified by the bracket on the right-hand side of the cell (sometimes these brackets may not appear because of the formatting and visualization options of the notebook or the cell). Each cell has a style. The style can be selected by clicking on the bracket and going to the "Format" menu and then "Style". The most styles are "Input", "Output", "Text" and others such as "Section", "Subsection" and so on. By default, the style of a cell is "Input" which is the style used for entering commands for the Kernel. When the Kernel returns a result, it appears in a cell of style "Output".

To create a new cell move the pointer in the notebook window until it becomes a horizontal I-beam. Click, and a cell insertion bar will appear so you can start typing. If you want to create a "Text" cell, choose "Format->Style->Text" or use the keyboard shortcut [ALT]-[7] before starting to type. To use a different style like section, the procedure is analogous: you just have to select the corresponding style.

Cells can be grouped together. This is essentially done by *Mathematica* although the default behaviour may be changed. For example, a "Section" will group all subsequent cells until a new "Section" (or maybe "Chapter") appears. When various cells are grouped, a larger bracket encloses all the other cells. To close a group of cells double-click the outermost cell bracket of the group. Double-click a closed group's cell bracket to open again.

Formatting text

The text displayed in any cell can be formatted in several different ways. You will typically format text only in "Text" cells, although this can be done with other styles as well. The way to do this is pretty much the same as in any text processor: select the text you want to format and choose the corresponding effect from the "Format" menu.

How to enter notation

You can enter mathematical formulas within the text. You can also write some of the commands in a more mathematical-like notation. We briefly explain how to do this.

The more straightforward way of entering mathematical notation is by means of palettes. If the "Basic Input" palette is not open, choose it with "File->Palettes- >BasicInput". The palette "Basic Typesetting" might also be useful for more sophisticated typing. The use of these palettes is quite obvious: just click on the symbol or object you want. Sometimes you will have a placeholder that will allow you to select an object before entering the notation; you can move to a different place holder with the [TAB] key or with the arrow keys. Note that, by means of this palette you have also gained access to more *Mathematica* commands.

Although palettes provide an easy way to enter notation, you might find it useful to learn by heart the shortcuts for some of the most common mathematical operations. For example, for entering the power of a number use [CTRL]-[∧] instead of "∧"; this way you can type 2^3 as 2[CTRL]-[∧]3 instead of $2^\wedge3$, for instance. For fractions use [CTRL]-[/] instead of "/", and for subscripts use [CTRL-[_]. An alias for the imaginary unit I is [ESC]-[ii]-[ESC], and for the Euler number E we can also use [ESC]-[ee]-[ESC]. Greek letters are entered by typing the roman character preceded and followed by [ESC]. For example [ESC]-[a]-[ESC] gives α and [ESC]-[p]-[ESC] gives π, which incidentally is another notation for Pi. It is also useful to remember that most of the symbols have a LaTeX alias. For example you could enter α as [ESC]-[\alpha]-[ESC] or $\oplus$ as [ESC]-[\oplus]-[ESC].

To type a formula into a text cell: place the cursor in the text cell where you want to enter the formula. Press [CTRL]-[(]. A placeholder box appears in the cell. Enter the formula. Press [CTRL]-[)] to end the formula. Note that [CTRL]-[(] and [CTRL]-[)] have a similar effect to $...$ in LaTeX.

BASIC PROGRAMMING WITH *MATHEMATICA*

When handling mathematical expressions the following functions are often useful: Expand (expands out products and positive integer powers), Factor (factors a polynomial over the integers) and Simplify (performs a sequence of algebraic and other transformations and returns the simplest form it finds; be careful when you use this function because the term 'simplest' may be something you do not expect). There are other related functions. See [4, Section 1.4.5] for details. These are a few examples of how the above functions work.

$\mathbf{In[1]} := \mathtt{Expand}[(-1+x^4)(1+x^4)(1+x^8)]$
$\mathbf{Out[1]} = -1+x^{16}$

$\mathbf{In[2]} := \mathtt{Factor}[-1+x^{16}]$
$\mathbf{Out[2]} = (-1+x)(1+x)(1+x^2)(1+x^4)(1+x^8)$

$\mathbf{In[3]} := \mathtt{Simplify}[(-1+x)(1+x)(1+x^2)(1+x^4)(1+x^8)-(-1+x^{16})]$
$\mathbf{Out[3]} = 0$

Variables and functions

Defining variables

In the above calculations, x is considered as a symbol. Any word can be a symbol unless it clashes with the name of a built-in function. However, all built-in functions have their first character capitalised, so you will not provoke clashes with built-in functions if your variables are written in lower case. Remember that *Mathematica* is a case-sensitive language. The symbols you use can have almost any name. There is no limit on the length of their names. One constraint, however, is that variable names can never start with numbers. For example, $x2$ could be a symbol, but $2x$ means $2*x$.

When you do long calculations, it is often convenient to give names to your intermediate results. Just as in standard mathematics, or in other computer languages, you can do this by introducing named variables.

This sets the value of the variable x to be 5.

$\mathbf{In[4]} := \mathtt{x=5}$
$\mathbf{Out[4]} = 5$

Now, the value of x is used for the calculations. Whenever x appears, Mathematica now replaces it with the value 5.

$\mathbf{In[5]} := (-1+x^4)(1+x^4)(1+x^8)$
$\mathbf{Out[5]} = 152587890624$

This assigns a new value to x.

$\mathbf{In[6]} := \mathtt{x=7+4}$

218

Out[6]= 11

In[7]:= $(-1+x^4)(1+x^4)(1+x^8)$
Out[7]= 45949729863572160

If you want a variable to work again as a variable, you have to delete its value. This is done with Clear.

In[8]:= Clear[x]

In[9]:= $(-1+x^4)(1+x^4)(1+x^8)$
Out[9]= $(-1+x^4)(1+x^4)(1+x^8)$

It is very important to take into account that values you assign to variables are permanent. Once you have assigned a value to a particular variable, the value will be kept until you explicitly remove it. The value will, of course, disappear if you start a whole new *Mathematica* session. Forgetting about definitions you made earlier is the most common cause of mistakes when using *Mathematica*. To avoid mistakes, you should *remove values you have defined as soon as you have finished using them.*

You can type formulas involving variables in Mathematica almost exactly as you would in mathematics. There are a few important points to watch, however: x y means x times y, xy with no space is the variable with name xy, 5x means 5 times x, x^2y means (x^2) y, not x^(2y).

Defining functions

So far we have seen built-in functions of *Mathematica*. Some of them operate just as numeric functions (for example, Sin), others do more complicated operations (such as $\sum_{\square=\square}^{\square} \square$) and others perform procedures (for example, Expand). You can define your own functions in *Mathematica*.

The syntax to define a function in *Mathematica* is "f [x1_, x2_, ...]=*expression involving the variables x1, x2,...*". Instead of the assignment operator "=" you may want to use ":=". We explain later the difference. For the moment we just define a few basic functions.

This defines the function $f(x) = x^2$. Incidentally, note the semicolon to finish the sentence. This prevents *Mathematica* from printing the result of the calculation.

In[10]:= f[x_]=x^2;

Now you can use this function as any other built-in function.

In[11]:= N[f[π]]
Out[11]= 9.8696

In[12]:= Expand[f[3x+1]]
Out[12]= $1+6x+9x^2$

The names like f that you use for functions in *Mathematica* are just symbols. Because of this, you should make sure to avoid using names that begin with capital letters, to prevent confusion with built-in *Mathematica* functions. You should also make sure that you have not used the names for anything else earlier in your session.

Again, if you want to delete the definition of a function, use `Clear`.

`In[13] := Clear[f]`

Now, there is no definition for f.

`In[14] := f[x]`
`Out[14]= f[x]`

A function can have more than one argument.

`In[15] := f[x_,y_,z_]=x y Cos[z];`

You can mix numerical values, symbols and expressions.

`In[16] := f[3,(x+1)², π]`
`Out[16]= −3(1+x)²`

Again, before starting another use of f, we clear the previous definition.

`In[17] := Clear[f]`

It is a common beginners' mistake to forget the underscores in the definition of a function. In that case you could have an unexpected behaviour.

The assignment operators = and :=

When defining mathematical functions, the operator "=" works well. However, if your functions are more like procedures, that is, they perform manipulations of their arguments instead of just mathematical calculations, the operator "=" does not work.

This is an attempt to define a procedure that takes a value, adds one, squares the result and expands it.

`In[18] := f[x_]=Expand[(1+x)²];`

But it does not work:

`In[19] := f[a+1]`
`Out[19]= 1+2(1+a)+(1+a)²`

You can look at the definitions that *Mathematica* saves in memory using "?".

`In[20] := ?f`
`Global`f`
`f[x_]=1+2x+x²`

The assignment operator "=" always evaluates the right-hand side of the assignment before saving the definition in memory. For this reason, when *Mathematica* evaluates `Expand[(1+x)^2]`, the result is $1+2x+x²$, and this is the actual expression that is stored in memory. To prevent *Mathematica* from evaluating the right hand side of the assignment use ":=".

```
In[21]:= Clear[f]
```

```
In[22]:= f[x_]:=Expand[(1+x)^2];
```

```
In[23]:= ?f
Global`f
f[x_]:=Expand[(1+x)^2]
```

Now, each time you evaluate f, *Mathematica* will perform the whole procedure.

```
In[24]:= f[a+1]
Out[24]= 4+4a+a^2
```

As you can see from the example above, both "=" and ":=" can be useful in defining functions, but they have different meanings, and you must be careful about which one to use in a particular case. One rule of thumb is the following. If you think of an assignment as giving the final value of an expression, use the "=" operator. If instead you think of the assignment as specifying a command for finding the value, use the ":=" operator. If in doubt, it is usually better to use the ":=" operator rather than the "=" one.

Lists

In doing calculations, it is often convenient to collect together several objects, and treat them as a single entity. Lists give you a way to make collections of objects in *Mathematica*. In particular, they are used to represent vectors, matrices and tensors. A list such as $\{3,5,1\}$ is a collection of three objects. But in many ways, you can treat the whole list as a single object. You can, for example, do arithmetic on the whole list at once, or assign the whole list to be the value of a variable.

This assigns v to be a list. Note that the elements of the list can be of completely different nature.

```
In[25]:= v={3,2x,π+1};
```

This a basic arithmetic operation

```
In[26]:= 2v+{1,2,3}
Out[26]= {7,2+4x,3+2(1+π)}
```

Basic operations with lists

You can create list by hand, but there are other methods to create lists from rules or functions. A very handy function for doing so is Table. The command Table[*expr*,{i,n1,n2,step}] evaluates *expr* for the different values of *i* starting at $n1$, finishing at $n2$ and using the step *step*. If you omit *step*, it is taken to be 1.

This gives a table of the values of i^2, with i running from 1 to 6.

```
In[27]:= Table[i^2,{i,1,6}]
Out[27]= {1,4,9,16,25,36}
```

You can also make tables that involve several parameters. Take a moment to examine how the indexes run.

```
In[28]:= Table[x[i,j],{i,1,4},{j,1,5}]
Out[28]= {{x[1,1],x[1,2],x[1,3],x[1,4],x[1,5]},
{x[2,1],x[2,2],x[2,3],x[2,4],x[2,5]},
{x[3,1],x[3,2],x[3,3],x[3,4],x[3,5]},
{x[4,1],x[4,2],x[4,3],x[4,4],x[4,5]}}
```

The table in this example is a list of lists. The elements of the outer list correspond to successive values of i. The elements of each inner list correspond to successive values of j, with i fixed.

This gives a list of four pseudo-random numbers. Table re-evaluates Random[] for each element in the list, so that you get a different pseudo-random number. So, in Table[*expr*,{i,n1,n2,step}] the expression *expr* is evaluated each time i takes a different value.

```
In[29]:= Table[Random[],{i,1,4}]
Out[29]= {0.588668,0.936042,0.411855,0.889148}
```

To access an element of a list use the operator "[[ ]]". For example, list[[i]] will access the ith element of list.

```
In[30]:= v=Table[i,{i,1,4}]
Out[30]= {1,2,3,4}
```

```
In[31]:= v[[3]]
Out[31]= 3
```

If you have a multidimensional list, you can access the element $(i1,i2,\ldots)$ by typing [[i1,i2,...]].

```
In[32]:= m=Table[i j,{i,1,3},{j,1,4}]
Out[32]= {{1,2,3,4},{2,4,6,8},{3,6,9,12}}
```

```
In[33]:= m[[1,2]]
Out[33]= 2
```

In a 2-dimensional list, each element is another list, and you can access it using "[[ ]]". For example, this gives the second row of m.

```
In[34]:= m[[2]]
Out[34]= {2,4,6,8}
```

Another useful command is Length, that gives the length of a list.

```
In[35]:= Length[v]
```
Out[35]= 4

The length of m is 3, the number of rows.

```
In[36]:= Length[m]
```
Out[36]= 3

However, each row has length 4, the number of columns of m.

```
In[37]:= Length[m[[1]]]
```
Out[37]= 4

Vectors and matrices

Vectors and matrices in *Mathematica* are simply represented by lists and by lists of lists, respectively.

This represents the vector (x, y, z).

```
In[38]:= v={x,y,z};
```

This is a 3×3 matrix.

```
In[39]:= m=Table[a[i,j],{i,3},{j,3}]
```
Out[39]= $\{\{a[1,1], a[1,2], a[1,3]\}, \{a[2,1], a[2,2], a[2,3]\}, \{a[3,1], a[3,2], a[3,3]\}\}$

The sum and product by scalars is denoted in the usual way

```
In[40]:= 2v
```
Out[40]= $\{2x, 2y, 2z\}$

```
In[41]:= v+{1,2,3}
```
Out[41]= $\{1+x, 2+y, 3+z\}$

```
In[42]:= 2m+Table[i j,{i,3},{j,3}]
```
Out[42]= $\{\{1+2a[1,1], 2+2a[1,2], 3+2a[1,3]\},$
$\{2+2a[2,1], 4+2a[2,2], 6+2a[2,3]\},$
$\{3+2a[3,1], 6+2a[3,2], 9+2a[3,3]\}\}$

The product of matrices and vectors is performed by the operator ".".

```
In[43]:= m.v
```
Out[43]= $\{xa[1,1]+ya[1,2]+za[1,3], xa[2,1]+ya[2,2]+za[2,3],$
$xa[3,1]+ya[3,2]+za[3,3]\}$

```
In[44]:= v.v
```
Out[44]= $x^2 + y^2 + z^2$

```
In[45]:= v.m
```
Out[45]= $\{xa[1,1]+ya[2,1]+za[3,1],xa[1,2]+ya[2,2]+za[3,2],$
$xa[1,3]+ya[2,3]+za[3,3]\}$

Be aware that depending on the side the vector v is, *Mathematica* will treat v as a column vector or a row vector. In particular v.w will represent the scalar product of v and w. Because of the way Mathematica uses lists to represent vectors and matrices, you never have to distinguish between row and column vectors.

To display matrices in a more mathematical way, use the function MatrixForm.

```
In[46]:= MatrixForm[m]
```
Out[46]//MatrixForm=

$$\begin{pmatrix} a[1,1] & a[1,2] & a[1,3] \\ a[2,1] & a[2,2] & a[2,3] \\ a[3,1] & a[3,2] & a[3,3] \end{pmatrix}$$

```
In[47]:= MatrixForm[m.v]
```
Out[47]//MatrixForm=

$$\begin{pmatrix} xa[1,1]+ya[1,2]+za[1,3] \\ xa[2,1]+ya[2,2]+za[2,3] \\ xa[3,1]+ya[3,2]+za[3,3] \end{pmatrix}$$

If you want to enter a matrix in this way, a simple way of doing so is by selecting "Create Table/Matrix/Palette" in the "Input" menu.

The table below shows some of the mathematical operations that *Mathematica* can perform on matrices.

c m	multiply by a scalar
a.b	matrix product
Inverse[m]	matrix inverse
Det[m]	determinant
Tr[m]	trace
Transpose[m]	transpose
Eigenvalues[m]	eigenvalues

To finish this section it is important to take the following fact into account. Whenever you want to multiply two matrices you have to use the operator ".". A blank space will *not* do. Instead, a blank space (or "*") will multiply matrices element-wise. Although this might cause some problem at first, it is convenient that *Mathematica* has this capability, which is known as *listability*. We discuss this in the following section.

Functional programming

It usually happens that one has to apply a function or procedure to a list of expressions or to each of the expressions themselves. Unlike classical programming languages,

Mathematica provides a way of treating lists of expressions at once. This implies that programmes will be more readable and that these functions are applied more efficiently. Although it might seem at first that functional programming is not so useful, it turns out that it is one of the most powerful tools provided by *Mathematica*.

Applying functions to lists

In an expression like f[{a, b, c}] you are giving a list as the argument to a function. This is different from typing f[a,b,c]. When you need to apply a function directly to the elements of a list you have to use Apply.

This makes each element of the list an argument of the function f.

```
In[1]:= Apply[f,{a,b,c}]
Out[1]= f[a,b,c]
```

There is a shortcut for Apply: Apply[f,{a1,a2,...}] can be written as f@@{a1,a2,...}. In what follows we will use this notation. However, Apply can have more than two arguments (basically, you can determine how deep the function will be applied) but we will not discuss that situation. See the online help for further details [4, Section 2.2.3].

```
In[2]:= f@@{a,b,c}
Out[2]= f[a,b,c]
```

We are going to have a short digression on how *Mathematica* represent expressions internally.

Each *Mathematica* expression has an internal form. It is often useful to know the particular internal form of commonly used functions. For example, the "+" operator has the internal name Plus, whereas the "*" operator has the internal name Times. You can see the internal form of an expression using FullForm.

```
In[3]:= FullForm[(a+b)c+d]
Out[3]//FullForm= Plus[Times[Plus[a, b], c], d]
```

```
In[4]:= FullForm[{a,b,c}]
Out[4]//FullForm= List[a, b, c]
```

So, internally, all *Mathematica* operators work exactly like functions. This allows you to type efficient code for certain operations on lists. For example, if you want to add up all the elements of a list, you can just type

```
In[5]:= Plus@@{a,b,c}
Out[5]= a+b+c
```

If you wan to multiply then, type

```
In[6]:= Times@@{a,b,c}
Out[6]= a b c
```

Although it will not be relevant in this notes, if you are going to manipulate the internal forms of expressions you will need to get familiar with the way *Mathematica* transforms those expressions internally. For example, there is no function for the "-" operator. Instead, an expression like $a - b$ is represented as $a + (-b)$:

```
In[7]:= FullForm[a-b]
Out[7]//FullForm= Plus[a, Times[-1, b]]
```

Also, note that for efficiency reasons, *Mathematica* might modify the ordering of an expression. The rule of thumb is that, whenever possible, *Mathematica* will arrange expressions in lexico-graphic order:

```
In[8]:= FullForm[c+b+a]
Out[8]//FullForm= Plus[a, b, c]
```

As an example of the function `Apply`, we provide a function to calculate the mean of the elements of a list.

```
In[9]:= mean[l_]:=(Plus@@l)/Length[l];
```

```
In[10]:= mean[{a,b,c}]
```
$$\text{Out[10]}= \frac{1}{3}(a+b+c)$$

```
In[11]:= mean[Table[i,{i,1,10}]]
```
$$\text{Out[11]}= \frac{11}{2}$$

Applying functions to elements of lists

Instead of performing an operation on the whole list, we are now interested in applying a function to *each* element of the list. You can do this using `Map`.

```
In[12]:= Map[f,{a,b,c}]
```
$$\text{Out[12]}= \{f[a],f[b],f[c]\}$$

There is also a shortcut for `Map`: `Map[f,{a1,a2,...}]` can be written as `f/@{a1,a2,...}`. In what follows we will use this notation. Also, `Map` can have more than two arguments. See [4, Section 2.2.4] for details.

```
In[13]:= f/@{a,b,c}
```
$$\text{Out[13]}= \{f[a],f[b],f[c]\}$$

Another example. We define a function that takes the first element of a list.

```
In[14]:= f[l_]:=l[[1]];
```

This can be used to take the first column of a matrix

$$\text{In[15]} := \text{f}/@\begin{pmatrix} a & b \\ c & d \end{pmatrix}$$
$$\text{Out[15]} = \{a,c\}$$

Some built-in functions have the attribute *listable* which means that you do not have to use Map in order to apply the same function to all the members of a list. Times is one of those functions, so this is why

$\text{In[16]} := \lambda\{a,b,c\}$
$\text{Out[16]} = \{a\lambda,b\lambda,c\lambda\}$

works like

$\text{In[17]} := \text{f}[\text{x\_}]=\lambda \text{ x}; \text{f}/@\{a,b,c\}$
$\text{Out[17]} = \{a\lambda,b\lambda,c\lambda\}$

More surprisingly: this also works for Plus.

$\text{In[18]} := \lambda+\{a,b,c\}$
$\text{Out[18]} = \{a+\lambda,b+\lambda,c+\lambda\}$

You might want to find out about the attribute Listable in the online help [4, Section 2.6.3].

Pure functions

When you use functional operations such as Apply and Map, you always have to specify a function to apply. In all the examples above, we have used the "name" of a function to specify it. Pure functions can be applied to arguments without having to define explicit names for the functions.

We begin with an example. The following lines define the same function.

$\text{In[19]} := \text{f}[\text{x\_}]=x^2;$

$\text{In[20]} := \text{g}=\text{Function}[\{\text{x}\},x^2];$

$\text{In[21]} := \text{h}=\#^2\&;$

$\text{In[22]} := \{\text{f}[a],\text{g}[a],\text{h}[a]\}$
$\text{Out[22]} = \{a^2,a^2,a^2\}$

So, a definition such as f[x1_,x2_,...]=*expression* can also be typed as f=Function[{x1,x2,...},*expression*], where in both cases, *expression* will ultimately contain the variables x1,x2,...

The last format is similar to the use of Function, but in that case the variables are compulsorily given the names #1,#2,... Note that the expression must end with the operator &, which has very low precedence so that in most cases it is not necessary to use brackets. Since the names of arguments in functions are irrelevant, *Mathematica* allows you to avoid using explicit names for the arguments of pure functions, and instead to specify the arguments by giving "slot numbers" #n (or Slot[n]). In a *Mathematica* pure function, #n stands for the *n*th argument you supply. # stands for the first argument. When you use short forms for pure functions, it is very important that you do not forget the ampersand. If you leave the ampersand out, *Mathematica* will not know that the expression you give is to be used as a pure function.

Coming back to example where we extracted the first column of a matrix, we can write that definition in the more concise way:

$$\texttt{In[23]:= \#[[1]]\&/@}\begin{pmatrix} a & b \\ c & d \end{pmatrix}$$

$$\texttt{Out[23]=} \{a, c\}$$

You can use a pure function the same way you would use any other function. For example, there was no need to define g or h before if we just wanted to evaluate at a.

$$\texttt{In[24]:= }\{\texttt{f[a],Function[\{x\},}x^2\texttt{][a],}\#^2\texttt{\&[a]}\}$$
$$\texttt{Out[24]=} \{a^2, a^2, a^2\}$$

If you are going to use a particular function repeatedly, then you can define the function using f[x_]:=*body*, and refer to the function by its name f. On the other hand, if you only intend to use a function once, you will probably find it better to give the function in pure function form, without ever naming it.

```
In[25]:= Clear[f,g,h]
```

Transformation rules

Sometimes, you will want to substitute the value of one variable for some number or another expression, but you do not want to give that value to the particular variable you want to evaluate. You can do this by using a rule. A rule is like an assignment, but it only works for the expression you apply the rule. The basic syntax of a rule is expr/.x->*value* which means that you substitute *x* for *value* for every occurrence of *x* in *expr*. The operator "/." is a shortcut for the function ReplaceAll, and the operator "->" can also be represented as "→" or Rule.

For example, you want to check that $x = 1$ is a solution of $x^2 - 3x + 2 = 0$.

```
In[26]:= x²-3x+2/.x→1
Out[26]= 0
```

But the value of *x* has not changed.

```
In[27]:= x
Out[27]= x
```

If you want to apply more than one rule at once, you have to use a list of rules.

$$\texttt{In[28]:= x y Cos[z]/.}\{x\to 3, y\to (x+1)^2, z\to \pi\}$$

`Out[28]=` $-3(1+x)^2$

Rules can be more complicated. You can give rules for functions as well. Essentially, everything that works for "=" works as well with rules, but without the permanent effect of a definition.

`In[29]:=` f[x]+(a+f[3+y])^2/.f[x_]→ x^2
`Out[29]=` $x^2 + \left((y+3)^2 + a\right)^2$

You can define rules for built-in functions:

`In[30]:=` $x^2 + \left(1+y^2\right)^2/.x\_^2 → 5x$
`Out[30]=` $5x + 5\left(y^2 + 1\right)$

Note that, in the above entry, the rule has been applied for the outer square of $\left(1+y^2\right)^2$ but not for y^2. The operator tries the rule for every part of the expression, but once the rule matches, it is no longer applied to any subexpression. Thus, when you use `expr/.rules`, each rule is tried in turn on each part of *expr*. As soon as a rule applies, the appropriate transformation is made, and the resulting part is returned. If you want to apply a rule more than once you will have to use "//.". We discus this latter.

Here is another example. The rule for x^3 is tried first; if it does not apply, the rule for x^{n-} is used.

`In[31]:=` $\left\{x^2, x^3, x^4\right\}/.\left\{x^3 → u, x^{n-} → p[n]\right\}$
`Out[31]=` $\{p[2], u, p[4]\}$

Be aware that the order in which you give the rules is relevant.

`In[32]:=` $\left\{x^2, x^3, x^4\right\}/.\left\{x^{n-} → p[n], x^3 → u\right\}$
`Out[32]=` $\{p[2], p[3], p[4]\}$

Rules are important not only because they provide a way of transforming expression, but also because functions such as `Solve`, `DSolve`, `NSolve` and `NDSolve` return lists whose elements are lists of rules, each representing a solution.

`In[33]:=` $\mathtt{Solve}\left[x^3 - 5x^2 + 2x + 8 == 0, x\right]$
`Out[33]=` $\{\{x → -1\}, \{x → 2\}, \{x → 4\}\}$

`In[34]:=` $\mathtt{DSolve}[\{x'[t] == ax[t], x[0] == x0\}, x, t]$
`Out[34]=` $\{\{x → \mathtt{Function}[t, \mathrm{E}^{at}x0]\}\}$

For example, you can use the above solution as follows

`In[35]:=` $\mathtt{FullSimplify}[x[\lambda + 1] - x[\lambda - 1]/.$
$\mathtt{DSolve}[\{x'[t] == ax[t], x[0] == x0\}, x, t][[1]]]$
`Out[35]=` $2\mathrm{E}^{a\lambda}x0\,\mathtt{Sinh}[a]$

Sometimes you may need to go on applying rules over and over again, until the expression you are working on no longer changes. You can do this using the repeated replacement operation `expr//.rules` (or `ReplaceRepeated[expr, rules]`). The only difference in syntax is that you have to replace "/." by "//.".

229

Now, your rule will be applied twice.

```
In[36]:= x^2 + (1+y^2)^2 //.x_^2 → 5x
Out[36]= 5x + 5(1+5y)
```

Here the rule is applied only once.

```
In[37]:= Log[a b c d]/.Log[x_ y_]→ Log[x]+Log[y]
Out[37]= Log[a]+Log[b c d]
```

With the repeated replacement operator, the rule is applied repeatedly, until the result no longer changes.

```
In[38]:= Log[a b c d]//.Log[x_ y_]→ Log[x]+Log[y]
Out[38]= Log[a]+Log[b]+Log[c]+Log[d]
```

When you use "//.", Mathematica repeatedly passes through your expression, trying each of the rules given. It goes on doing this until it gets the same result on two successive passes. If you give a set of rules that is circular, then "//." can keep on getting different results forever. In practice, *Mathematica* has mechanisms to stop when a rule provokes an infinite loop, but it could take a while before this happens. You can always stop by explicitly interrupting Mathematica. Needless to say that, you should never use rules that are circular or endless.

For example, this will not work.

```
In[39]:= x//.x→ x+1
ReplaceRepeated::rrlim : Exiting after x scanned 65536
times. More...
Out[39]= 65536+x
```

One last issue about rules. Instead of the operator "->" when defining a rule you can use ":>" or ":→". The difference between "->" and ":>" is pretty much the same as the difference between "=" and ":=".

An advantage of applying rules instead of defining functions is that rules are not permanent, so there is no need to use Clear after the calculations we have done in this section.

Modules and programmes

We begin with an example. This defines a function exprod which constructs a product of n terms, then expands it out. Note that the operator ":=" is important in this case.

```
In[40]:= exprod[n_]:=Expand[∏_{i=1}^{n}(x+i)];
```

Every time you use the function, it will execute the Product and Expand operations.

```
In[41]:= {exprod[3],exprod[4],exprod[5]}
```

Out[41]= $\{6 + 11x + 6x^2 + x^3, 24 + 50x + 35x^2 + 10x^3 + x^4, 120 + 274x + 225x^2 + 85x^3 + 15x^4 + x^5\}$

The functions you define in *Mathematica* are essentially procedures that execute the commands you give. You can have several steps in your procedures, separated by semicolons ";". A semicolon after a command produces no output, but the command is still executed. The result you get from the whole function is simply the last expression in the procedure.

This takes the coefficient of x^i of the polynomial produced by `exprod`. We use an auxiliary variable `pol` to store the polynomial. Notice that you have to put parentheses around the procedure when you define it like this.

```
In[42]:= cex[n_,i_]:=(pol=exprod[n]; Coefficient[pol,x^i]);
```

This runs the procedure.

```
In[43]:= cex[5,3]
Out[43]= 85
```

A very important issue in the above procedure is that the variable `pol` now has a value.

```
In[44]:= pol
Out[44]= 120 + 274x + 225x^2 + 85x^3 + 15x^4 + x^5
```

This is potentially dangerous because you could have been using this variable before for other purposes, and this way your previous value is deleted. It is rather obvious in this case that you are using the variable `pol`, but if you were using code from someone else, you would not want other programmers' code to use your variables!

When you write procedures in *Mathematica*, it is important to make variables you use inside the procedures local, so that they do not interfere with things outside the procedures. You can do this by setting up your procedures as modules, in which you give a list of variables to be treated as local. The syntax of a module is `Module[{a,b,...}, proc]`, where a, b,... are the local variables and *proc* is the sequence of operations performed. The alternative `Module[{a=a0,b=b0,...}, proc]` gives the initial values *a0, b0,...* to the local variables. The value returned by a module is the output of the last command in *proc*.

```
In[45]:= Clear[cex,pol]
```

This is the right way of defining the previous function:

```
In[46]:= cex[n_,i_]:=Module[{pol=exprod[n]},
                    Coefficient[pol,x^i]];
```

Now, your function works without interfering with the variable `pol`.

```
In[47]:= cex[5,3]
Out[47]= 85
```

```
In[48]:= ?pol
```

Actually, this definition of `exprod` is not right. The problem is that `i` is not a *local* variable. Consider the following situation.

```
In[49]:= exprod[3]
```
$$\text{Out}[49]= 6+11x+6x^2+x^3$$

```
In[50]:= x=i;
```

```
In[51]:= exprod[3]
```
$$\text{Out}[51]=48$$

You would expect a behaviour like this:

$$\text{In}[52]:= \text{Expand}\left[\prod_{i=1}^{3}(y+i)\right]/.\,y\rightarrow i$$
$$\text{Out}[52]= 6+11i+6i^2+i^3$$

But with `x=i` this is what you actually evaluated:

$$\text{In}[53]:= \text{Expand}\left[\prod_{i=1}^{3}(i+i)\right]$$
$$\text{Out}[53]=48$$

There is very little chance that you would like your programme to do something like this. In order to have the right behaviour you must localise the variable `i`.

```
In[54]:= Clear[exprod]
```

$$\text{In}[55]:= \text{exprod}[n\_]:=\text{Module}\left[\{i\},\text{Expand}\left[\prod_{i=1}^{n}(x+i)\right]\right];$$

```
In[56]:= exprod[3]
```
$$\text{Out}[56]= 6+11i+6i^2+i^3$$

The reason for this working is that, although the local variable is named after `i` inside the module, *Mathematica* actually changes this name to something else. See [4, Section 2.7.3] for a more detailed explanation.

Actually, in the last definition something like

$$\text{In}[57]:= \text{exprod}[n\_][x\_]:=\text{Module}\left[\{i\},\text{Expand}\left[\prod_{i=1}^{n}(x+i)\right]\right];$$

would be preferred, as it would also isolate the variable `x`.

```
In[58]:= Clear[exprod, cex, x]
```

An example of functional programming

We are going to define a function that calculates the norm of a vector of real numbers (of course you could do that with the built-in functions `Norm` or `Dot`, but it is our intention to provide a more general discussion).

A Fortran77 or C programmer would be tempted to type something like this (we do not explain this code, although it should be obvious if you are a programmer).

```
In[59]:= norm1[v_]:=Module[{i=1,aux=0,n=Length[v]},
           While[i≤ n,  aux+=v[[i++]]∧2];
           Sqrt[aux]];
```

This is a bad programming style in *Mathematica* for various reasons. Firstly, it is not clear what the code does. One needs to read it through in order to understand what is going on. Secondly it does not take advantage of the power of *Mathematica*: the code is too restrictive and *Mathematica* has no room to improve its efficiency.

If you are a mathematician, this looks like what you would write in your notes.

$$In[60]:= \quad norm2[v\_]:=\sqrt{\sum_{i=1}^{Length[v]} v[[i]]^2};$$

This sentence uses some of the power of *Mathematica* to perform this kind of operations. It is a very direct and clear code, but it is in the nature of the function Sum to perform a "do" loop to carry out this kind of calculations. Although much better than the previous codes, it still has some of their drawbacks.

Actually, It would be preferable to write this in a slightly different way, as we saw before (the variable i should be localised to prevent harmful side effects):

```
In[61]:= norm3[v_]:=Module[{i},
                    Sqrt[Sum[v[[i]]∧2, {i,Length[v]}]]];
```

A C++ programmer, a LISP programmer or a *Mathematica* programmer should type something like this.

```
In[62]:= norm4[v_]:=Sqrt[Plus@@ (#²&/@v) ];
```

Here, we use the Map function to take the square of all the components. Then we use Apply and the Plus function to add all the squared components. Finally we take square root as usual. You should aim to use functional programming to produce concise code like this. Whether this is more efficient than the above piece of code depends on the programming language, but this programming style is more elegant, powerful, easier to maintain and could lead to more efficient implementations. In this particular case, norm4 is several times faster than norm1!

There is further improvement that one can do in the definition.

```
In[63]:= norm5[v_]:=Sqrt[Plus@@v∧2];
```

Since the function Power is listable, *Mathematica* understands that by squaring a list what you actually want to do is to square all its components. Hence, you can use v^2 instead of #²&/@v. The effect is the same, but v^2 is optimised so that the process is even quicker and more efficient.

```
In[64]:= norm[v_]:=Sqrt[Total[v∧2]];
```

New in *Mathematica* 5: the function Total adds the components of a vector, so

`Total[v]` is equivalent to `Plus@@v`. However, the performance of this function is even better than `Apply` combined with `Plus`.

In this section we discussed several ways of defining the norm of a vector, comparing traditional programming, mathematical functions and functional programming. As you have seen, the are several reasons why functional programming supersedes other programming paradigms; this is the kind of programming that you should aim to use.

Apart from `Map` and `Apply`, there are other functions that are considered as functional programming. Listability and `Total` are examples of them. If you decide to use *Mathematica* for your programming, read [4, sections 2.1, 2.2, 2.3, 2.5 and 2.6]. Another good reference is [3].

```
In[65]:= Clear["norm*"]
```

GRAPHICS

2D plots

Plotting a function $f : [a,b] \to \mathbb{R}$ is something very simple in *Mathematica*. One just has to use a command like `Plot[f,{x,a,b}]`. Here `f` is of course an expression depending on x that evaluates to a real number for all $x \in [a,b]$.

```
In[1]:= Plot[Sin[x],{x,0,2π}];
```

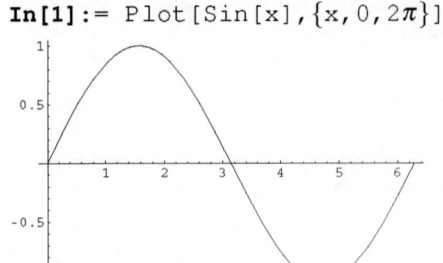

You can plot functions that have singularities. *Mathematica* will try to choose appropriate scales.

```
In[2]:= Plot[Tan[x],{x,0,2π}];
```

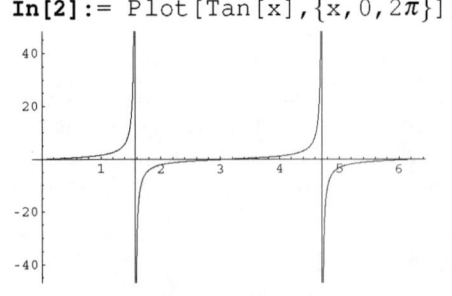

We can plot several functions at a time using a list as the first argument. For example,

In[3]:= `Plot[{Sin[x],Sin[2x],Sin[3x]},{x,0,2π}];`

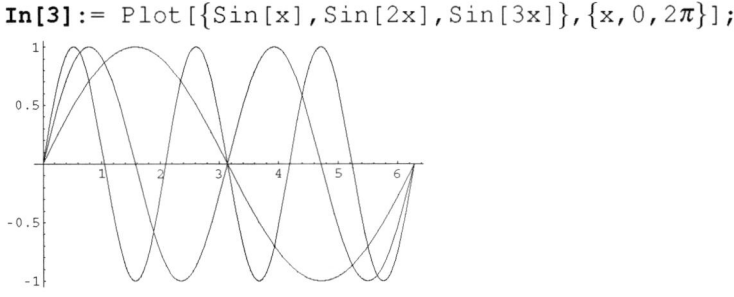

To get smooth curves, *Mathematica* has to evaluate the functions you plot at a large number of points. As a result, it is important that you set things up so that each function evaluation is as quick as possible. When you plot an object, say f, as a function of x, *Mathematica* first works out what values of x are needed, and only then evaluates f with those values of x. This has the advantage that *Mathematica* only tries to evaluate f for specific numerical values of x; it does not matter whether sensible values are defined for f when x is symbolic. There are, however, some cases in which it is much better to have *Mathematica* evaluate f before it starts to make the plot. A typical case is when f is actually a command that generates a table of functions. You want to have *Mathematica* first produce the table, and then evaluate the functions, rather than trying to produce the table afresh for each value of x. You can do this by typing `Plot[Evaluate[f],{x,`x_{min}`,`x_{max}`}]`. For example:

In[4]:= `Plot[Evaluate[Table[BesselJ[n,x],{n,4}]],`
`{x,0,10}];`

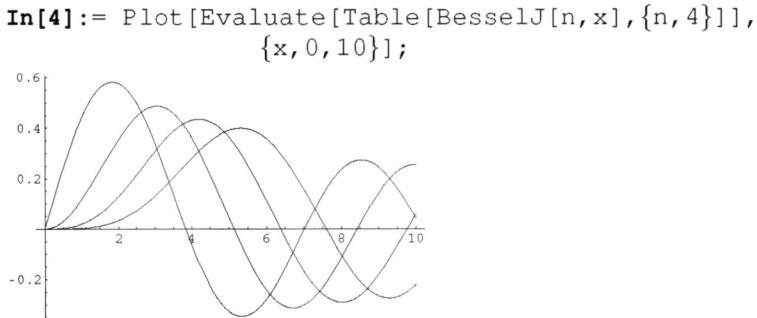

In most cases, the simple commands above will work fine. Sometimes, however, you may want to change the default way *Mathematica* chooses to plot a function. We introduce here the basic options. A more complicated syntax of `Plot` that allows you to include options would be `Plot[Evaluate[f],{x,`x_{min}`,`x_{max}`}`, *option1*→*value1*, *option2*→*value2*, ...]`. The following table shows some of the options you can change. Any option for which you do not give an explicit rule is taken to have its default value.

You can find in the Help Browser a more thorough discussion.

option name	default value	
`AspectRatio`	`1/GoldenRatio`	the height-to-width ratio for the plot; `Automatic` sets it from the absolute x and y coordinates
`DisplayFunction`	`$DisplayFunction`	how to display graphics; `Identity` causes no display
`PlotPoints`	`25`	the minimum number of points at which to sample the function
`PlotRange`	`Automatic`	the range of coordinates to include in the plot; `All` includes all points; with $\{y1, y2\}$ the explicit limits are given
`PlotStyle`	`Automatic`	specifies that all objects are to be generated with the given graphics directive, or list of graphics directives; for example, to draw lines with different colours use something like `RGBColor[r,g,b], ...`

When plotting several functions at a time it is particularly interesting to plot each function in a different colour to visualise the difference.

```
In[5]:=Plot[{Sin[x],Sin[2x],Sin[3x]},{x,0,2π},PlotStyle→
       {RGBColor[1,0,0],RGBColor[0,1,0],RGBColor[0,0,1]}];
```

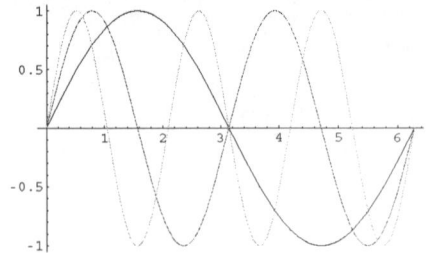

There are many options that you can change for `Plot`, and things that you can customise for each option (for example, with `PlotStyle` you can specify not only the colour but other properties such as the thickness or whether the lines should be dashed). We content ourselves with these basic options.

Also, note that some functions like `DSolve` and `NDSolve` give solutions as rules. Hence, you have to transform the entry to a plot in a suitable way.

```
In[6]:= DSolve[{y"[x]==-y[x],y[0]==y0,y'[0]==v0},y,x]
Out[6]= {{y→Function[{x},y0 Cos[x]+v0 Sin[x]]}}
```

This is the way to plot the solution of the previous differential equation.

```
In[7]:= Plot[Evaluate[y[x]/.DSolve[{y"[x]==-y[x],
    y[0]==y0,y'[0]==v0},y,x]/.{y0→0,v0→1}], {x,-2π ,2π}];
```

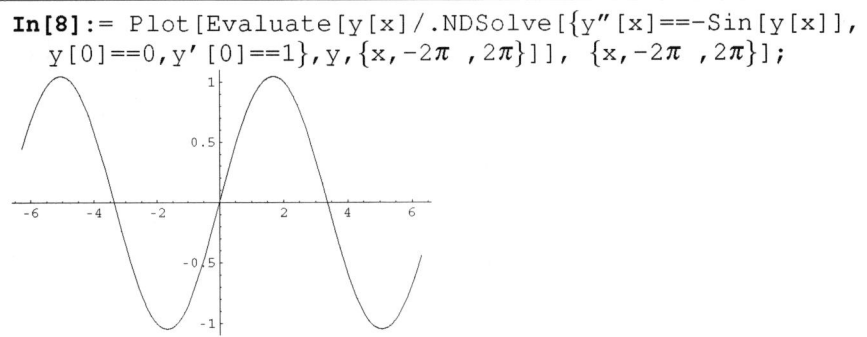

The same procedure works for numerical solutions. Note that in this case it is necessary to provide the numerical initial conditions already in NDSolve.

```
In[8]:= Plot[Evaluate[y[x]/.NDSolve[{y"[x]==-Sin[y[x]],
    y[0]==0,y'[0]==1},y,{x,-2π ,2π}]], {x,-2π ,2π}];
```

Now we can compare the pendulum equation with its linearisation.

```
In[9]:= Plot[Evaluate[y[x]/.
    {{DSolve[{y"[x]==-y[x],y[0]==0,y'[0]==1},y,x]},
     {NDSolve[{y"[x]==-Sin[y[x]],y[0]==0,y'[0]==1},
        y,{x,-2π,2π}]}}],
    {x,-2π,2π},
    PlotStyle→ {RGBColor[1,0,0],RGBColor[0,1,0]}];
```

3D plots

To plot a function $f : [a_1, b_1] \times [a_2, b_2] \to \mathbb{R}$, we use Plot3D instead of Plot, but there is not much difference with what we have just discussed.

```
In[10]:= Plot3D[Sin[x y],{x,0,3},{y,0,3}];
```

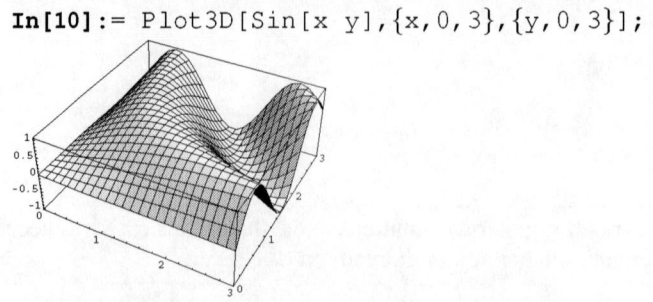

As with two dimensional plots, you can move and resize your plots. Unfortunately, you cannot dynamically rotate plots to watch them from different points of view in all versions of *Mathematica* up to version 5. In version 6, it is now possible to rotate 3d graphics dynamically. Also, note that there are many applications (some of them free) that you can use to manipulate graphics produced with *Mathematica*. In the author's opinion, *JavaView* is particularly useful and easy to handle.

Mathematica 4 and 5 do have an experimental package that allows you to rotate 3d plots dynamically. To load this package write <<RealTime3D ';. To restore the initial configuration use <<Default3D ';.

We introduce the basic 3d options in the following table.

option name	default value	
AspectRatio	Automatic	the ratio among the axes of the plot;
DisplayFunction	$DisplayFunction	how to display graphics; Identity causes no display
Lighting	True	whether to use ambient light
Mesh	True	whether to draw the lines of polygons in the 3d plot
PlotPoints	25	the minimum number of points at which to sample the function
PlotRange	Automatic	the range of coordinates to include in the plot; All includes all points; you can give explicit limits
ViewPoint	{1.3,-2.4,2.0}	specifies the point of view from where you look at the graphic

This plots the previous graph with more sample points, no polygons and from a different point of view.

```
In[11]:= Plot3D[Sin[x y],{x,0,3},{y,0,3},
    PlotPoints→50,Mesh→False,ViewPoint→{2,1,1}];
```

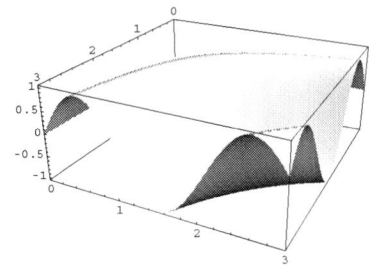

An introduction to curves and surfaces

In this section, we start using *Mathematica* for understanding several concepts of differential geometry. A theorem of Whitney states that every differentiable manifold can be embedded in a Euclidean space of sufficiently large dimension. So, we can consider a manifold, at least locally, as the image of a smooth map, $\phi : U \to \mathbb{R}^n$, where U is an open set in certain Euclidean space. In the case of curves, U will be an interval of $\mathbb{R}$ and $n = 2, 3$. In the case of a surface, U will be an open rectangle of $\mathbb{R}^2$ and $n = 3$.

We are interested in the set $\phi(U) \subset \mathbb{R}^n$, not the graph of ϕ. For that reason, we need to introduce parametric plots that are capable of plotting these sets.

Parametric plots

The syntax of a parametric plot is similar to the plots we have seen so far. The command `ParametricPlot[{`f_x, f_y`}, {t, `t_{min}`, {`t_{max}`}]` plots a curve in $\mathbb{R}^2$ with x and y coordinates f_x and f_y generated as a function of t. This function can also take options, which are essentially the same as for `Plot`.

For example, this plots a circle.

`In[12]:= ParametricPlot[{Cos[t],Sin[t]},{t,0,2`π`}];`

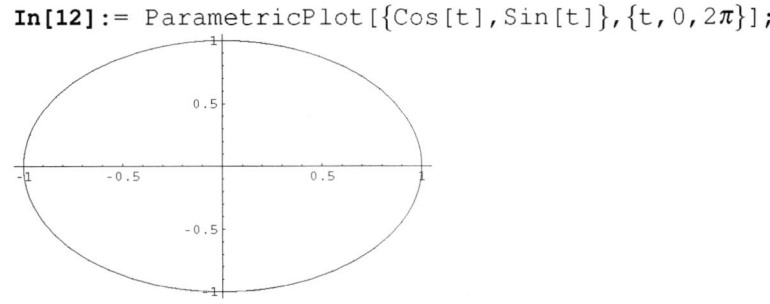

But because, *Mathematica* uses `1/GoldenRatio` as the ratio between axes, the above picture looks more like an ellipse. To sort this out, we set the same scale for both axes as follows:

```
In[13]:= ParametricPlot[{Cos[t],Sin[t]},{t,0,2π},
    AspectRatio→Automatic];
```

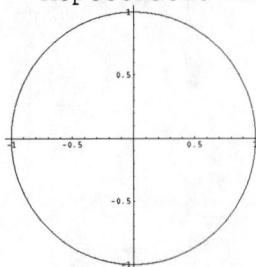

For three dimensional graphics, we use `ParametricPlot3D`. The command `ParametricPlot[{$f_x, f_y, f_z$}, {t, $t_{min}$, {$t_{max}$}]` is the direct analog in three dimensions of `ParametricPlot` in two dimensions. In both cases, *Mathematica* effectively generates a sequence of points by varying the parameter t, then forms a curve by joining these points.

This makes a parametric plot of a helical curve. Varying t produces circular motion in the x, y plane, and linear motion in the z direction.

```
In[14]:= ParametricPlot3D[{Cos[t],Sin[t],t/3},{t,0,4π}];
```

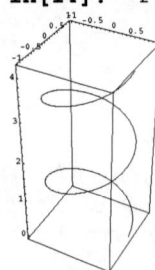

Note that, these parametric plots just plot parametrised curves, so you could have self-intersections, cusps, points of discontinuity and so on. For example, the following plot shows a four-leafed rose with self-intersections.

```
In[15]:= ParametricPlot[Sin[2t]{Cos[t],Sin[t]}, {t,0,2π},
    AspectRatio→Automatic,Axes→ None];
```

The function `ParametricPlot3D` can also be used to plot surfaces. Again, it is your duty to make sure that the parametric plot you produce does not have self intersections or any strange behaviour. The syntax to plot a parametrised surface is `ParametricPlot[{`f_x, f_y, f_z`}, {u, `u_{min}`, {`u_{max}`}, {v, `v_{min}`, {`v_{max}`}].` The surface is formed from a collection of quadrilaterals whose corners have coordinates corresponding to the values of the f_i when u and v take on values in a regular grid.

`In[16]:=ParametricPlot3D[{x,y,Sin[x y]},{x,0,3},{y,0,3}];`

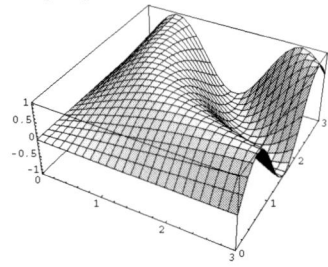

Using the helical curve, we can produce an helicoid.

`In[17]:= ParametricPlot3D[{u Cos[t],u Sin[t],t/3},`
`  {t,0,4`π`},{u,-1,1}];`

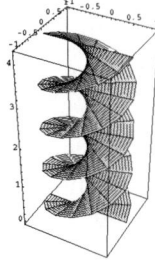

Again, the options you can use with `ParametricPlot3D` are essentially the same as with `Plot3D`.

`In[18]:= ParametricPlot3D[{u Cos[t],u Sin[t],t/3},`
`  {t,0,4`π`},{u,-1,1},`
`  Boxed`$\rightarrow$`False,Axes`$\rightarrow$`False, PlotPoints`$\rightarrow$`{60,25}];`

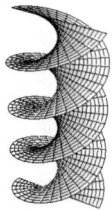

When you draw surfaces with `ParametricPlot3D`, the exact choice of parametrisation is crucial. You should avoid parametrisations in which all or part of your surface
is covered more than once. Such multiple coverings often lead to discontinuities in the
mesh drawn on the surface, and may make `ParametricPlot3D` take much longer to
render the surface.

A final observation is that you can combine plots of the same dimension. To achieve
that, use the command `Show`. Let us take these previous two plots (we do not print them
again). We store then in one variable for further use:

```
In[19]:= g1=Plot3D[Sin[x y],{x,0,3},{y,0,3}];
```

```
In[20]:= g2=ParametricPlot3D[{u Cos[t],u Sin[t],t/3},
    {t,0,4π},{u,-1,1},
    Boxed→False,Axes→False,PlotPoints→{60,25}];
```

```
In[21]:= Show[g1,g2];
```

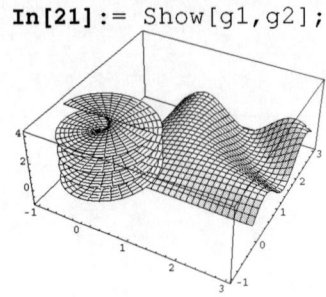

The function `Show` takes the same options as the usual plots, with a few differences.
Here is an example (there is no need to store plots in variables before combination). The
use of `DisplayFunction` prevents the plotting of the intermediate graphics.

```
In[22]:= Show[Plot3D[Sin[x y],{x,0,3},{y,0,3},
    DisplayFunction→Identity],
    ParametricPlot3D[{u Cos[t],u Sin[t],t/3},
    {t,0,4π},{u,-1,1},
    PlotPoints→{60,25},DisplayFunction→Identity],
    Boxed→False,Axes→False,
    DisplayFunction→$DisplayFunction];
```

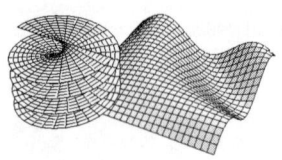

Curves

In this section we illustrate some of the basic geometric concepts associated with curves. Many of the concepts defined here are well known and there is a vast bibliography to explore curves using software like *Mathematica*. We do not aim to be thorough: we just want to show a few examples.

```
In[1]:= helix[t_]={Cos[t],Sin[t],t/3};
```

```
In[2]:= ParametricPlot3D[helix[t],{t,0,4π}];
```

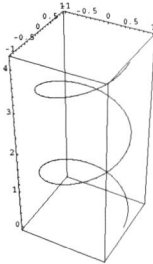

We quickly remind the basic formulas and definitions of the basic geometric objects associated with regular curves in $\mathbb{R}^3$. Let $\alpha : t \in I \to \alpha(t) \in \mathbb{R}^3$ be a regular curve, that is, $\alpha'(t) \neq 0$ for all t. Such a regular curve can be parametrised by arc length $s(t) = \int_{t_0}^{t} \|\alpha'(t)\| dt$. With respect to the arc length parameter (whose derivatives are denoted by a dot), we define the *tangent vector* as $T(s) = \dot{\alpha}(s)$, the *curvature* as $\kappa(s) = \|\dot{T}(s)\|$, the *normal vector* as $N(s) = \dot{T}(s)/\kappa(s)$, the *binormal vector* as $B(s) = T(s) \times N(s)$, and the *torsion* as $\tau(s) = \langle \dot{B}(s), N(s) \rangle$ whenever $\ddot{\alpha}(s) \neq 0$.

Although the arc length parameter always exists, but it can be very difficult (or impossible) to calculate. Hence, we work with the parameter t. The formulas are well known (see for example [1]): $\kappa(t) = \|\alpha'(t) \times \alpha''(t)\|/\|\alpha'(t)\|^3$, $T(t) = \alpha'(t)/\|\alpha'(t)\|$, $N(t) = T'(t)/(\kappa(t)\|\alpha'(t)\|)$, $B(t) = T(t) \times N(t)$, $\tau(t) = -\det(\alpha'(t), \alpha''(t), \alpha'''(t))/\|\alpha'(t) \times \alpha''(t)\|^2$.

```
In[3]:= frenetT[curve_][t_]:=
  curve'[t]/Sqrt[Total[curve'[t]∧2]]];
```

```
In[4]:= frenetN[curve_][t_]:=
  frenetT[curve]'[t]/
    (curvature[curve][t] Sqrt[Total[curve'[t]∧2]]);
```

```
In[5]:= frenetB[curve_][t_]:=
  Cross[frenetT[curve][t],frenetN[curve][t]];
```

```
In[6]:= curvature[curve_][t_]:=
  Sqrt[Total[Cross[curve'[t],curve"[t]]∧2]]/
    (Total[curve'[t]∧2]∧(3/2));
```

243

```
In[7]:= torsion[curve_][t_]:=
    -Det[{curve'[t],curve"[t],curve"'[t]}]/
        Total[Cross[curve'[t],curve"[t]]^2];
```

As an example, we calculate these geometric objects for the helix above. Note the use of `Simplify`.

```
In[8]:= Simplify[curvature[helix][t]]
Out[8]= 9/10
```

```
In[9]:= Simplify[torsion[helix][t]]
Out[9]= -3/10
```

```
In[10]:= Simplify[frenetT[helix][t]]
Out[10]= {-3Sin[t]/√10), 3Cos[t]/√10), 1/√10)}
```

```
In[11]:= Simplify[frenetN[helix][t]]
Out[11]= {-Cos[t],-Sin[t],0}
```

```
In[12]:= Simplify[frenetB[helix][t]]
Out[12]= {Sin[t]/√10), -Cos[t]/√10), 3/√10)}
```

Sometimes, a plot helps more than the exact formulas you get using the symbolic capability of *Mathematica*. For example, we write here a function to plot the Frenet-Serret frame at a point of a curve. This uses the formulas for the Frenet-Serret frames above. We use the command Line to plot the corresponding vectors of the Frenet-Serret frame. In this case, Line[{p,q}] just represents a line from *p* to *q*. Once you have a line, wrap it using Graphics3D to produce the actual graphic. Latter, we will use Show to render the result on screen.

```
In[13]:= frenet[curve_][t_]:=
    Module[{c=curve[t], v1=frenetT[curve][t],
        v2=frenetN[curve][t],v3=frenetB[curve][t],
        style=Sequence[Thickness[0.015],RGBColor[0,0,1]]},
    Graphics3D[{style,{Line[{c,c+v1}],Line[{c,c+v2}],
        Line[{c,c+v3}]}}]];
```

This plots a few Frenet-Serret frames on the curve.

```
In[14]:= Show[ParametricPlot3D[helix[t],{t,0,4π},
    DisplayFunction→Identity],
    Table[frenet[helix][t],{t,0,4π,π}],
    DisplayFunction→ $DisplayFunction, Boxed→False,
    Axes→False];
```

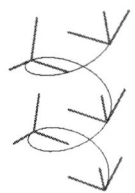

There is a better way to visualise the Frenet-Serret frame. If you plot a series of graphics you can animate them by double-clicking one of the graphics. What we do here is to plot the curve together with a few Frenet-Serret frames in several points of the curve. The animation produces the effect of the frame moving along the curve.

```
In[15]:= Table[Show[ParametricPlot3D[helix[t],{t,0,4π},
    DisplayFunction→Identity],
  frenet[helix][s],
  DisplayFunction→ $DisplayFunction,Boxed→False,
  Axes→False, PlotRange→{{-1.5,1.5},{-1.5,1.5},{0,5}}],
  {s,0,4π,π/10}];
```

This is a small modification that plots the normal and the binormal vectors so that they have as length the curvature and the torsion.

```
In[16]:= plotCurvatures[curve_][t_]:=
Module[{c=curve[t],v1=frenetT[curve][t],
    v2=frenetN[curve][t],v3=frenetB[curve][t],
    styleFrenet=Sequence[Thickness[.015],RGBColor[0,0,1]],
    styleCurvature=Sequence[Thickness[.02],RGBColor[1,0,0]],
    styleTorsion=Sequence[Thickness[.02],RGBColor[0,1,0]]},
Graphics3D[{{styleFrenet,{Line[{c,c+v1}],
    Line[{c,c+v2}],Line[{c,c+v3}]]}},
    {styleCurvature,{Line[{c,c+curvature[curve][t]v2}]}},
    {styleTorsion,{Line[{c,c+torsion[curve][t]v3}]}}}]];
```

This produces a plot as before.

```
In[17]:= Show[ParametricPlot3D[helix[t],{t,0,4π},
    DisplayFunction→Identity],
    Table[plotCurvatures[helix][t],{t,0,4π,π}],
    DisplayFunction→$DisplayFunction,
    Boxed→False,Axes→False];
```

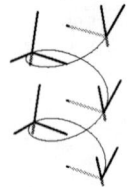

Surfaces

In this section we start our study of manifolds by the first non trivial step: surfaces. As usual, we will consider parametrised surfaces. It will be the task of the user to check whether the parametrisation produces an embedded surface or not (that is, one has to check whether there are self intersections, singular points and so on). We will illustrate some of the geometric objects that can be constructed on a surface and give the code to calculate them.

Here it is the definition of a torus.

```
In[18]:= torus[R_,r_][u_,v_]=
    {(R+r Cos[u])Cos[v],(R+r Cos[u])Sin[v],r Sin[u]};
```

The first obvious thing we can do with a surface is to plot it.

```
In[19]:= ParametricPlot3D[torus[2,1][u,v],
   {u,0,2π},{v,0,2π}];
```

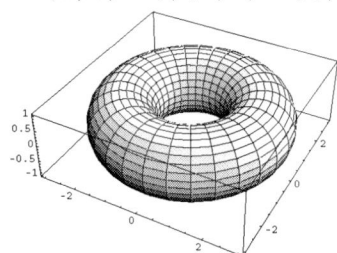

This is a Möbius strip.

```
In[20]:= moebiusStrip[a_][u_,v_]=
   a{Cos[u]+v Cos[u/2]Cos[u],Sin[u]+v Cos[u/2]Sin[u],
     vSin[u/2]};
```

```
In[21]:= ParametricPlot3D[moebiusStrip[3][u,v],
   {u,0,2π},{v,-0.3,0.3},PlotPoints→{25,7}];
```

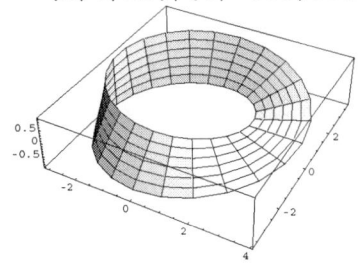

Most of the geometry of a surface in $\mathbb{R}^3$ is encoded in the Gauss map. This is defined by $x_3 = (x_1 \times x_2)/\|x_1 \times x_2\|$ where $x_i = \partial x/\partial u_i$ and $x(u_1,u_2)$ is the parametrization. The code is straightforward. The only interesting point here is the way the derivatives are calculated: Derivative[$i_1, i_2, \ldots$][f][$x_1, x_2, \ldots$] represents $\frac{\partial^{i_1+i_2+\cdots} f}{\partial x_1^{i_1} \partial x_2^{i_2} \cdots}$.

```
In[22]:= gaussMap[surface_][u_,v_]:=Module[
   {x3=Cross[Derivative[1,0][surface][u,v],
     Derivative[0,1][surface][u,v]]},
   x3/Sqrt[x3.x3]];
```

This calculates the Gauss map of the torus. An interesting point here is that Simplify does not cancel square roots with squares. This is because *Mathematica* assumes that variables are complex numbers, so this simplification cannot be performed in the general case. To expand powers use PowerExpand.

```
In[23]:= PowerExpand[Simplify[gaussMap[torus[R,r]][u,v]]]
Out[23]= {-Cos[u] Cos[v], -Cos[u] Sin[v], -Sin[u]}
```

247

This plots the Gauss map along one of the generating circles of the torus.

```
In[24]:= Show[Graphics3D[
    {Thickness[0.01], Table[Line[{torus[2,1][u,0],
        torus[2,1][u,0]-gaussMap[torus[2,1]][u,0]}],
      {u,0,2π,π/4}]}],
    ParametricPlot3D[torus[2,1][u,v],{u,0,2π},{v,0,2π},
      DisplayFunction→Identity],
    DisplayFunction→ $DisplayFunction,Boxed→False];
```

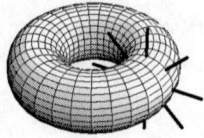

A more interesting plot is the corresponding Gauss map of the Möbius strip.

```
In[25]:= Show[Graphics3D[
  {Thickness[0.01],Table[Line[{moebiusStrip[3][u,0],
      moebiusStrip[3][u,0]-gaussMap[moebiusStrip[3]][u,0]}],
    {u,0,2π,π/24}]}],
  ParametricPlot3D[moebiusStrip[3][u,v],
    {u,0,2π},{v,-0.3,0.3},
    PlotPoints→{25,7},DisplayFunction→Identity],
  DisplayFunction→ $DisplayFunction,Boxed→False,
  ViewPoint→{3, 1,1}];
```

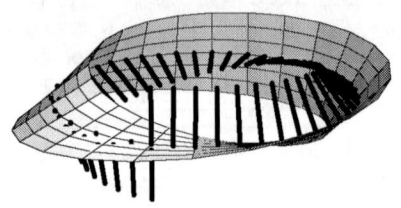

An animation is probably more representative of the fact that the normal vector is not continuous along the surface. This is because of the well known fact that the Möbius strip is non-orientable. For space reasons, we delete the output.

```
In[26]:= Module[{g=ParametricPlot3D[moebiusStrip[3][u,v],
    {u,0,2π},{v,-0.3,0.3},PlotPoints→{25,7},
    DisplayFunction→Identity]},
  Table[Show[g,Graphics3D[
```

```
{Thickness[0.01],Line[{moebiusStrip[3][u,0],
    moebiusStrip[3][u,0]-gaussMap[moebiusStrip[3]][u,0]}]}],
  DisplayFunction→ $DisplayFunction,Boxed→False,
  Axes→False,ViewPoint→ {3,1,1}],{u,0,2π,π/24}]];
```

The first intrinsic geometric object of a surface is the *metric* or first fundamental form. Its calculation is straightforward from the Euclidean metric.

```
In[27]:= metric[surface_][u_,v_]:=Module[{i,j,dx},
    dx=Table[Derivative[KroneckerDelta[i,1],
      KroneckerDelta[i,2]][surface][u,v],{i,2}];
    Table[dx[[i]].dx[[j]],{i,2},{j,2}]];
```

For example, these are the metric of the torus and the Möbius strip.

```
In[28]:= MatrixForm[Simplify[metric[torus[R,r]][u,v]]]
Out[28]//MatrixForm=
```

$$\begin{pmatrix} r^2 & 0 \\ 0 & (R+rCos[u])^2 \end{pmatrix}$$

```
In[29]:=MatrixForm[Simplify[metric[moebiusStrip[a]][u,v]]]
Out[29]//MatrixForm=
```

$$\begin{pmatrix} \frac{1}{4}a^2\left(2Cos[u]v^2+3v^2+8Cos\left[\frac{u}{2}\right]v+4\right) & 0 \\ 0 & a^2 \end{pmatrix}$$

The differential of the Gauss map contains also very important geometric information of a surface. It essentially encodes the second fundamental form. Its trace is the *mean curvature* and its determinant the *Gauss curvature*.

```
In[30]:= gaussianCurvature[surface_][u_,v_]:=
    Module[{x11=Derivative[2,0][surface][u,v],
      x12=Derivative[1,1][surface][u,v],
      x22=Derivative[0,2][surface][u,v],
      x3=gaussMap[surface][u,v]},
    Det[{{x11.x3,x12.x3},{x12.x3,x22.x3}}]
      /Det[metric[surface][u,v]]];
```

```
In[31]:= meanCurvature[surface_][u_,v_]:=
    Module[{x11=Derivative[2,0][surface][u,v],
      x12=Derivative[1,1][surface][u,v],
      x22=Derivative[0,2][surface][u,v],
      x3=gaussMap[surface][u,v],gAux=metric[surface][u,v]},
    (x11.x3 gAux[[2,2]]-2 x12.x3 gAux[[1,2]]
      +x22.x3 gAux[[1,1]])/Det[gAux]/2];
```

For example, these are the Gaussian and mean curvature of the torus.

```
In[32]:= Simplify[gaussianCurvature[torus[R,r]][u,v]]
```

$$\text{Out[32]}= \frac{Cos[u]}{rR + r^2 Cos[u]}$$

```
In[33]:= PowerExpand[Simplify[
    meanCurvature[torus[R,r]][u,v]]]
```

$$\text{Out[33]}= \frac{R + 2rCos[u]}{rR + r^2 Cos[u]}$$

The following cell plots the torus coloured by its Gaussian curvature. First, we define the function `curvatureColour`. It uses the function `Hue`: its first parameter gives one colour of the rainbow and the second its saturation. We produce a tone of red (if the curvature is positive) or blue (if the curvature is negative). The closer the tone is to white, the more flat the surface is at that point. In the parametric plot we turn off the lighting so that *Mathematica* does not use ambient light to colour the surface. The actual colour of the surface is given by appending a fourth coordinate to the parametrisation. This fourth coordinate is a `FaceForm` command whose parameter is the colour defined above. Take some time to understand (or look it up) how the option `Lighting` works.

```
In[34]:= curvatureColour[k_?Positive]=
    Hue[0,2N[ArcTan[k]/π],1];
curvatureColour[k_?NonPositive]=
    Hue[.7,2N[ArcTan[-k]/π],1];
ParametricPlot3D[Evaluate[Append[torus[2,1][u,v],
    FaceForm[curvatureColour[
    gaussianCurvature[torus[2,1]][u,v]]]]],
    {u,0,2π},{v,0,2π},Lighting→False];
```

As a final example, we show how to plot geodesics of a surface. This is very easy using the *Mathematica* functions. First, we give the definition of the Christoffel symbols.

```
In[35]:= christoffelSymbols[surface_][u_,v_]:=
    Module[{i,j,k,l,dg,gInv=Inverse[metric[surface][u,v]]},
    dg={Table[Derivative[1,0][metric[surface]][u,v][[j,k]],
        {j,2},{k,2}],
    Table[Derivative[0,1][metric[surface]][u,v][[j,k]],
        {j,2},{k,2}]};
    Table[Sum[gInv[[k,l]](dg[[j,i,k]]+dg[[i,j,k]]
        -dg[[k,i,j]]),{l,2}]/2,
        {i,2},{j,2},{k,2}]];
```

For example:

```
In[36]:= Simplify[christoffelSymbols[torus[R,r]][u,v]]
```

$$\text{Out[36]}= \{0,0\},\{0,-\frac{rSin[u]}{R+rCos[u]}\},\{0,-\frac{rSin[u]}{R+rCos[u]}\},\{\frac{(R+rCos[u])Sin[u]}{r},0\}$$

Recall that the geodesic equation is given by

$$\frac{d^2x_k}{dt^2}+\sum_{i,j}\Gamma^k_{ij}\frac{dx_i}{dt}\frac{dx_j}{dt}=0, \quad k=1,2.$$

By appending the initial conditions, we can use NDSolve to numerically solve the geodesic equation. The command would look something like this:

```
In[37]:= geodesic[t_]=torus[2,1]@@(#[t]&/@
(Table[x[i],{i,2}]/.
  NDSolve[Flatten[{Table[x[k]"[t]
    +Sum[christoffelSymbols[torus[2,1]]
      [x[1][t],x[2][t]][[i,j,k]]
    x[i]'[t]x[j]'[t],{i,2},{j,2}]==0,{k,2}],
    x[1][0]==0,x[2][0]==0,
    x[1]'[0]==1,x[2]'[0]==0.11395}],
  Table[x[i],{i,2}],{t,0,100}][[1]]])));
```

The above initial conditions produce (almost) a periodic geodesic.

```
In[38]:= ParametricPlot3D[geodesic[t],{t,0,100},
PlotPoints→1000];
```

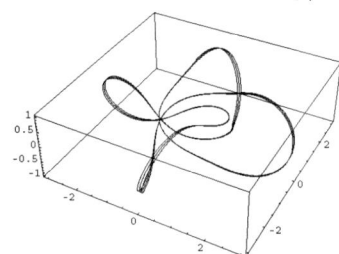

REFERENCES

1. M. P. do Carmo, *Differential geometry of curves and surfaces*, Prentice-Hall, Inc., Englewood Cliffs, N.J., 1976.
2. A. Gray, *Modern differential geometry of curves and surfaces with Mathematica*, Second edition, CRC, Boca Raton, FL, 1998.
3. R. Maeder, *Programming in Mathematica (3rd Edition)*, Addison-Wesley Professional, 1996.
4. S. Wolfram, *The Mathematica Book, Fifth Edition*, Wolfram Media, Inc., 2003.

List of Participants

Albujer Brotons, Alma Luisa
Universidad de Murcia, Spain
albujer@um.es

Alvarez Dios, Jose A.
Univ. de Santiago de Compostela, Spain
jaaldios@usc.es

Arroyo, Josu
Universidad del País Vasco, Spain
josujon@ehu.es

Blow, Matthew Lewis
Oxford University, United Kingdom
blow@thphys.ox.ac.uk

Caballero, Magdalena
Universidad de Granada, Spain
mccmm@ugr.es

Castro-Villarreal, Pavel
Univ. Nacional Autonoma, Mexico
pcastro@nucleares.unam.mx

Cordero, Luis A.
Univ. de Santiago de Compostela, Spain
cordero@zmat.usc.es

Daglas, Hert
University of Pennsylvania, USA
hertdaglas@yahoo.com

Fernández, Marisa
Universidad del País Vasco, Spain
marisa.fernandez@ehu.es

Ferrández, Angel
Universidad de Murcia, Spain
aferr@um.es

Garcia de Andrade, Luiz
UERJ, Brasil
garciluiz@gmail.com

Alejo Plana Miguel Ángel
Universidad del País Vasco, Spain
miguelangel_alejoplana@yahoo.es

Alvarez Lopez, Jesus Antonio
Univ. Santiago de Compostela, Spain
jalvarez@usc.es

Barros, Manuel
Universidad de Granada, Spain
mbarros@ugr.es

Brozos Vázquez, Miguel
Univ. de Santiago de Compostela, Spain
mbrozos@usc.es

Calviño Louzao, Esteban
Univ. de Santiago de Compostela, Spain
estebcl@usc.es

Conde Pena, Eduardo
Univ. de Santiago de Compostela, Spain
eduardo.conde@rai.usc.es

Cortés Ayaso, Alexandre Andrés
Univ. de Santiago de Compostela, Spain
alancoia@usc.es

Díaz Ramos, José Carlos
University College Cork, Ireland
jc.diazramos@ucc.ie

Fernández Sanmartín, Iván Alejandro
Univ. de Santiago de Compostela, Spain
orange21@gmail.com

Garay, Óscar J.
Universidad del País Vasco, Spain
oscarj.garay@ehu.es

García Río, Eduardo
Univ. de Santiago de Compostela, Spain
xtedugr@usc.es

Gómez Tato, Antonio
Univ. de Santiago de Compostela, Spain
agtato@usc.es

Haji Badali, Ali
Tabriz University, Iran
ahajibadali@gmail.com

Hervella, Luis M.
Univ. de Santiago de Compostela, Spain
dumaso@usc.es

Lipowsky, Reinhard
Max Planck Institute, Germany
Reinhard.Lipowsky@mpikg-golm.mpg.de

Nieto Roig, Juan J.
Univ. de Santiago de Compostela, Spain
amnieto@usc.es

Palmer, Bennett
Idaho State University, USA
palmbenn@isu.edu

Santangelo, Christian
University of Massachusetts, USA
santancd@physics.upenn.edu

Tang, Tsing-Young
Imperial College, England
tsing-young.tang02@imperial.ac.uk

Vázquez Lorenzo, Ramón
Univ. de Santiago de Compostela, Spain
ravazlor@usc.es

Verpoort, Steven
K.U.Leuven, Belgium
steven.verpoort@wis.kuleuven.be

Guven, Jemal
Univ. Nacional Autonoma, Mexico
jemal@nucleares.unam.mx

Herrera Fernández, Jonatan
Universidad de Málaga, Spain
jherrera@agt.cie.uma.es

Kierfeld, Jan
Max Planck Institute, Germany
Jan.Kierfelduni@dortmund.de

Matsumoto, Elisabetta
University of Pennsylvania, USA
ematsumo@sas.upenn.edu

Oubiña, José A.
Univ. de Santiago de Compostela, Spain
jaoubina@usc.es

Rodrigo Fernández, César
Academia Militar Lisboa, Portugal
crodrigo@us.es

Singer, David
Case Western Reserve University, USA
das5@case.edu

Vázquez Abal, M. Elena
Univ. de Santiago de Compostela, Spain
meva@zmat.usc.es

Vázquez Montejo, Pablo Agustín
Univ. Nacional Autónoma, Mexico
vazqmont@nucleares.unam.mx

Vilariño Fernández, Silvia
Univ. de Santiago de Compostela, Spain
svfernan@usc.es

AUTHOR INDEX

A

Arroyo, J., 186

B

Baczynski, K., 151
Barros, M., 71

D

Díaz-Ramos, J. C., 213

G

Gutjahr, P., 151

K

Kierfeld, J., 151

L

Lipowsky, R., 151

P

Palmer, B., 33

S

Santangelo, C., 114
Singer, D. A., 3